N-Heterocyclic Carbenes in Synthesis

Edited by
Steven P. Nolan

Related Titles

A. Berkessel, H. Gröger

Asymmetric Organocatalysis
From Biomimetic Concepts to Applications
in Asymmetric Synthesis

2005. ISBN 3-527-30517-3

A. de Meijere, F. Diederich (Eds.)

Metal-Catalyzed Cross-Coupling Reactions

2004. ISBN 3-527-30518-1

R. H. Grubbs (Ed.)

Handbook of Metathesis
3 Volumes

2003. ISBN 3-527-30616-1

N-Heterocyclic Carbenes in Synthesis

Edited by
Steven P. Nolan

WILEY-VCH Verlag GmbH & Co. KGaA

The Editor:

Prof. Steven P. Nolan
Department of Chemistry
University of New Orleans
2000 Lakeshore Drive
New Orleans, LA 70148
USA

All books published by Wiley-VCH are carefully produced. Nevertheless, authors, editors, and publisher do not warrant the information contained in these books, including this book, to be free of errors. Readers are advised to keep in mind that statements, data, illustrations, procedural details or other items may inadvertently be inaccurate.

Library of Congress Card No.: applied for
British Library Cataloguing-in-Publication Data
A catalogue record for this book is available from the British Library.

Bibliographic information published by the Deutsche Nationalbibliothek
The Deutsche Nationalbibliothek lists this publication in the Deutsche Nationalbibliografie; detailed bibliographic data are available in the Internet at http://dnb.d-nb.de

© 2006 WILEY-VCH Verlag GmbH & Co. KGaA, Weinheim

All rights reserved (including those of translation into other languages). No part of this book may be reproduced in any form – nor transmitted or translated into machine language without written permission from the publishers. Registered names, trademarks, etc. used in this book, even when not specifically marked as such, are not to be considered unprotected by law.

Typesetting Kühn & Weyh, Satz und Medien, Freiburg
Printing betz-druck GmbH, Darmstadt
Bookbinding Litges & Dopf Buchbinderei GmbH, Heppenheim

Printed in the Federal Republic of Germany
Printed on acid-free paper

ISBN-13: 978-3-527-31400-3
ISBN-10: 3-527-31400-8

Contents

Preface *XI*

List of Contributors *XIII*

1 N-Heterocyclic Carbene–Ruthenium Complexes in Olefin Metathesis *1*
Samuel Beligny and Siegfried Blechert

1.1 Introduction *1*
1.2 N-Heterocyclic Carbene–Ruthenium Complexes *2*
1.2.1 Introduction of N-Heterocyclic Carbenes *2*
1.3 Second-generation NHC-Ru Catalysts *6*
1.3.1 Variations on the NHC Group *7*
1.3.2 Variation on the Benzylidene Group *8*
1.3.3 Phosphine-free NHC-Ruthenium Complexes *9*
1.3.4 Variation of the Anionic Ligands *13*
1.3.5 14-Electron NHC-Ruthenium Complexes *13*
1.4 Enantioselective Ruthenium Olefin Catalysts *13*
1.4.1 Grubbs II Analogues *14*
1.4.2 Phosphine-free Chiral NHC-Ruthenium Complexes *14*
1.4.2.1 First-generation Catalysts *14*
1.4.2.2 Second-generation Chiral Ru Complexes *15*
1.5 Solid Supported NHC-Ru Complexes *16*
1.5.1 Immobilization via the NHC Ligand *17*
1.5.2 Attachment Through the Anionic Ligand *18*
1.5.3 Attachment Through the Alkylidene Moiety *18*
1.5.4 Homogenous Catalysts *20*
1.5.5 Ionic Liquids *20*
1.6 Conclusion and Outlook *22*
References *22*

N-Heterocyclic Carbenes in Synthesis. Edited by Steven P. Nolan
Copyright © 2006 WILEY-VCH Verlag GmbH & Co. KGaA, Weinheim
ISBN: 3-527-31400-8

2	**Ruthenium N-Heterocyclic Carbene Complexes in Organic Transformations (Excluding Metathesis)** *27*
	Suzanne Burling, Belinda M. Paine, and Michael K. Whittlesey

2.1	Introduction *27*
2.2	Hydrogenation and Hydrosilylation Reactions *27*
2.3	Isomerization *34*
2.4	Other Reactivity *37*
2.5	Tandem Reactions [29] *42*
2.5.1	Metathesis and Hydrogenation *42*
2.5.2	Metathesis and Isomerization *44*
2.5.3	Tandem Reactions not Involving Metathesis *50*
2.6	Conclusions *51*
	References *52*

3	**Cross-coupling Reactions Catalyzed by Palladium N-Heterocyclic Carbene Complexes** *55*
	Natalie M. Scott and Steven P. Nolan

3.1	Introduction *55*
3.2	Palladium(0) NHC Complexes *56*
3.3	Palladium(II) N-Heterocyclic Carbene Complexes *58*
3.4	Palladium/NHC Complexes as Catalysts *59*
3.4.1	C–N Bond-forming Reactions: the Hartwig–Buchwald Reaction *59*
3.4.2	C–C Bond-forming Reactions: α-Arylation of Ketones *63*
3.4.3	Suzuki–Miyaura Cross-coupling of Aryl Chlorides with Arylboronic Acids *64*
3.4.4	C–H Bond-forming Reactions: Dehalogenation of Aryl Halides *67*
3.4.5	C–C Bond-forming Reactions: Hydroarylation of Alkynes *69*
3.5	Conclusion *70*
	References *70*

4	**Pd-NHC Complexes as Catalysts in Telomerization and Aryl Amination Reactions** *73*
	David J. Nielsen and Kingsley J. Cavell

4.1	Introduction *73*
4.2	Telomerization *74*
4.2.1	Definition and Background *74*
4.2.1.1	Commercial Viability of the Telomerization Reaction *75*
4.2.2	Catalyst Design: Ligand Selection *76*
4.2.3	Mechanism of the Pd-catalyzed Telomerization of Buta-1,3-diene with Methanol *77*
4.2.4	Pd-(NHC) Complexes as Telomerization Catalysts *83*
4.2.5	Telomerization in Imidazolium-based Ionic Liquids *90*

4.3	Buchwald–Hartwig Amination Reactions Catalyzed by Pd(NHC) Complexes *92*
4.3.1	Introduction *92*
4.3.2	Mechanism of Aryl Amination *93*
4.3.3	Palladium-NHC Systems as Catalysts for Aryl Amination *95*
4.3.3.1	Application of Preformed Pd(0/II)(NHC) Complexes *95*
4.3.3.2	*In situ* Pd Imidazolium Catalyst Systems *98*
4.4	Conclusions *100*
	References 100

5	**Metal-mediated and -catalyzed Oxidations Using N-Heterocyclic Carbene Ligands** *103*
	Mitchell J. Schultz and Matthew S. Sigman
5.1	Introduction *103*
5.2	Metal–NHC-mediated Activation of Molecular Oxygen *103*
5.2.1	Co *103*
5.2.2	Ni *105*
5.2.3	Pd *107*
5.3	Metal-catalyzed Oxidations, Pd *108*
5.3.1	Methane Oxidation *108*
5.3.2	Alcohol Oxidation *108*
5.3.3	Wacker-type Oxidations *112*
5.3.4	Oxidative Carbonylation *113*
5.4	Ir-catalyzed Oppenauer Oxidation of Alcohols *115*
5.5	Conclusion *117*
	References 117

6	**Efficient and Selective Hydrosilylation of Alkenes and Alkynes Catalyzed by Novel N-Heterocyclic Carbene Pt^0 Complexes** *119*
	Guillaume Berthon-Gelloz and István E. Markó
6.1	Introduction *119*
6.2	Initial Results *120*
6.3	Synthesis, Structure and Reactivity of (NHC)Pt(dvtms) Complexes *123*
6.3.1	(Alkyl-NHC)Pt(dvtms) Complexes *123*
6.3.2	(Aryl-NHC)Pt(dvtms) Complexes *131*
6.3.3	(Benzimidazolyl-NHC)Pt(dvtms) Complexes *134*
6.4	Kinetic and Mechanistic Studies *137*
6.5	Hydrosilylation of Alkynes *150*
6.6	Summary *158*
	References 158

7 Ni-NHC Mediated Catalysis 163
Janis Louie

7.1 Introduction 163
7.2 Rearrangement Reactions 163
7.2.1 Rearrangement Reactions of Vinyl Cyclopropanes 163
7.2.2 Rearrangement Reactions of Cyclopropylen-Ynes 164
7.3 Cycloaddition Reactions 167
7.3.1 Cycloaddition of Diynes and Carbon Dioxide 167
7.3.2 Cycloaddition of Unsaturated Hydrocarbons and Carbonyl Substrates 169
7.3.3 Cycloaddition of Diynes and Isocyanates 170
7.3.4 Cycloaddition of Diynes and Nitriles 172
7.4 Reductive Coupling Reactions 174
7.4.1 Reductive Coupling Reactions: No added Reductant 174
7.4.2 Reductive Coupling Reactions in the Presence of a Reductant 175
7.5 Oligomerization and Polymerization 178
7.6 Hydrogenation 181
7.7 Conclusions 181
References 181

8 Asymmetric Catalysis with Metal N-Heterocyclic Carbene Complexes 183
Marc Mauduit and Hervé Clavier

8.1 Introduction 183
8.2 Concept, Design and Synthesis of Chiral NHC Complexes 185
8.2.1 Synthesis of Ligand Precursors 185
8.2.2 Synthesis of NHC Complexes 186
8.2.3 Concept and Design of Chiral NHCs 187
8.3 Asymmetric Hydrogenation 193
8.4 Asymmetric 1,4-Addition 199
8.4.1 Copper-NHC Complexes 200
8.4.2 Rhodium-NHC Complexes 203
8.4.3 Palladium-NHC Complexes 205
8.5 Asymmetric 1,2-Addition 205
8.6 Asymmetric Hydrosilylation 207
8.6.1 Rhodium-NHC Complexes 207
8.6.2 Ruthenium-NHC Complexes 211
8.7 Asymmetric Olefin Metathesis 211
8.7.1 Asymmetric Ring-closing Metathesis 212
8.7.2 Asymmetric Ring-opening Metathesis/Cross Metathesis 213
8.8 Allylic Substitution Reaction 215
8.8.1 Palladium Catalysis 215
8.8.2 Copper Catalysis 216

8.9	Asymmetric α-Arylation *217*
8.10	Palladium-catalyzed Kinetic Resolution *218*
8.11	Conclusion and Outlook *219*
	References 220

9 Chelate and Pincer Carbene Complexes   *223*
Guillermina Rivera and Robert H. Crabtree

9.1	Introduction *223*
9.2	Design Strategy *224*
9.2.1	Bite Angle *227*
9.2.2	Tripod Ligands *227*
9.3	Synthetic Strategies *228*
9.4	Failure to Chelate *231*
9.5	Ligand Properties *233*
9.5.1	Types of Ligand *234*
9.5.1.1	C,C Chelates *234*
9.5.1.2	N,C Chelates and Pincers *234*
9.5.1.3	P,C Chelates and Pincers *235*
9.6	Catalysis *235*
9.6.1	Medicinal Applications *236*
9.7	Conclusions *237*
	References 237

10 The Quest for Longevity and Stability of Iridium-based Hydrogenation Catalysts: N-Heterocyclic Carbenes and Crabtree's Catalyst   *241*
Leslie D. Vazquez-Serrano and Jillian M. Buriak

10.1	Introduction: Rhodium and Iridium-based Hydrogenation Catalysts *241*
10.2	Building upon Crabtree's Catalyst with N-Heterocyclic Carbenes *243*
10.3	Chiral Iridium N-Heterocyclic Catalysts *250*
10.4	Conclusions *253*
	References 253

11 Cu-, Ag-, and Au-NHC Complexes in Catalysis   *257*
Pedro J. Pérez and M. Mar Díaz-Requejo

11.1	Introduction *257*
11.2	Copper *258*
11.2.1	Conjugate Additions *258*
11.2.2	Reduction of Carbonyl Compounds *261*
11.2.3	Enantioselective Allylic Alkylations *265*
11.3	Silver *268*
11.3.1	Synthesis of 1,2-bis(Boronate) Esters *268*
11.3.2	NHC-Ag as Carbene Delivery Agents *269*

11.4	Gold	*270*
11.5	Cu-, Ag-, and Au-NHC Complexes as Catalysts for Carbene Transfer Reactions from Ethyl Diazoacetate	*271*
	References	*274*

12 N-Heterocyclic Carbenes as Organic Catalysts *275*
Andrew P. Dove, Russell C. Pratt, Bas G. G. Lohmeijer, Hongbo Li, Erik C. Hagberg, Robert M. Waymouth, and James L. Hedrick

12.1	Introduction	*275*
12.2	*In situ* Generation of Free Carbenes	*276*
12.3	Small Molecule Transformations	*278*
12.3.1	Benzoin and Formoin Condensation	*278*
12.3.2	Michael-Stetter Reaction	*281*
12.3.2.1	Stetter Reaction: Addition of Acyl Intermediate to α,β-Unsaturated Aldehydes	*281*
12.3.3	α,β-Unsaturated Aldehydes as Homoenolate Equivalents	*283*
12.3.4	Conversion of α-Substituted Aldehydes into Esters	*284*
12.3.5	Transesterification	*285*
12.3.6	Nucleophilic Aromatic Substitution	*287*
12.4	Living Ring-opening Polymerization	*288*
12.4.1	Imidazol-2-ylidenes	*289*
12.4.2	Imidazolin-2-ylidenes	*291*
12.4.3	1,2,4-Triazol-5-ylidenes	*293*
12.4.4	Thiazol-2-ylidenes	*294*
	References	*294*

Subject Index *297*

Preface

This project was begun after a symposium organized at the fall 2004 ACS meeting in Philadelphia. The occasion appeared timely to gather some of the most important developments in the use of N-heterocyclic carbenes (NHCs) in catalysis and synthesis and find a home for them. This project has developed quite rapidly, as has the area, and we hope the reader will find the emerging science useful and stimulating.

The seminal work of Arduengo is key to the development of NHCs as ligands in transition metal chemistry and as reactive entities. The importance of this development, involving the isolation and characterization by X-ray diffraction of a free carbene, cannot be overstated. It opened the door to the most recent developments in the use of NHCs as catalyst modifiers and organic catalysts. The earlier work of Michael Lappert is also of note as the chemistry developed in Sussex was the first concerted effort to use NHCs as ligands for transition metal systems. Even earlier developments by Wanzlick at the Technical University in Berlin are at the origin of the ideas associated with the possible intermediacy and existence of a "stable carbene".

The work described in this book is, as with most science, not done in a vacuum but "built on the shoulders of giants". Ranging from developments of catalysts that have become known as "second-generation olefin metathesis catalysts" to the use of NHCs in palladium cross-coupling to the role of NHCs in organic catalysis, the areas described in the following chapters are quite diverse. We hope the work described here will encourage others to investigate this area and to proceed in new and exciting directions.

Ottawa, Canada *Steven P. Nolan*
August 2006

N-Heterocyclic Carbenes in Synthesis. Edited by Steven P. Nolan
Copyright © 2006 WILEY-VCH Verlag GmbH & Co. KGaA, Weinheim
ISBN: 3-527-31400-8

List of Contributors

Samuel Beligny
Institut für Organische Chemie
Technische Universität Berlin
Straße des 17. Juni 135
10623 Berlin
Germany

Guillaume Berthon-Gelloz
Université Catholique de Louvain
Département de Chimie
Place Louis Pasteur, 1
1348 Louvain-la-Neuve
Belgium

Siegfried Blechert
Institut für Organische Chemie
Technische Universität Berlin
Straße des 17. Juni 135
10623 Berlin
Germany

Jillian M. Buriak
Department of Chemistry
University of Alberta
and the National Institute for
Nanotechnology (NINT)
National Research Council
Edmonton, Alberta T6G 2G2
Canada

Suzanne Burling
Department of Chemistry
University of Bath
Claverton Down
Bath, BA2 7AY
United Kingdom

Kingsley J. Cavell
School of Chemistry
Cardiff University
Park Place
Cardiff, CF10 3AT
United Kingdom

Hervé Clavier
Equipe Synthèse Organique et
Systèmes Organisés
UMR CNRS 6226
"Sciences Chimiques de Rennes"
Ecole Nationale Supérieure de
Chimie de Rennes
Avenue du Général Leclerc
35700 Rennes
France

Robert H. Crabtree
Department of Chemistry
Yale University
P.O. Box 208107
New Haven, CT 06520-8107
USA

N-Heterocyclic Carbenes in Synthesis. Edited by Steven P. Nolan
Copyright © 2006 WILEY-VCH Verlag GmbH & Co. KGaA, Weinheim
ISBN: 3-527-31400-8

M. Mar Díaz-Requejo
Departamento de Química y
Ciencia de Materiales
Universidad de Huelva
Campus de El Carmen s/n
21007 Huelva
Spain

Andrew P. Dove
IBM Almaden Research Center
650 Harry Road
San Jose, CA 95120
USA

Erik C. Hagberg
IBM Almaden Research Center
650 Harry Road
San Jose, CA 95120
USA

James L. Hedrick
IBM Almaden Research Center
650 Harry Road
San Jose, CA 95120
USA

Hongbo Li
IBM Almaden Research Center
650 Harry Road
San Jose, CA 95120
USA

Bas G. G. Lohmeijer
IBM Almaden Research Center
650 Harry Road
San Jose, CA 95120
USA

Janis Louie
Department of Chemistry
University of Utah
315 South 1400 East
Salt Lake City, UT 84112
USA

István E. Markó
Université Catholique de Louvain
Département de Chimie
Place Louis Pasteur,1
1348 Louvain-la-Neuve
Belgium

Marc Mauduit
Equipe Synthèse Organique et
Systèmes Organisés
UMR CNRS 6226
"Sciences Chimiques de Rennes"
Ecole Nationale Supérieure de
Chimie de Rennes
Avenue du Général Leclerc
35700 Rennes
France

David J. Nielsen
School of Chemistry
Cardiff University
Park Place
Cardiff, CF10 3AT
United Kingdom

Steven P. Nolan
Department of Chemistry
University of New Orleans
New Orleans, LA 70148
USA

Belinda M. Paine
Department of Chemistry
University of Bath
Claverton Down
Bath, BA2 7AY
United Kingdom

Pedro J. Pérez
Departamento de Química y
Ciencia de Materiales
Universidad de Huelva
Campus de El Carmen s/n
21007 Huelva
Spain

Russell C. Pratt
IBM Almaden Research Center
650 Harry Road
San Jose, CA 95120
USA

Guillermina Rivera
Departamento de Química
FES-Cuautitlán UNAM
Apartado Postal 25
Cuautitlán Izcalli
Edo. de México, 54740
México

Mitchell J. Schultz
Department of Chemistry
University of Utah
Salt Lake City, UT 84112
USA

Natalie M. Scott
Department of Chemistry
University of New Orleans
New Orleans, LA 70148
USA

Matthew S. Sigman
Department of Chemistry
University of Utah
Salt Lake City, UT 84112
USA

Leslie D. Vazquez-Serrano
Owens Corning
Science & Technology Center
2790 Columbus Road, Route 16
Granville, OH 43023
USA

Robert M. Waymouth
Department of Chemistry
Stanford University
Stanford, CA 94305
USA

Michael K. Whittlesey
Department of Chemistry
University of Bath
Claverton Down
Bath, BA2 7AY
United Kingdom

1
N-Heterocyclic Carbene–Ruthenium Complexes in Olefin Metathesis

Samuel Beligny and Siegfried Blechert

1.1
Introduction

Metal-catalyzed olefin metathesis has established itself as a powerful tool for carbon–carbon bond formation in organic chemistry [1]. The development of catalysts since the initial discoveries of the early 1990s has been tremendous: molybdenum [2], tungsten [3] and ruthenium catalysts have proved to be very fruitful metals for this reaction (Fig. 1.1).

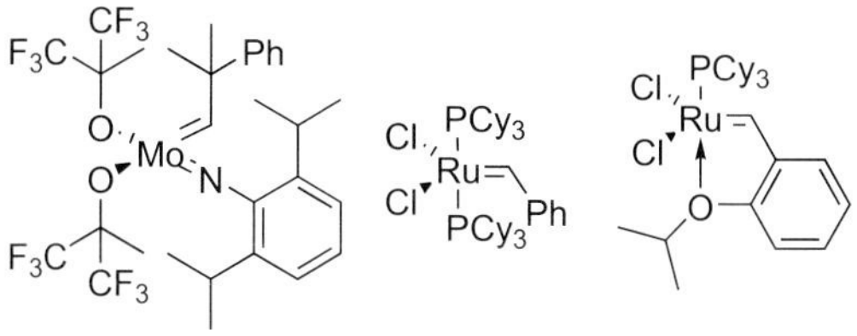

Fig. 1.1 Metathesis catalysts.

Ruthenium based olefin metathesis catalysts have been the focus of great attention. The first major breakthrough for ruthenium-catalyzed metathesis was from the work of Grubbs, which developed catalyst **2**, known as Grubbs I catalyst [4], which is less reactive than the Schrock molybdenum-based alkylidene complexes but has greater functional group tolerance and simplified handling characteristics. However, these species still show relatively low thermal stability and suffer significant decomposition at elevated temperatures through P–C bond degradation [5]. Hoveyda and coworkers have serendipitously discovered catalyst **3** [6], which contains an internal metal–oxygen chelate. This Ru-carbene complex offers excellent stability to air and moisture and can be recycled in high yield by silica-gel column chromatography. The ability of catalyst **3** to be recycled is based on a release–return mechanism. Considerable evidence that this mechanism is at least partially

N-Heterocyclic Carbenes in Synthesis. Edited by Steven P. Nolan
Copyright © 2006 WILEY-VCH Verlag GmbH & Co. KGaA, Weinheim
ISBN: 3-527-31400-8

supported has been given recently [7]. After the first turnover, the styrene moiety is released from the ruthenium core but can return at the end of the sequence. However, despite this progress, ruthenium complexes **2** and **3** do not generally allow the formation of tri- and tetra-substituted double bonds by ring-closing metathesis (RCM); only Schrock's tetra-coordinated alkylidene species, such as **1**, can promote such reactions efficiently.

1.2
N-Heterocyclic Carbene–Ruthenium Complexes

1.2.1
Introduction of N-Heterocyclic Carbenes

Intimate understanding of the mechanism of the metathesis reaction promoted by ruthenium complexes was crucial for the development of more efficient catalysts. The mechanism of olefin metathesis promoted by **2** and its analogues has been the subject of extended theoretical [8] and experimental [9] studies. There is consensus on the mechanism depicted in Scheme 1.1. Phosphine dissociation was critical to the process and a low ratio of phosphine reassociation to the ruthenium species was necessary for high activity.

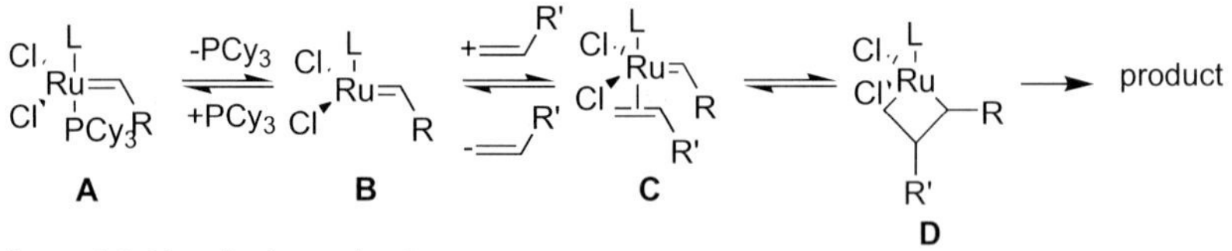

Scheme 1.1 Metathesis mechanism.

In addition, catalyst activity is directly related to the electron-donating ability of the phosphine ligands [1h]. The steric bulk of the ligand may also play an important role, contributing to phosphine dissociation by destabilizing the crowded bis(phosphine) olefin complex. An understanding of the mechanism has made clear that a highly active but unstable 14-electron mono(phosphine) intermediate **B** is formed during the catalytic cycle. To have more stable and active catalysts it was necessary to incorporate more basic and sterically demanding ligands than PCy$_3$. N-Heterocyclic carbenes (NHC) were perfect candidates.

The second breakthrough in ruthenium catalysts was the introduction of NHCs as ligands to the ruthenium complex. The use of nucleophilic NHCs is an attractive alternative to phosphine ligands since they are relatively easy to prepare. NHCs are strong σ-donors but poor π-acceptor ligands and bind strongly to the metal center with little tendency to dissociate from it. Solution calorimetry has shown that the NHC ligand binds by approximately 5 kcal mol^{-1} more than PCy$_3$ to ruthenium [10]. Herrmann reported the first such complex [11]. Both PCy$_3$ moieties were replaced by N,N'-disubstituted 2,3-dihydro-1H-imidazol-2-ylidene units

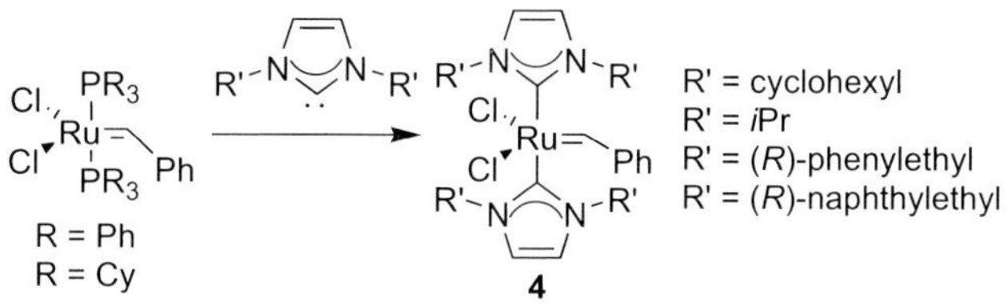

1.2 N-Heterocyclic Carbene–Ruthenium Complexes

Scheme 1.2 The first NHC-Ru complexes (reported by Herrmann [11]).

to give ruthenium complex **4** (Scheme 1.2). The product is stable but the catalytic activity was not considerably improved.

The lack of improved catalytic activity is due to the strong bonding between the NHC and the ruthenium core, which renders the dissociative pathway less likely and leads to a low concentration of the catalytically active 14-electron species in solution. However, the combination of a strongly binding, electron-donating NHC ligand with a more labile ligand should afford the desired effect, leading to a more active and more stable species. Both the 14-electron catalyst species **B** and the 16-electron olefin complex should be stabilized by the NHC ligand due to its strong σ-donor ability. Three groups independently and almost simultaneously reported the synthesis and catalytic properties of such ruthenium complexes (Fig. 1.2) [10, 12, 13].

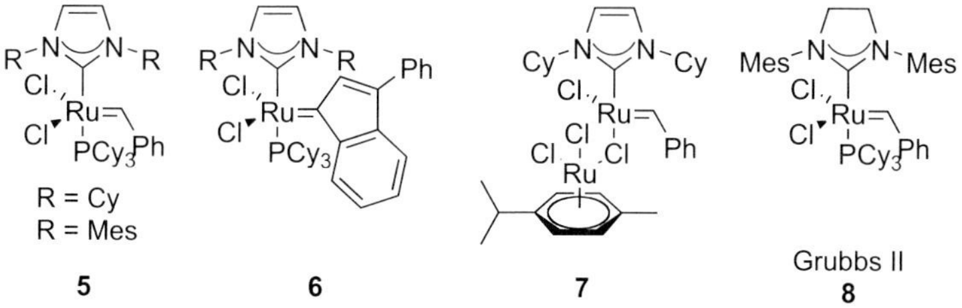

Fig. 1.2 Initial second-generation ruthenium complexes for metathesis.

As expected, this new generation of catalysts proved to be very stable, combined with greater reactivity than the Grubbs I catalyst, which opened new possibilities for organic synthesis. The NHC-ruthenium complexes are stable to air and have reactivity that can even surpass, in some cases, that of molybdenum catalyst **1**. Formation of tri and even tetra substituted double bonds, which were generally only possible using Schrock's catalyst, were now possible using NHC-ruthenium complexes (Table 1.1).

Differences in reactivity with Grubbs I catalyst and analogues were not only noticeable in terms of rates of reaction but also in terms of E/Z selectivity in RCM. Fürstner and coworkers, during the total synthesis of herbarumin I and II, discovered that the use of the ruthenium indenylidene complex **9** leads only to the lactone with E-geometry. Whereas catalyst **5** favors the corresponding Z-geometry with good selectivity [14]. This selectivity reflects kinetic control versus thermodynamic control; catalyst **9** is not active enough to equilibrate the E-isomer of the lactone to its more thermodynamically favored Z-isomer (Scheme 1.3).

Table 1.1 Comparison of reactivity of first- and second-generation Ru and Mo catalysts.

Entry	Substrate	Product	Time (min)	Yield (%)		
				1	2	8
1	 E,E-diene	 cyclopentene E,E	10	Quant.	Quant.	Quant.
2	 E,E,Me-diene	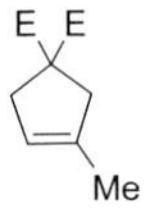 cyclopentene E,E,Me	10	Quant.	20	Quant.
3	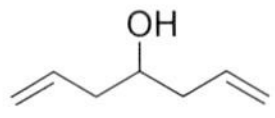 OH-diene	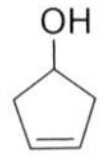 cyclopentenol	10	0	0	Quant.
4	 E,E,t-Bu-diene	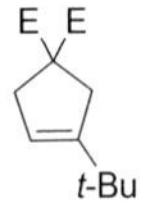 cyclopentene E,E,t-Bu	60	37	0	Quant.
5	 Me,E,E,Me-diene	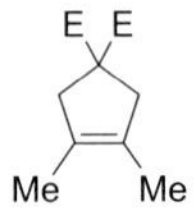 cyclopentene E,E,Me,Me	24 h	93	0	31
6	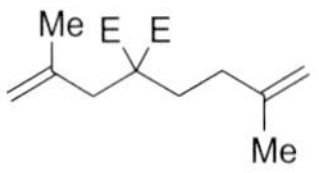 Me,E,E,Me-triene	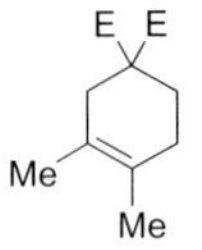 cyclohexene E,E,Me,Me	90	52	0	90

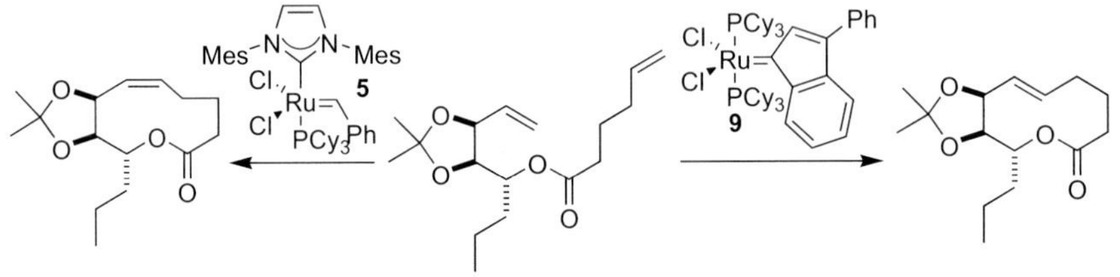

Scheme 1.3 Differences in E/Z selectivity between first and second-generation Ru complexes in RCM.

The NHC-ruthenium complex **8**, commonly called Grubbs II, was the most active catalyst of these early second-generation complexes [15]. Due to the absence of a π-system in the NHC the carbene is not stabilized by resonance. This makes the carbene more basic than the unsaturated analogue and this higher basicity translates into an increased activity of the resulting ruthenium complex. Nolan and coworkers have directly compared the NHC ligand SIMes to its unsaturated analogue IMes with respect to steric bulk and electron donor activity with calorimetric and structural investigations [16]. In view of the relatively important difference of reactivity between **5** and **8**, surprisingly minor differences in donor ability were found.

The increased reactivity of this new generation of ruthenium catalysts was highlighted in the total synthesis of epothilones by Sinha's group [17]. The synthesis of epothilones via a C9–C10 disconnection was first explored in Danishefsky's group. Unfortunately, the attempted connection of C9–C10 by RCM using either the Grubbs I catalyst **2** or Schrock's molybdenum catalyst **1** was unsuccessful [18]. However, Sinha showed that this strategy was viable using the Grubbs II catalyst **8**. The ring-closed product was obtained in 89% yield. The mixture of geometric isomers was of no consequence since the double bond was subsequently hydrogenated (Scheme 1.4).

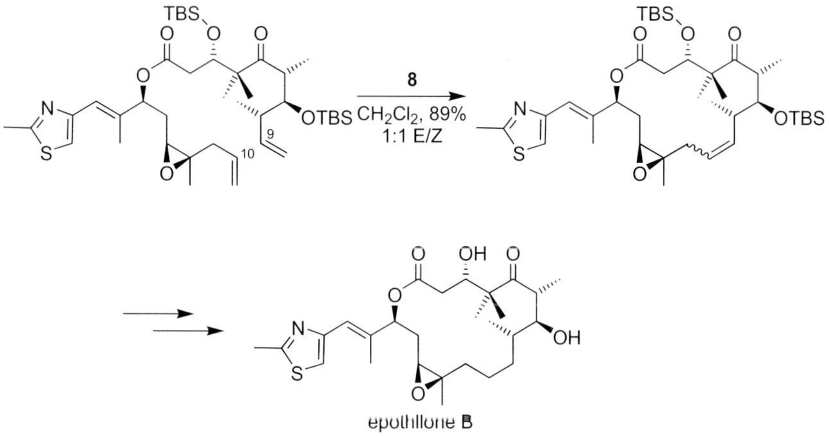

Scheme 1.4 Epothilone B synthesis.

The Grubbs II catalyst was also the first catalyst to enable the formation of tri-substituted alkenes by cross-metathesis (CM) [19]. This was of prime importance since tri-substituted carbon–carbon double bonds are a recurring motif in a wide array of organic molecules. Grubbs and coworkers at Caltech reported the formation of tri-substituted double bonds in good yield with moderate to excellent E-selectivity [12c, 20]. The CM of α,β-unsaturated compounds (ester, aldehydes and ketones) and simple terminal olefins in the presence of **8** (5 mol%) was remarkably efficient. This particular reactivity was used by Spessard and Stoltz towards the total synthesis of garsubellin A, a potential Alzheimer therapeutic [21]. The CM between the bicyclo[3.3.1]nonane core and 2-methylbut-2-ene highlighted this reactivity as it gave the CM product in 88% yield (Scheme 1.5).

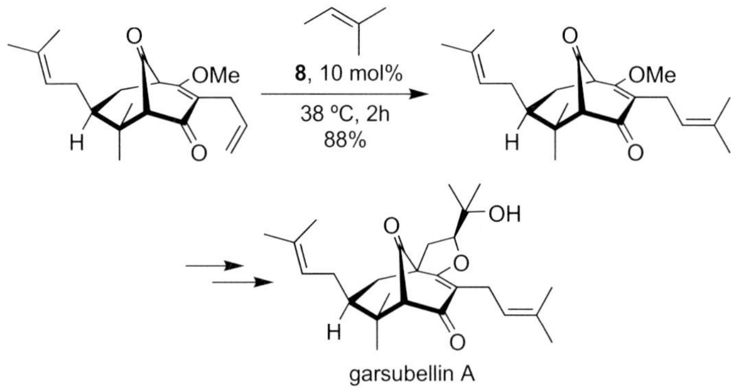

Scheme 1.5 Towards the synthesis of garsubellin A.

Mechanistic studies have shown an interesting feature of this new family of catalysts. Initially, the improved catalytic properties were ascribed to the exacerbated ability of the phosphine moiety to dissociate owing to the presence of the bulky NHC. Conversely, phosphine dissociation from **8** was two orders of magnitude slower than from **2**, which makes the Grubbs II catalyst a slower initiator than Grubbs I [9b, 22]. However, **8** showed an increased preference for coordination of olefinic substrates relative to phosphines compared to the Grubbs I catalyst. This is certainly due to the increased σ-donor character of NHCs in comparison to phosphines [1i]. Hence, Grubbs II catalyst **8** remains longer in the catalytic cycle even if it initiates slower. The strong donor ability of NHCs leads to overall faster rates of catalysis and enables the metathesis of olefins for which **2** was ineffective.

1.3
Second-generation NHC-Ru Catalysts

These new vistas of reactivities prompted an impressive amount of research towards the development of new NHC-ruthenium catalysts for metathesis reactions. Research was directed towards the use of new types of NHCs and also to variations of moieties around the ruthenium core.

1.3.1
Variations on the NHC Group

Several "second generation" metathesis catalysts have been prepared from Grubbs I catalyst 2 and various NHCs (Table 1.2). The influence of the N-substituent on both imidazol-2-ylidene and 5,5-dihydroimidazol-2-ylidene has been studied by different groups [10b, 23–27]. The SIMes analogue bearing two 2,6-diisopropylphenyl groups displayed even greater activity than 8 for the metathesis of terminal olefins [23]. Other analogues generally displayed lower reactivity. Substitution on the backbone of the NHC ligand with two chlorides (entry 1) afforded little change

Table 1.2 Variation of the NHC.

Entry	Catalyst	Catalytic activity	Ref.
1	**10**	RCM, enyne metathesis	[29]
2	**11**, **12**	RCM, enyne metathesis	[29]
3	**13**	RCM	[29, 30]
4	**14**	–	[31]
5	**15**	RCM, ROMP	[32]

in reactivity compared with catalyst 5. Fürstner and coworkers also showed that asymmetrically substituted NHC-ruthenium (entry 2) complexes promote the formation of tetra-substituted double bonds by RCM in moderate to good yields. Complex 11, bearing a pendant terminal olefin, was shown to form tethered carbene 12, which potentially could regenerate once the substrate is subjected to metathesis and has been completely consumed.

The NHC complex using the triazol-5-ylidene carbene developed by Enders [28] exhibits good catalyst activity; however, its limited lifetime in solution does not enable the reaction to reach completion in demanding cases. Adamantyl-substituted NHC-Ru complex (entry 4) was a poor metathesis catalyst, most likely because of the steric hindrance of the trans position to the benzylidene moiety by the adamantyl group. Catalyst 15 (entry 5), bearing a six-membered NHC, was synthesized by the Grubbs group. This catalyst showed limited reactivity for RCM and ROMP (ring-opening metathesis polymerization) compared with its five-membered NHC-Ru complex analogues.

To date, the SIMes ligand is still the ligand of choice as it affords the most potent NHC-ruthenium catalyst for olefin metathesis.

1.3.2
Variation on the Benzylidene Group

The effect of the variation or the replacement of the benzylidene group has also been studied. The Grubbs group have prepared a series of NHC-Ru complexes with electron-donating groups on the carbene carbon [33]. These carbenes are often referred to as Fischer-type carbenes. Complexes **16–19** (Fig. 1.3) were prepared from the reaction between the Grubbs I catalyst and an excess of the corresponding vinylic compound followed by treatment with the free IMes carbene in benzene.

Analogue **20** was prepared directly by treatment of Grubbs II catalyst **8** with an excess of ethyl vinyl ether. These complexes initiated the ROMP of norbornene and norbornene derivatives and gave the corresponding polymer in quantitative yield. However, polymerization was significantly slower than with the parent NHC-Ru complexes **5** and **8**. They also promoted the RCM of diethyl diallylmalonate in good yield. The rates of RCM and ROMP suggest that the reactivity follows a general trend: E = C > N > S > O.

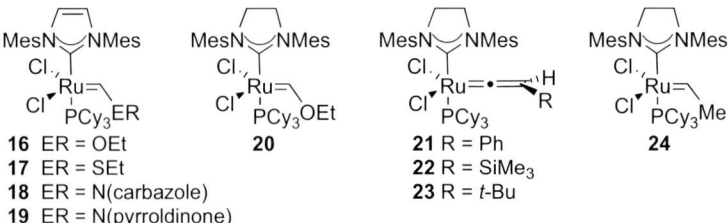

Fig. 1.3 Variation on the benzylidene group.

The benzylidene group has also been replaced by vinylidene groups [34]. Complexes **21–23** display good metathesis activity for the ROMP and RCM, yet the reactivity is still inferior to the benzylidene analogues.

To install a linear alkyl end group on ROMP polymers, NHC-Ru complex **24** was prepared from Grubbs II (**8**) and but-2-ene gas [35]. These complexes are again slightly less active than the parent benzylidenes but are suitable for ROMP and acyclic diene metathesis (ADMET).

1.3.3
Phosphine-free NHC-Ruthenium Complexes

Tremendous efforts have been made to obtain phosphine-free NHC-ruthenium complexes. The first breakthrough was reported almost simultaneously by the Hoveyda [36] and the Blechert [37] groups and was based on the Hoveyda–Grubbs and the Grubbs II catalyst. This new catalyst (**25**) is now one of the most widely used ruthenium catalysts for metathesis reactions, alongside both Grubbs I and II and the Hoveyda–Grubbs catalyst, and is prepared from the Grubbs II catalyst and 2-isopropoxystyrene (Scheme 1.6).

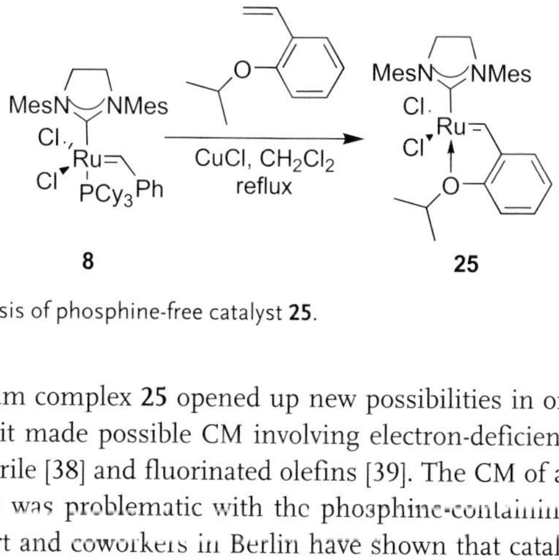

Scheme 1.6 Synthesis of phosphine-free catalyst **25**.

NHC-ruthenium complex **25** opened up new possibilities in organic synthesis. Most noticeably it made possible CM involving electron-deficient olefin partners such as acrylonitrile [38] and fluorinated olefins [39]. The CM of acrylonitrile with terminal alkenes was problematic with the phosphine-containing catalyst **8** [40]; however, Blechert and coworkers in Berlin have shown that catalyst **25** promoted such reactions in high yield and with good to excellent Z-selectivity (Scheme 1.7).

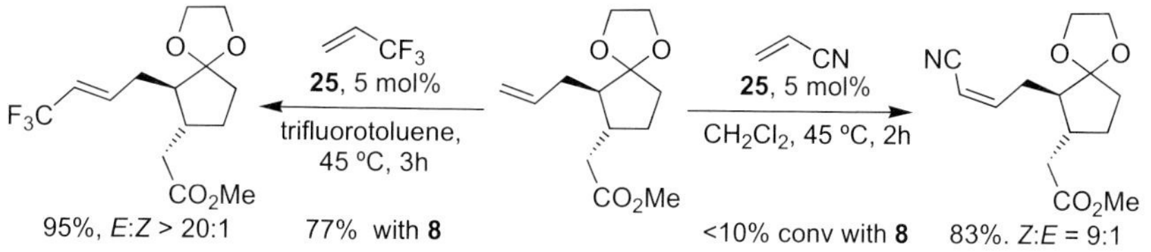

Scheme 1.7 Reactivity of phosphine-free catalyst **25**.

Catalyst **25** also made possible the efficient synthesis of biologically interesting molecules. Hoveyda et al. have reported the enantioselective total synthesis of erogorgiaene, an inhibitor of *Mycobacterium tuberculosis* [41]. This synthesis involves two metathesis steps: an enyne metathesis and a CM. Both catalysts **8** and **25** promote the enyne metathesis; however, the Grubbs II catalyst led to the formation of side products and a lower reaction rate in the CM step with methyl vinyl ketone (MVK) and only **25** gave the desired product in good yield and with excellent E-selectivity (Scheme 1.8).

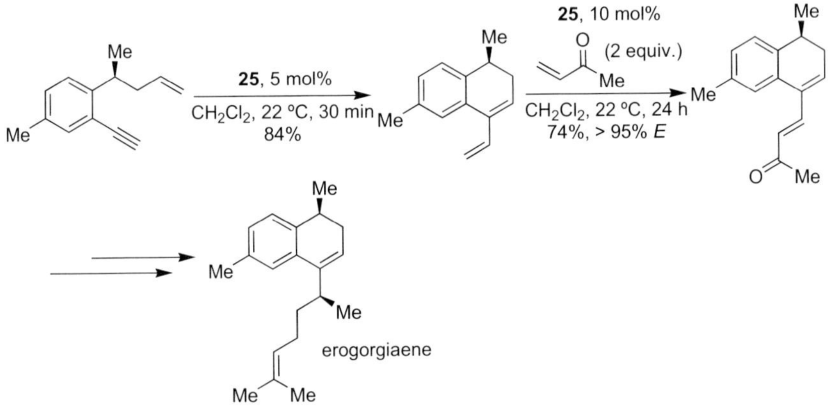

Scheme 1.8 Synthesis of erogorgiaene.

The activity and reactivity profile of catalyst **25** is greatly affected by the released phosphine, which is able to intercept and deactivate the 14-electron active species [22, 42].

To improve further the reactivity of phosphine-free NHC–Ru complexes, the groups of Blechert and Grela both embarked on systematic studies on the effect of substitution on the 2-isopropoxystyrene ligand. Blechert's group have shown that increased steric hindrance adjacent to the chelating isopropoxy group is crucial for increasing the catalytic activity. Replacing the benzylidene ligand in **25** with BINOL- or biphenyl-based ligands results in a large improvement in initiation (Table 1.3). These catalysts, especially **27**, were shown to initiate significantly more rapidly than **8** and **25**. Formation of the 14-electron active species is, presumably, facilitated by the increased bulk of the ligand, which helps dissociation. Analogue **28**, which displays a similar reactivity profile as **27** (entry 3), is particularly interesting since its synthesis is more facile than the other analogues, starting with *o*-vanillin. Systematic studies on the effect of substituents on the styrene showed that decreased electron density on both the chelating oxygen and the Ru=C bond had a significant effect on the rate of acceleration [48]. Reassociation to the metal center, which deactivates the catalyst, is also suppressed. Grela and coworkers have developed catalyst **29**, derived from inexpensive α-asarone, which showed catalytic activity comparable to the parent catalyst **25** [46, 49]. They also

Table 1.3 Phosphine-free NHC-Ru complexes.

Entry	Catalyst	Entry	Catalyst
1 [43]	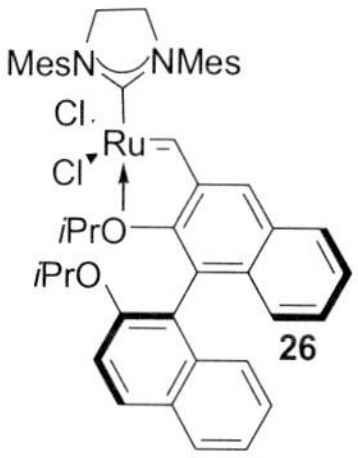 **26**	4 [46]	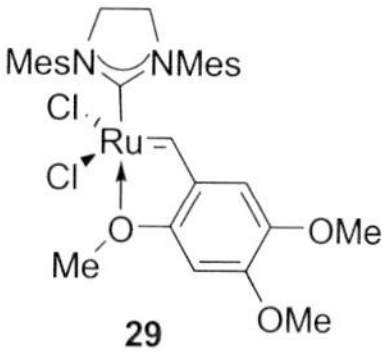 **29**
2 [44]	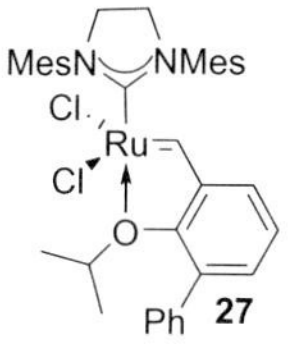 **27**	5 [47]	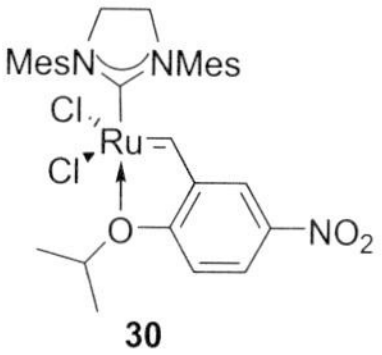 **30**
3 [45]	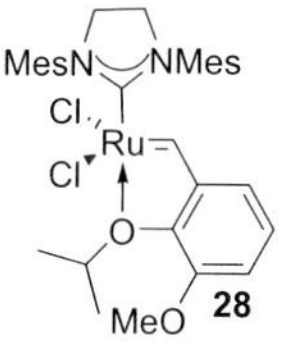 **28**		

synthesized catalyst **30**, which contains the electron-withdrawing group (EWG) NO$_2$ [47]. It is assumed that the NO$_2$ group weakens the iPrO→Ru bond and therefore renders the initiation more facile. This catalyst, which showed enhanced activity, has been used in the total synthesis of (−)-securinine and (+)-viroallosecurinine [50]. This example illustrates the potency of complex **30** as it promotes the tandem enyne-RCM of a dienyne system, enabling the formation of three rings of the core of securinine in excellent yield (Scheme 1.9).

The Grela group then embarked on a program to study the effect of combining an EWG, to decrease the electronic density of the styrene moiety, and steric bulk close to the chelating isopropoxy substituent, in the hope of combining the effects

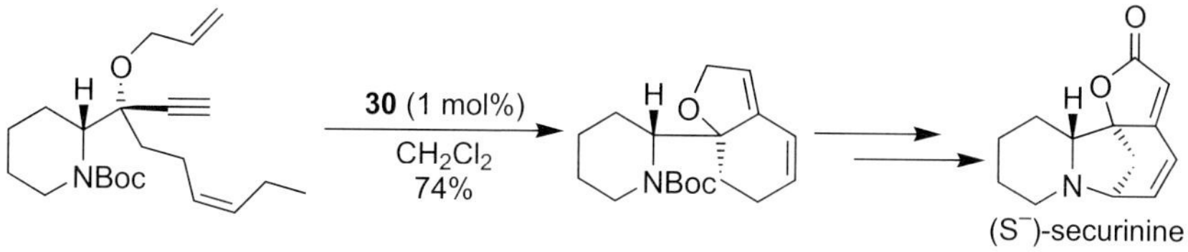

Scheme 1.9 Synthesis of (−)-securinine.

shown in **27** and **30** to increase still further the catalytic activity [51]. Unfortunately, combination of those two modes of activation, steric and electronic, resulted in a significant decrease in stability.

Grubbs and coworkers have prepared phosphine-free catalysts **31** [42] and **32** [52] (Fig. 1.4). Catalyst **31** was initially developed to promote CM with acrylonitrile. This catalyst is easily obtained in good yield from treatment of Grubbs II catalyst **8** with an excess of 3-bromopyridine and has been shown to be a very fast initiator. It initiates at least six orders of magnitude faster than **8**. Presumably, dissociation of the electron-deficient 3-bromopyridine is extremely rapid and the rebinding is slow, which contributes to an excellent turnover. This catalyst was also found to be an excellent promoter of living polymerization, not only with norbornene but also with oxo-norbornene derivatives, which do not undergo living polymerization with other catalysts [53].

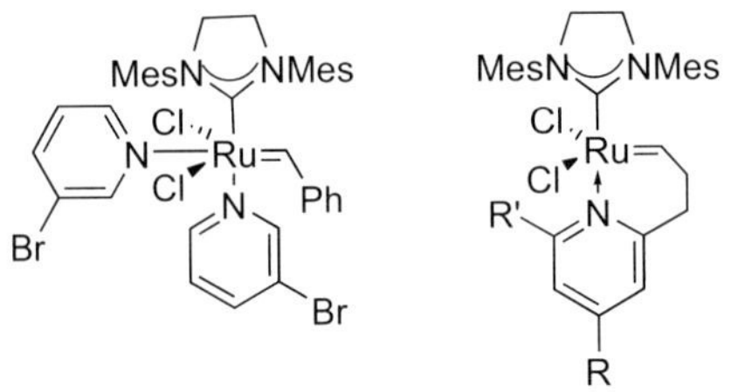

Fig. 1.4 Phosphine-free NHC-Ru complexes.

In contrast, phosphine-free catalyst **32** was a much slower catalyst than **8**.

Recently, Grubbs [54] and Buchmeiser [55] have also prepared NHC-ruthenium complexes **33** and **34** (Fig. 1.5). Catalyst **33** is the first NHC-Ru complex bearing a four-membered cyclic NHC and has been synthesized in moderate yield and showed slow reactivity towards olefin metathesis. Presumably, this arises from the less basic character of the NHC, which makes it a lesser α-donor than the SIMes NHC. Buchmeiser and coworkers disclosed the preparation of catalyst **34** based on tetrahydropyrimidin-2-ylidenes. This catalyst and its analogue with two chlorides have been shown to be very potent catalysts for RCM and ROCM.

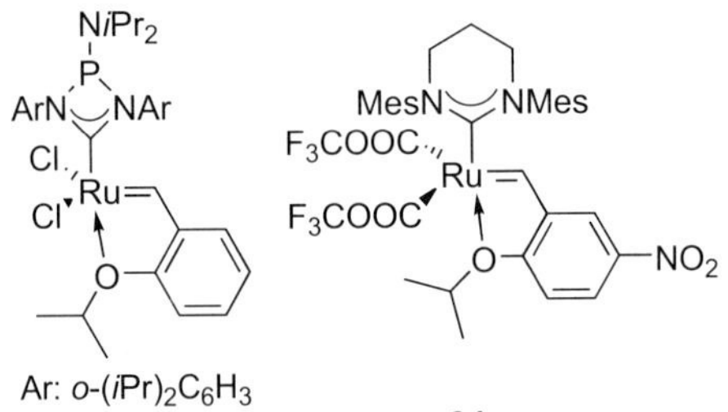

Fig. 1.5 More phosphine-free NHC-Ru complexes.

1.3.4
Variation of the Anionic Ligands

Exchange of the anionic chlorine ligand has been studied at length in the Buchmeiser and Fogg laboratories [56, 57]. The chlorine ions were substituted by strongly electron-withdrawing ligands. Buchmeiser has prepared analogues of catalyst **25** and of Grela's variation **30** with $CF_3(CF_2)_xCOO$ ($x = 0$–2) and showed that these catalysts displayed great activity, enabling the cyclopolymerization of diethyldipropargylmalonate, for which the chloride analogues were inactive. These catalysts also proved to be highly stable.

1.3.5
14-Electron NHC-Ruthenium Complexes

To have a fast initiating catalyst, Piers and coworkers prepared a 14-electron ruthenium catalyst [58]. This compound is isoelectronic with the active species in the olefin metathesis and possesses a vacant coordination site in direct analogy to it. This catalyst (**35**) is prepared from the corresponding Grubbs II catalyst **8**. It is reasonably stable and was found to be extremely potent. The rate for RCM is qualitatively comparable to the best Blechert catalyst (**27**) (Fig. 1.6).

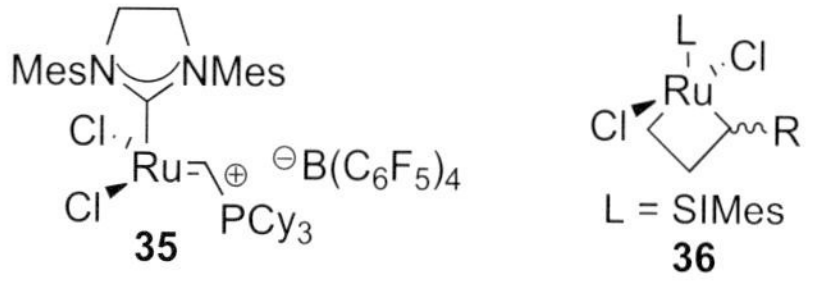

Fig. 1.6 Fourteen-electron NHC-Ru complexes.

The reason for such high reactivity is that there is no need for the ruthenium complex to dissociate a ligand to enter the catalytic cycle and that the initiation binding of the C=C substrate to the ruthenium core is now energetically more favorable. This catalyst has also enabled direct observation of the long postulated 14-electron ruthenocyclobutane metathesis intermediate **36** [59].

1.4
Enantioselective Ruthenium Olefin Catalysts

Reactions requiring high e.e. are still the domain of molybdenum-based Schrock catalysts [60]. However, asymmetric Ru-catalysts are starting to show promising results.

1.4.1
Grubbs II Analogues

The Grubbs group reported an early example of a chiral NHC-ruthenium complex for the enantioselective metathesis reaction [61]. Ruthenium complex **37** was shown to promote desymmetrization of achiral trienes in up to 82% conversion with 90% e.e. in the presence of NaI, which gives the iodine analogue *in situ* (Scheme 1.10). The asymmetric induction is clearly dependant on the degree of substitution of the olefins [62]. The stereocenters of the NHC are too remote from the reaction center to have any noticeable effect. The mesityl groups were replaced by o-substituted aryl groups to afford a steric effect, which was expected to transfer the stereochemistry of the ligand nearer the metal center by placing the aryl group in an arrangement anti to the substituent on the imidazole ring. Crystal studies showed that the NHC ligand was approximately C_2-symmetric. The increased steric bulk generated by replacement of the two chloride atoms with iodine also plays an important role in the enantioselectivity.

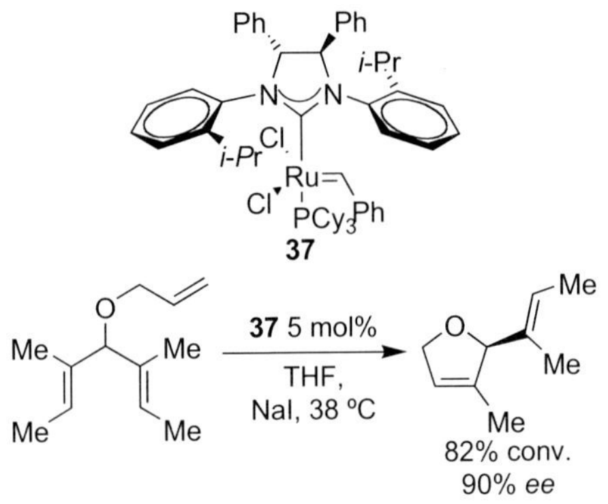

Scheme 1.10 Grubbs chiral NHC-Ru complex.

1.4.2
Phosphine-free Chiral NHC-Ruthenium Complexes

1.4.2.1 First-generation Catalysts
Hoveyda and coworkers have developed a series of chiral catalysts for enantioselective olefin metathesis. These catalysts are particularly efficient for asymmetric ring-opening/cross metathesis reactions. The first generation of such catalysts was prepared in 2002 [63]. The optically pure phosphine-free complex **38**, isolated as a single enantiomer, is air-stable and can be purified by silica-gel chromatography. It was prepared from an unsymmetrical NHC and the triphenylphosphine analogue of the Hoveyda I catalyst (Scheme 1.11). The Hoveyda group chose a bidentate chiral imidazolidene on the hypothesis that such a ligand would induce

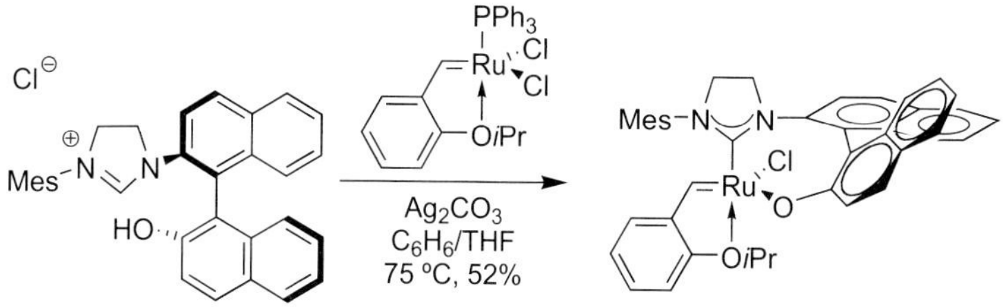

Scheme 1.11 Synthesis of Hoveyda's chiral NHC-Ru complex.

enantioselectivity more efficiently than a monodentate ligand. The fact that one of the chloride ions was substituted for an aryl oxide ligand was not too much of a concern in terms of the effects on activity since there have been some reports of active Ru catalyst bearing a bidentate phenolic base [64]. However, catalyst **38** is less active than its achiral parent **25**.

The combination of replacing a chlorine ion with a less electronegative phenoxide and the increased steric bulk due to the presence of the binaphthyl group is probably responsible for this loss of reactivity. Yet, catalyst **38** showed excellent selectivity in AROM/CM (Scheme 1.12). The catalyst was efficient even when the reaction was performed in air and with non-distilled solvents. The catalyst was recovered in excellent yields and could be reused without significant loss of activity.

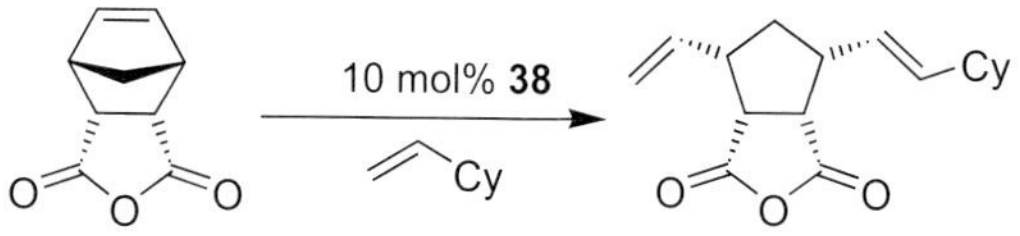

Scheme 1.12 AROM/CM with catalyst **38**.

1.4.2.2 Second-generation Chiral Ru Complexes

To compensate the loss of reactivity resulting from the bulk of the binaphthyl ligand and the replacement of a chlorine ion by an aryl oxide group, Hoveyda's group studied the effect of sterics and electronic alteration on the parent catalyst **38**. Modifications of the benzylidene by installing the EWG NO_2, an electron-donating OMe, or a bulky phenyl group were performed to see if the effect observed by Blechert and Grela on the achiral analogue (Section 2.2.3) was translated to the present class of chiral Ru catalysts (Fig. 1.7) [65]. Enantiomerically pure catalysts **39d** and **39e** were prepared to study the influence on the catalytic activity of reduced electron donation of the aryl oxide oxygen to the Ru core.

Compounds **39c** and **39d** were the more potent catalysts as their reactivity levels are three orders of magnitude higher than catalyst **37**. Catalyst **39d** promotes AROM/CM of **41** in good yield with excellent e.e. (Scheme 1.13). Catalyst **38** leads to less than 10% conversion and chiral Mo catalysts result in rapid polymerization

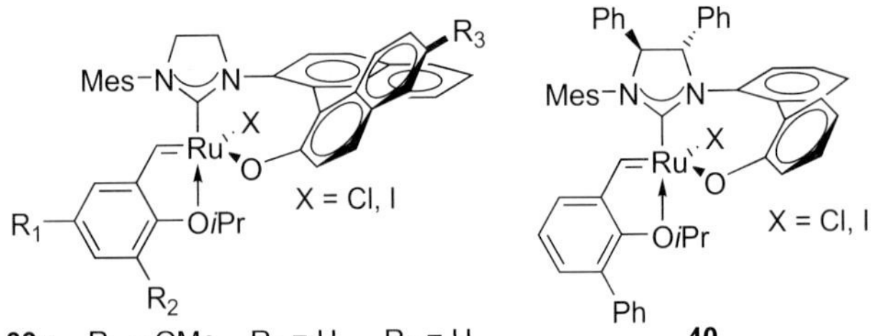

39a	R_1 = OMe	R_2 = H	R_3 = H
39b	R_1 = NO_2	R_2 = H	R_3 = H
39c	R_1 = H	R_2 = Ph	R_3 = H
39d	R_1 = H	R_2 = H	R_3 = CF_3
39e	R_1 = H	R_2 = Ph	R_3 = CF_3

Fig. 1.7 Second-generation chiral Ru complexes.

[66]. Chiral complex **39d** also promoted ARCM in good yield with good e.e. However, notably, chiral Mo-based catalysts are still generally the complexes of choice for such reactions.

The synthesis of these catalysts is lengthy. However, a new, more readily available chiral bidentate NHC-Ru complex has recently been reported by the Hoveyda group [67]. The synthesis is considerably shorter than that of the parent catalyst **39**. The chloride version is not stable on silica but can be prepared *in situ*. The iodine analogue **40**, though, is more stable. Previous studies have shown the effect of substitution of the chloride group by an iodine on the reactivity and the stability of **39** [68]. The iodine analogues are generally less active than their chlorine counterparts; however, they give the product with higher enantioselectivity.

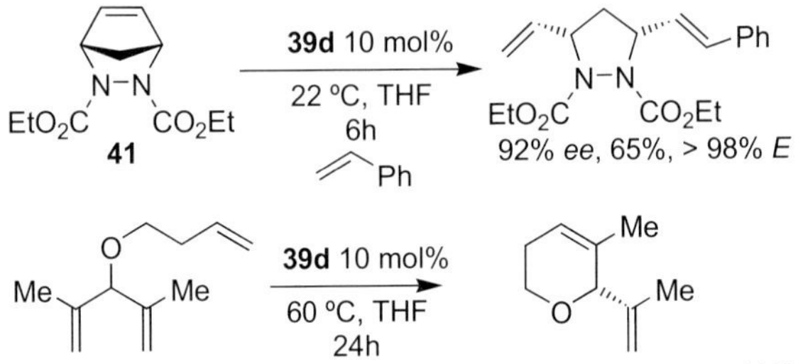

Scheme 1.13 Reactivity of the second-generation chiral Ru complexes.

1.5
Solid Supported NHC-Ru Complexes

There has been an increased demand for supported versions of modern catalysts in recent years and NHC-ruthenium complexes for metathesis are no exception. There are various reasons for this interest: to reduce metal contamination, especially in medicinal chemistry, the possibility of recovering the catalyst to reuse it is

1.5 Solid Supported NHC-Ru Complexes

also very important in terms of cost and, finally, solid supported catalysts offer access to high-throughput chemistry and continuous flow reactors. There are two classes of solid supported catalysts:

- Heterogeneous catalysts, which are covalently attached to an insoluble polymer support; filtration enables the recovery of the catalyst.
- Homogenous catalysts, which are covalently attached to a soluble polymer support. Addition of solvent selectively precipitates the polymer supported catalyst and filtration enables the recovery of the catalyst.

These classes of catalysts can be further classified by the type of attachment:

- attachment to the permanently bound ancillary NHC ligand,
- attachment through the anionic ligand directly to the Ru metal,
- attachment to the alkylidene moiety.

1.5.1
Immobilization via the NHC Ligand

The first such catalyst was reported by Blechert and coworkers (Table 1.4, entry 1). This Merrifield-supported version of the Grubbs II catalyst was successful in RCM and enyne metathesis and was easy to handle. The same group also reported a supported version of the phosphine-free catalyst **25** bearing the same attached NHC [69]. The supported version of the Grubbs II catalyst **42** showed excellent reactivity for RCM but, disappointingly, proved to be a mediocre promoter for

Table 1.4 Immobilization via the NHC ligand.

Entry	Catalyst	Entry	Catalyst
1 [70]	**42**	3 [72]	R = Me, R = Ph **44**
2 [71]	R = adamantyl, R = Mes **43**	4 [73]	**45**

CM. This is probably because the 14-electron active species remained immobilized and suffers diffusion problems. This factor is not important for RCM as only one substrate is involved but had a noticeable effect on CM when two substrates were involved. Buchmeiser's group have worked on the development of non-porous supports suitable for continuous flow experiments (entry 2). These catalysts showed high activity for RCM and ROMP and the cis/trans ratio of the polymer is the same as that obtained with homogeneous systems. These monolithic systems can be used as cartridges for combinatorial chemistry; the products obtained are also virtually ruthenium free, with a ruthenium content of only ≤70 ppm. A silica-based version of **43**, also prepared by the same group, gave good results in RCM.

Fürstner and coworkers have prepared a solid supported version of catalyst **11**, which they previously developed in their laboratory (entry 3). This solid supported catalyst is attached to the silica gel support via one nitrogen of the NHC and exhibits similar reactivity for RCM as the homogeneous non-attached version. A monolith-immobilized version of this catalyst was also prepared by Fürstner and Buchmeiser (entry 4). It showed moderate activity in metathesis reactions.

1.5.2
Attachment Through the Anionic Ligand

The Buchmeiser group has devoted a lot of research to this type of solid supported catalyst. Work was conducted toward monolith- (Table 1.5, entry 1) and silica- (entry 2) supported catalysts. These catalysts displayed high RCM activity, demonstrating high turnover numbers (TON) at elevated temperatures, and also led to a unprecedented low content of ruthenium in the RCM products at <70 ppb. These catalysts also showed high activity for enyne and ring-opening cross metathesis, giving the final products in high yields.

Table 1.5 Attachment through the anionic ligand.

Entry	Catalyst	Entry	Catalyst
1 [74]	AgOOC **46** Mes-N-Ru(Cy₃P)=CHPh Cl Mes	2 [55, 75]	F₃COOC-Ru **47** L = SIMes

1.5.3
Attachment Through the Alkylidene Moiety

The most versatile and most widely used method for attaching catalysts to the solid support has been through the alkylidene moiety. This is due to the ease of functionalization. After one catalytic cycle, the catalyst is detached from the solid support – the active species is then in solution. This is one reason why this type of solid supported catalyst exhibits activities comparable to homogenous catalysts. However, for this technology to be valid and viable as a solid support technology, the catalysts have to reattach to the solid support at the end of the sequence (see Section 1.1 and [7]).

The first such catalysts were prepared by the Nolan group (Table 1.6, entry 1). The catalysts were attached to poly-divinylbenzene (poly-DVB), a macroporous resin polymer. The catalysts have been shown to be recoverable and to display an activity comparable to their homogenous analogues for the RCM of unsubstituted dienes. However, they performed poorly with substrates that have different co-ordination modes competitive with ruthenium recapture by DVB, such as highly hindered substrates. Catalysts **49–52** (entries 2–5), analogous to catalyst **25**, all showed high activity for RCM and CM with 5 mol% loading. They are usually highly recyclable, generally 5–6×, and up to 15× under inert conditions for the Hoveyda version **51**. This particular catalyst has also been shown to display high reactivity for ROM-CM and ring rearrangement metathesis (RRM).

Table 1.6 Attachment through the alkylidene moiety.

Entry	Catalyst	Entry	Catalyst
1 [76]	**48**, L = IMes, SIMes	4 [78]	**51**
2 [69]	**49**	5 [79]	**52**
3 [77]	**50**		

1.5.4
Homogenous Catalysts

Homogenous versions of this type of catalyst have also been prepared. The advantage of homogenous supported catalysts is that their reactivity profile corresponds to their unattached analogues. They usually display higher overall activity than their heterogeneous counterparts. However, much solvent waste is generated due to the techniques used for their recovery. A change of solvent polarity is necessary to precipitate the catalyst. Catalyst **53** (Table 1.7, entry 1) has been shown to be efficient for RCM, ROM-CM and RRM, including the synthesis of tri-substituted double bonds with only 1 mol% of loading. It can be reused up to eight times without loss of activity and gave the final products with very low ruthenium contamination (0.0004% in the first four cycles). Catalyst **54** (entry 2) promotes RCM and EYM and can lead to the formation of tri-substituted double bonds. However, high loadings of catalyst are necessary (10 mol%) and it can be recycled only twice before there is a noticeable loss of activity. Catalyst **55** was shown to be highly recyclable as it could be used 17× in RCM and was particularly active as it can lead to tetra-substituted double bonds. The light fluorous catalyst **56** displayed activities similar to its non-fluorous analogue **25** (entry 4). It can be recovered either by fluorous solid-phase extraction or by filtration when it is initially added on fluorous silica gel. It can be used up to five times without significant loss of activity.

1.5.5
Ionic Liquids

An alternative to solid supported reagents that is environmentally friendly, and with the potential for recyclability, is to use ionic liquids. Recently, Mauduit and Guillemin in Rennes have developed the phosphine-free NHC ruthenium complex **57** (Fig. 1.8) to perform metathesis in a BMI·PF_6/toluene biphasic medium [84]. High reactivity was observed in RCM. The catalyst was reused up to eight times without significant loss of reactivity and the final product had ruthenium contamination as low as 1 ppm.

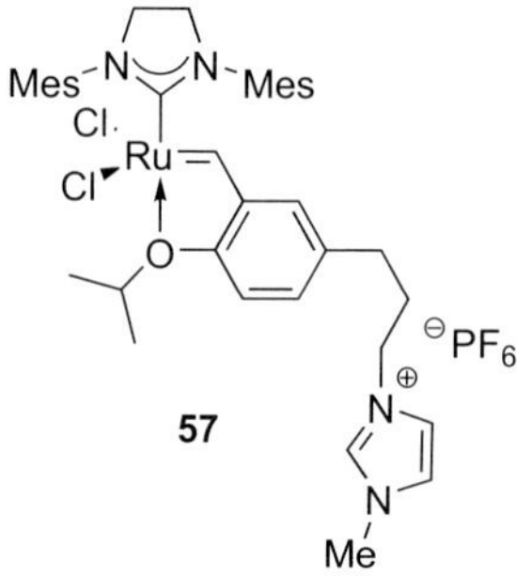

Fig. 1.8 NHC-Ru complex for reactions in ionic liquids.

Table 1.7 Homogenous catalysts.

Entry	Catalyst
1 [80]	**53** (x:y:z = 1:9:30)
2 [81]	**54**
3 [82]	**55**
4 [83]	**56**

1.6
Conclusion and Outlook

In a relatively short period, an impressive amount of research has been done towards the synthesis of potent NHC-ruthenium complexes for olefin metathesis. The advent of such metal catalysts has had a tremendous impact in organic synthesis, enabling transformations not possible before. No single catalyst performs better than all the other catalysts in all possible reactions. Phosphine-free catalysts are generally the catalysts of choice for CM whereas the phosphine analogues are more potent in the formation of tetra-substituted double bonds. We can be sure that further improvements will arise in the near future with perhaps the introduction of novel NHCs.

References

1 For reviews on catalytic olefin metathesis, see: (a) Grubbs, R. H., Miller, S. J., Fu, G. C. *Acc. Chem. Res.* **1995**, *28*, 446. (b) Schmalz, H.-G. *Angew. Chem. Int. Ed.* **1995**, *34*, 1833. (c) Schuster, M., Blechert, S. *Angew. Chem. Int. Ed.* **1997**, *36*, 2036. (d) Fürstner, A. *Top. Catal.* **1997**, *4*, 285. (e) Armstrong, S. K. *J. Chem. Soc., Perkin Trans. 1*, **1998**, 371. (f) Grubbs, R. H., Chang, S. *Tetrahedron* **1998**, *54*, 4413. (g) Fürstner, A. *Angew. Chem. Int. Ed.* **2000**, *39*, 3012. (h) Trnka, T. M., Grubbs, R. H. *Acc. Chem. Res.* **2001**, *34*, 18. (i) *Handbook of Olefin Metathesis*, ed. Grubbs, R. H., VCH-Wiley, Weinheim, **2003**. (j) Schrock, R. R., Hoveyda, A. H. *Angew. Chem. Int. Ed.* **2003**, *42*, 4592. (k) Grubbs, R. H. *Tetrahedron* **2004**, *60*, 7117.

2 Schrock, R. R., Murdzek, J. S., Bazan, G. C., Robbins, J., DiMare, M., O'Regan, M. *J. Am. Chem. Soc.* **1990**, *112*, 3875.

3 Tsang, W. C. P., Hultzsch, K. C., Alexander, J. B., Bonitatebus, P. J., Schrock, R. R., Hoveyda, A. H. *J. Am. Chem. Soc.* **2003**, *125*, 2652.

4 (a) Schwab, P., France, M. B., Ziller, J. W., Grubbs, R. H. *Angew. Chem. Int. Ed.* **1995**, *34*, 2039. (b) Schwab, P., Grubbs, R. H., Ziller, J. W. *J. Am. Chem. Soc.* **1996**, *118*, 100.

5 Collman, J. P., Hegedus, L. S., Norton, J. R., Finke, R. G. *Principles and Applications of Organotransition Metal Chemistry*, 2nd edn., University Science, Mill Valley, CA, **1987**.

6 Kingsbury, J. S., Harrity, J. P. A., Bonitatebus, P. J., Hoveyda, A. H. *J. Am. Chem. Soc.* **1999**, *121*, 791.

7 Kingsbury, J. S., Hoveyda, A. H. *J. Am. Chem. Soc.* **2005**, *127*, 4510.

8 (a) Adlart, C., Hinderling, C., Baumann, H., Chen, P. *J. Am. Chem. Soc.* **2000**, *122*, 8204. (b) Cavallo, L. *J. Am. Chem. Soc.* **2002**, *124*, 8965.

9 (a) Dias, E. L., Nguyen, S. T., Grubbs, R. H. *J. Am. Chem. Soc.* **1997**, *119*, 3887. (b) Sanford, M. S., Love, J. A., Grubbs, R. H. *J. Am. Chem. Soc.* **2001**, *123*, 6543.

10 (a) Huang, J., Stevens, E. D., Nolan, S. P., Peterson, J. L. *J. Am. Chem. Soc.* **1999**, *121*, 2674. (b) Huang, J., Schanz, H.-J., Stevens, E. D., Nolan, S. P. *Organometallics* **1999**, *18*, 5375.

11 Weskamp, T., Schattenmann, W. C., Spiegler, M., Herrmann, W. A. *Angew. Chem. Int. Ed.* **1998**, *37*, 2490.

12 (a) Scholl, M., Trnka, T. M., Morgan, J. P., Grubbs, R. H. *Tetrahedron Lett.* **1999**, *40*, 2247. (b) Scholl, M., Ding, S., Lee, C. W., Grubbs, R. H. *Org. Lett.* **1999**, *1*, 953. (c) Chatterjee, A. K., Morgan, J. P., Scholl, M., Grubbs, R. H. *J. Am. Chem. Soc.* **2000**, *122*, 3783.

13 (a) Ackermann, L., Fürstner, A., Weskamp, T., Kohl, F. J., Herrmann, W. A. *Tetrahedron Lett.* **1999**, *40*, 4787. (b) Weskamp, T., Kohl, F. J., Hieringer, W., Gleich, D., Herrmann, W. A. *Angew. Chem. Int. Ed.* **1999**, *38*, 2416. (c) Weskamp, T., Kohl, F. J., Herrmann, W. A. *J. Organomet. Chem.* **1999**, *582*, 362.

14 Fürstner, A., Radkowski, K., Wirtz, C., Goddard, R., Lehmann, C. W., Mynott, R. *J. Am. Chem. Soc.* **2002**, *124*, 7061.

15 For an improved synthesis of second-generation ruthenium catalysts, see: Jafarpour, L., Hillier, A. C., Nolan, S. P. *Organometallics* **2002**, *21*, 442.

16 Hillier, A. C., Sommer, W. J., Yong, B. S., Petersen, J. L., Cavallo, L., Nolan, S. P. *Organometallics* **2003**, *22*, 4322.

17 Sun, J., Sinha, S. C. *Angew. Chem. Int. Ed.* **2002**, *41*, 1381.

18 Meng, D., Bertinato, P., Balog, A., Su, D.-S., Kamenecka, T., Sorensen, E. J., Danishefsky, S. J. *J. Am. Chem. Soc.* **1997**, *119*, 10073.

19 For a review on olefin cross metathesis, see: Connon, S. J., Blechert, S. *Angew. Chem. Int. Ed.* **2003**, *42*, 1900.

20 Chatterjee, A. K., Grubbs, R. H. *Org. Lett.* **1999**, *1*, 1751.

21 Spessard, S. J., Stoltz, B. M. *Org. Lett.* **2002**, *4*, 1943.

22 Love, J. A., Sanford, M. S., Day, M. W., Grubbs, R. H. *J. Am. Chem. Soc.* **2003**, *125*, 10103.

23 Dinger, M. B., Mol, J. C. *Adv. Synth. Catal.* **2002**, *344*, 671.

24 Jafarpour, L., Nolan, S. P. *J. Organomet. Chem.* **2001**, *617–618*, 17.

25 Jafarpour, L., Stevens, E. D., Nolan, S. P. *J. Organomet. Chem.* **2000**, *606*, 49.

26 Fürstner, A., Ackermann, L., Gabor, B., Goddard, R., Lehmann, C. W., Mynott, R., Stelzer, F., Thiel, O. R. *Chem. Eur. J.* **2001**, *7*, 3236.

27 Fürstner, A., Krause, H., Ackermann, L., Lehmann, C. W. *Chem. Commun.* **2001**, 2240.

28 Enders, D., Breuer, K., Raabe, G., Runsink, J., Teles, J. H., Melder, J.-P., Ebel, K., Brode, S. *Angew. Chem. Int. Ed.* **1995**, *34*, 1021.

29 Fürstner, A., Ackermann, L., Gabor, B., Goddard, R., Lehmann, C. W., Mynott, R., Stelzer, F., Thiel, O. R. *Chem. Eur. J.* **2001**, *7*, 3236.

30 Trnka, T. M., Morgan, J. P., Sanford, M. S., Wilhelm, T. E., Scholl, M., Choi, T.-L., Ding, S., Day, M. W., Grubbs, R. H. *J. Am. Chem. Soc.* **2003**, *125*, 2546.

31 Dinger, M. B., Nieczypor, P., Mol, J. C. *Organometallics* **2003**, *22*, 5291.

32 Yun, J., Marinez, E. R., Grubbs, R. H. *Organometallics* **2004**, *23*, 4172.

33 Louie, J., Grubbs, R. H. *Organometallics* **2002**, *21*, 2153.

34 Opstal, T., Verpoort, F. *J. Mol. Cat. A: Chem.* **2003**, *200*, 49.

35 Lehman, Jr., S. E., Wagener, K. B. *Organometallics* **2005**, *24*, 1477.

36 Garber, S. B., Kingsbury, J. S., Gray, B. L., Hoveyda, A. H. *J. Am. Chem. Soc.* **2000**, *122*, 8168.

37 Gessler, S., Randl, S., Blechert, S. *Tetrahedron Lett.* **2000**, *41*, 9973.

38 Randl, S., Gessler, S., Wakamatsu, H., Blechert, S. *Synlett.* **2001**, 430.

39 Imhof, S., Randl, S., Blechert, S. *Chem. Commun.* **2001**, 1692.

40 For activation of the Grubbs II catalyst towards acrylonitrile by addition of CuCl, see: Rivard, M., Blechert, S. *Eur. J. Org. Chem.* **2003**, 2225.

41 Cesati III, R. R., de Armas, J., Hoveyda, A. H. *J. Am. Chem. Soc.* **2004**, *126*, 96.

42 Love, J. A., Morgan, J. P., Trnka, T. M., Grubbs, R. H. *Angew. Chem. Int. Ed.* **2002**, *41*, 4035.

43 Wakamatsu, H., Blechert, S. *Angew. Chem. Int. Ed.* **2002**, *41*, 794.

44 Wakamatsu, H., Blechert, S. *Angew. Chem. Int. Ed.* **2002**, *41*, 2403.

45 Buschmann, N., Wakamatsu, H., Blechert, S. *Synlett.* **2004**, 667.

46 Grela, K., Kim, M. *Eur. J. Org. Chem.* **2003**, 963.

47 Grela, K., Harutyunyan, S., Michrowska, A. *Angew. Chem. Int. Ed.* **2002**, *41*, 4038.

48 (a) Zaja, M., Connon, S. J., Dunne, A. M., Rivard, M., Buschmann, N., Jiricek, J., Blechert, S. *Tetrahedron* **2003**, *59*, 6545. (b) Connon, S. J., Rivard, M., Zaja, M., Blechert, S. *Adv. Synth. Catal.* **2003**, *345*, 572.

49 Krause, J. O., Zarka, M. T., Anders, U., Weberskirch, R., Nuyken, O., Buchmeiser, M. R. *Angew. Chem. Int. Ed.* **2003**, *42*, 5965.

50 (a) Honda, T., Namiki, H., Kaneda, K., Mizutani, H. *Org. Lett.* **2004**, *6*, 87. (b) Honda, T., Namiki, H., Watanabe, M., Mizutani, H. *Tetrahedron Lett.* **2004**, *45*, 5211.

51 Michrowska, A., Bujok, R., Harutyunyan, S., Sashuk, V., Dolgonos, G., Grela, K. *J. Am. Chem. Soc.* **2004**, *126*, 9318.

52 Ung, T., Hejl, A., Grubbs, R. H., Schrodi, Y. *Organometallics* **2004**, *23*, 5399.

53 Choi, T.-L., Grubbs, R. H. *Angew. Chem. Int. Ed.* **2003**, *42*, 1743.

54 Despagnet-Ayoub, E., Grubbs, R. H. *Organometallics* **2005**, *24*, 338.

55 Yang, L., Mayr, M., Wurst, K., Buchmeiser, M. R. *Chem. Eur. J.* **2004**, *10*, 5761.

56 Krause, J. O., Nuyken, O., Buchmeiser, M. R. *Chem. Eur. J.* **2004**, *10*, 2029.

57 Conrad, J. C., Amoroso, D., Czechura, P., Yap, G. P. A., Fogg, D. E. *Organometallics* **2003**, *22*, 3634.

58 Romero, P. E., Piers, W. E., McDonald, R. *Angew. Chem. Int. Ed.* **2004**, *43*, 6161.

59 Romero, P. E., Piers, W. E. *J. Am. Chem. Soc.* **2005**, *127*, 5032.

60 (a) Dolman, S. J., Schrock, R. R., Hoveyda, A. H. *Org. Lett.* **2003**, *5*, 4899. (b) Dolman, S. J., Sattely, E. S., Hoveyda, A. H., Schrock, R. R. *J. Am. Chem. Soc.* **2002**, *124*, 6991. (c) Tsang, W. C. P., Jernelius, J. A., Cortez, G. A., Weatherhead, G. S., Schrock, R. R., Hoveyda, A. H. *J. Am. Chem. Soc.* **2003**, *125*, 2591.

61 Seiders, T. J., Ward, D. W., Grubbs, R. H. *Org. Lett.* **2001**, *3*, 3225.

62 Costabile, C., Cavallo, L. *J. Am. Chem. Soc.* **2004**, *126*, 9592.

63 Van Veldhuizen, J. J., Garber, S. B., Kingsbury, J. S., Hoveyda, A. H. *J. Am. Chem. Soc.* **2002**, *124*, 4954.

64 Chang, S., Jones, L., Wang, C., Henling, L. M., Grubbs, R. H. *Organometallics* **1998**, *17*, 3460.

65 Van Veldhuizen, J. J., Gillingham, D. G., Garber, S. B., Kataoka, O., Hoveyda, A. H. *J. Am. Chem. Soc.* **2003**, *125*, 12502.

66 (a) Weatherhead, G. S., Ford, J. G., Alexanian, E. J., Schrock, R. R., Hoveyda, A. H. *J. Am. Chem. Soc.* **2000**, *122*, 1828. (b) La, D. S., Sattely, E. S., Ford, J. G., Schrock, R. R., Hoveyda, A. H. *J. Am. Chem. Soc.* **2001**, *123*, 7767. (c) Tsang, W. C. P., Jernelius, J. A., Cortez, G. A., Weatherhead, G. S., Schrock, R. R., Hoveyda, A. H. *J. Am. Chem. Soc.* **2003**, *125*, 2591.

67 Van Veldhuizen, J. J., Campbell, J. E., Giudici, R. E., Hoveyda, A. H. *J. Am. Chem. Soc.* **2005**, *127*, 6877.

68 Gillingham, D. G., Kataoka, O., Garber, S. B., Hoveyda, A. H. *J. Am. Chem. Soc.* **2004**, *126*, 12288.

69 Randl, S., Buschmann, N., Connon, S. J., Blechert, S. *Synlett.* **2001**, *10*, 1547.

70 Schürer, S. C., Gessler, S., Buschmann, N., Blechert, S. *Angew. Chem. Int. Ed.* **2000**, *39*, 3898.

71 (a) Mayr, M., Mayr, B., Buchmeiser, M. R. *Angew. Chem. Int. Ed.* **2001**, *40*, 3839. (b) Buchmeiser, M. R. *J. Mol. Cat. A: Chem.* **2002**, *190*, 145. (c) Mayr, M., Buchmeiser, M. R., Wurst, K. *Adv. Synth. Catal.* **2002**, *344*, 712.

72 Prühs, S., Lehmann, C. W., Fürstner, A. *Organometallics* **2004**, *23*, 280.

73 Mayr, M., Wang, D., Kröll, R., Schuler, N., Prühs, S., Fürstner, A., Buchmeiser, M. R. *Adv. Synth. Catal.* **2005**, *347*, 484.

74 (a) Krause, J. O., Lubbad, S., Nuyken, O., Buchmeiser, M. R. *Adv. Synth. Catal.* **2003**, *345*, 996. (b) Krause, J. O., Nuyken, O., Wurst, K., Buchmeiser, M. R. *Chem. Eur. J.* **2004**, *10*, 777.

75 (a) Krause, J. O., Nuyken, O., Wurst, K., Buchmeiser, M. R. *Chem. Eur. J.* **2004**, *10*, 777. (b) Halbach, T. S., Mix, S., Fischer, D., Maechling, S., Krause, J. O., Sievers, C., Blechert, S., Nuyken, O., Buchmeiser, M. R. *J. Org. Chem.* **2005**, *70*, 4687.

76 (a) Jafarpour, L., Nolan, S. P. *Org. Lett.* **2000**, *2*, 4075. (b) Jafarpour, L., Heck, C. Baylon, M.-P., Lee, A. M., Mioskowski, C., Nolan, S. P. *Organometallics* **2002**, *21*, 671.

77 Grela, K., Tryzmowski, M., Bieniek, M. *Tetrahedron Lett.* **2002**, *43*, 9055.

78 Kingsbury, J. S., Garber, S. B., Giftos, J. M., Gray, B. L., Okamoto, M. M., Farrer, R. A., Fourkas, J. T., Hoveyda, A. H. *Angew. Chem. Int. Ed.* **2001**, *40*, 4251.

79 Connon, S. J., Blechert, S. *Bioorg. Med. Chem. Lett.* **2002**, *12*, 1873.

80 Connon, S. J., Dunne, A. M., Blechert, S. *Angew. Chem. Int. Ed.* **2002**, *41*, 3835.

81 Varray, S., Lazaro, R., Martinez, J., Lamaty, F. *Organometallics* **2003**, *22*, 2426.

82 Yao, Q., Rodriguez, M. A. *Tetrahedron Lett.* **2004**, *45*, 2447.

83 Matsugi, M., Curran, D. P. *J. Org. Chem.* **2005**, *70*, 1636.

84 Clavier, H., Audic, N., Mauduit, M., Guillemin, J.-C. *Chem. Commun.* **2004**, 2282.

2
Ruthenium N-Heterocyclic Carbene Complexes in Organic Transformations (Excluding Metathesis)

Suzanne Burling, Belinda M. Paine, and Michael K. Whittlesey

2.1
Introduction

The development of N-heterocyclic carbene (NHC) based second-generation Grubbs metathesis catalysts has been at the forefront of developments in ruthenium NHC chemistry [1]. As a result of this emphasis, the use of Ru NHC complexes in other catalytic reactions is still in its infancy. The present chapter outlines the developments so far involving single-step reactions (hydrogenation, hydrosilylation, isomerization etc.) and also the progress in utilizing more than one type of reaction for tandem processes.

2.2
Hydrogenation and Hydrosilylation Reactions

Based on the success associated with ruthenium phosphine complexes in both direct and transfer hydrogenation chemistry, it is perhaps unsurprising that several research groups have utilized ruthenium NHC complexes for this type of reactivity.

The tricyclohexylphosphine complex [Ru(PCy$_3$)$_2$(CO)HCl] (**1**) proved to be more active than the mono-NHC species, [Ru(IMes)(PCy$_3$)(CO)HCl] (**2**), for the hydrogenation of hex-1-ene at ambient temperature (0.1 mol% catalyst, 4 atm H$_2$), although at higher temperature (100 °C) the two systems become comparable (TOF: **1**, 21 500 h^{-1}; **2**, 24 000 h^{-1}). In all cases, only trace amounts of alkene isomerization products were detected [2]. The activity of **2** can also be enhanced by the presence of 1–2 equivalents of HBF$_4$ (which favors formation of the reactive 14-electron species [Ru(IMes)(CO)HCl] through loss of [HPCy$_3$][BF$_4$]), resulting in improvements in turnover rates of between ca. two- and seven-fold for the hydrogenation of cyclooctene, hex-1-ene and allylbenzene (0.1 mol% cat, 1 atm H$_2$, 25 °C). The comparable activity of **1** and **2** in the hydrogenation of hexene is

N-Heterocyclic Carbenes in Synthesis. Edited by Steven P. Nolan
Copyright © 2006 WILEY-VCH Verlag GmbH & Co. KGaA, Weinheim
ISBN: 3-527-31400-8

1 R = H
5 R = Ph

2

3

4 R = Ph, R' = H
6 R = Cy, R' = Ph

perhaps somewhat surprising given that the more donating NHC ligand might be expected to favor alkene binding and H_2 oxidative addition [3].

Subsequent work by Fogg and Nolan has suggested that any enhancement that may actually arise due to the presence of an NHC ligand is offset by the poor lability of the PCy_3 group; indeed, they have shown that changing to the more labile PPh_3 yields compounds that will hydrogenate sterically encumbered alkenes. These compounds also exhibit competitive reactivity for the isomerization of terminal alkenes and polymerization of strained cyclic alkenes [4]. Thus, [Ru(NHC)(PPh$_3$)(CO)HCl] (NHC = IMes, **3**; NHC = SIMes, **4**) is able to reduce both cis- and trans-cyclododecene, whereas **2** (and indeed **1**) are only active for the less sterically hindered cis-isomer. Table 2.1 summarizes the isomerization activity

Table 2.1 Hydrogenation and isomerization activity of **2–4** with allylbenzene.[a]

Catalyst	$p(H_2)$ (atm)	PhCHMe$_2$ (%)	PhCHCHMe (% cis)	PhCHCHMe (% trans)	TOF (h^{-1})[b]
2	3.3	51	1	6	2040
	9.3	94	0	6	3760
3	3.3	60	0	40	2400
	9.3	89	0	11	3560
4	9.3	49	10	37	1960

a) Conditions: 0.05 mol% [Ru], 2 mmol allylbenzene, 80 °C, 30 min.
b) These values refer to the hydrogenation step.

2.2 Hydrogenation and Hydrosilylation Reactions

of **2–4** with allylbenzene; **3** affords 40% of *trans*-propenylbenzene in a short time even in the presence of 3 atm H_2.

Dinger and Mol have compared the activity of **1**, [Ru(PCy$_3$)$_2$(CO)(Ph)Cl] (**5**) and [Ru(SIMes)(PCy$_3$)(CO)(Ph)Cl] (**6**) in the hydrogenation of oct-1-ene [5]. Comparable data was found to that outlined above. Thus, **6** shows no activity under ambient conditions, but can be activated upon heating to 100 °C to yield comparable activity to that seen with **1**. However, at low catalyst loading (1:350 000 **6**:oct-1-ene), **6** showed poorer selectivity for conversion into octane than either **1** or **5**. A later report on the bulky saturated isopropyl NHC complex [Ru(SIPr)(PCy$_3$)(CO)HCl] (**7**) (formed by the decomposition of [Ru(SIPr)(PCy$_3$)(=CHPh)Cl$_2$] in the presence of primary alcohols and base) [6] showed that hydrogenation activity is compromised by the ability to isomerize alkenes.

The stoichiometric hydrogenation of alkenes has been reported for the 18-electron dihydride NHC complex [Ru(IMes)(PPh$_3$)$_2$(CO)H$_2$] (**8**) [7]. Addition of one equivalent of H$_2$C=CHSiMe$_3$ leads to quantitative formation of trimethylethylsilane and **9**, which results from C–H bond activation of an ortho-CH$_3$ group of the IMes ligand. Reaction of **9** with H$_2$ (1 atm) at room temperature or ^{i}PrOH at 50 °C resulted in the reformation of **8**, suggesting that the reversibility of the C–H activation chemistry allows **8** to be used for catalytic hydrogenation reactions [8]. Thus, H$_2$C=CHSiMe$_3$ is hydrogenated by ^{i}PrOH in the presence of 2 mol% **8** to give CH$_3$CH$_2$SiMe$_3$ in quantitative yield after 10 h at 70 °C; this reactivity has been further utilized for tandem dehydrogenation/hydrogenation reactions (Section 2.5.3).

The catalytic hydrogenation of C=C, as well as C=O bonds, can be conducted in aqueous solution with the water-soluble half-sandwich complexes **10** and **11**. Thus, acetone, acetophenone, cinnamaldehyde and benzylideneacetone (the latter two undergoing reduction of the C=C bond) react with both precursors under 10 atm H_2 at 80 °C, with significantly higher activity being found for the cationic compound **11**. With allyl alcohol as the substrate, both propanal and propanol are formed, although rate data suggest that the latter is formed not only by reduction of propanal but by a second, direct, isomerization pathway acting in concert [9].

Several bidentate as well as monodentate carbene ligands have been utilized on ruthenium to bring about transfer hydrogenation of ketones and imines. The pincer system **12**, based on 2,6-bis(butylimidazol-2-ylidene)pyridine, catalyzes the reduction of aryl and alkyl ketones by ^{i}PrOH/KOH at elevated temperature to afford the corresponding alcohols in quantitative yields. An important consideration in the catalysis is that the stability of **12** allows reactions to be carried out without special exclusion of air and moisture [10]. The related aryl pincer carbene complex **13** displays lower turnover numbers for the transfer hydrogenation of ketones than **12**, although lower temperatures, but higher catalyst loadings, have been employed [11]. Table 2.2 summarizes the activities of **12** and **13**. Most recently, the pincer carbene based cationic dimer **14** has been utilized for the indirect hydrogenation of cyclohexanone, acetophenone or benzophenone with ^{i}PrOH in the presence of KO^{t}Bu or KOH [12].

Table 2.2 Transfer hydrogenation of ketones by **12** and **13**.[a, b]

Substrate	Catalyst/ conditions	Loading (mol%)	Temperature (°C)	Time (h)	TON
Cyclohexanone	**12**[a]	0.07	80	3	8800
Cyclohexanone	**12**[a]	0.07	80	20	12 600
Cyclohexanone	**12**[a]	0.0007	80	20	126 000
Cyclohexanone	**13**[b]	0.01	55	20	8800
Acetophenone	**12**[a]	0.07	80	6	700
Acetophenone	**13**[b]	0.015	80	12	4000
Benzophenone	**12**[a]	0.07	80	3	1400

a) Conditions: 2 mmol substrate, 10 mL 0.1 M KOH in ^{i}PrOH.
b) Conditions: 50 mmol substrate, 4.5 mmol KO^{t}Bu in ^{i}PrOH.

2.2 Hydrogenation and Hydrosilylation Reactions | 31

To remove the need for base, the ruthenium hydride complex [Ru(IMes)-(PPh$_3$)$_2$(CO)H$_2$] (**8**) has been employed in the transfer hydrogenation of ketones with ^{i}PrOH at 50 °C with 2 mol% catalyst precursor [13]. Because **8** also oxidizes alcohols to the corresponding ketones in the presence of acetone as the hydrogen acceptor, equilibria determined by the relative oxidation potentials of the alcohol and ketone products dictate the overall percentage conversions (Fig. 2.1).

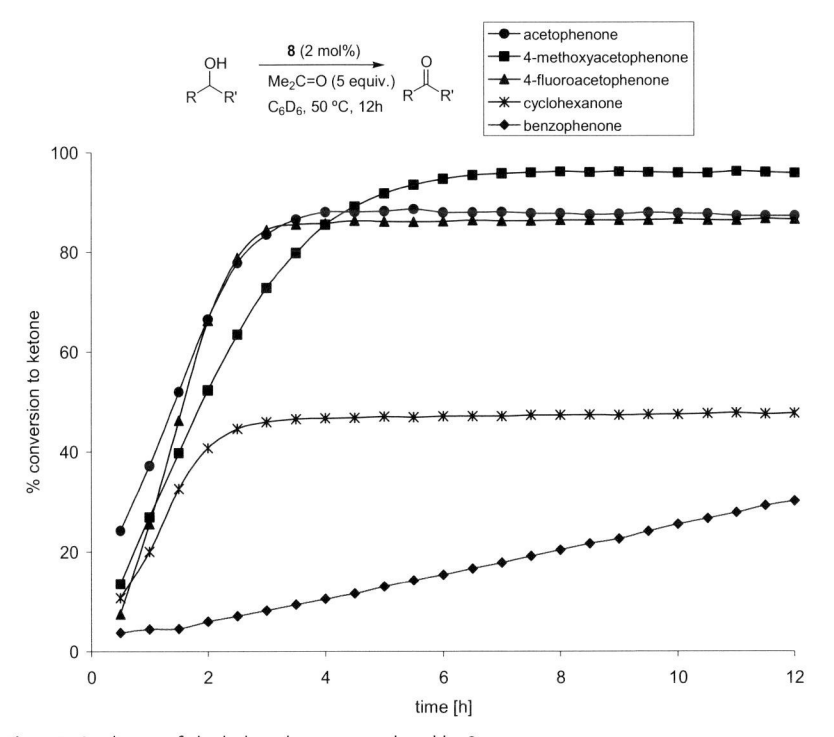

Fig. 2.1 Oxidation of alcohols to ketones catalyzed by **8**.

Quantitative conversion of the same range of ketones can be brought about through direct hydrogenation under 5 atm H$_2$ at 70 °C for 24 h with only 0.4 mol% catalyst loading. A preliminary indication that this reactivity can be extended from ketones to imines follows in that N-benzylideneaniline undergoes both transfer (2 mol% **8**, 5 equivalents ^{i}PrOH, 70 °C, 16 h) and direct (0.4 mol% **8**, 5 atm H$_2$, 70 °C, 24 h) hydrogenation through to the amine (Scheme 2.1).

Scheme 2.1 N-Benzylideneaniline hydrogenation by [Ru(IMes)(PPh$_3$)$_2$(CO)H$_2$] (**8**).

There have been two recent reports detailing hydrosilylation by ruthenium NHC complexes, in one case involving addition across the C≡C bond of terminal alkynes and secondly reporting the asymmetric addition of the Si–H bond to ketones. Thus, the Ru NHC alkylidene complexes [Ru(SIMes)(PCy$_3$)(=CHPh)Cl$_2$] (**15**) and [Ru(IMes)(PCy$_3$)(=CHPh)Cl$_2$] (**16**), as well as the 3-phenylindenylid-1-ene analogue **17**, afford (Z)-PhCH=CH(SiEt$_3$) with moderate conversion from PhC≡CH and Et$_3$SiH at elevated temperature (Scheme 2.2). In all cases, the (Z)-alkene **18** is formed in preference to the (a)-isomer **19**, while minor amounts of the alkyne dimer PhC≡CCH=CHPh (**20**) are also formed [14].

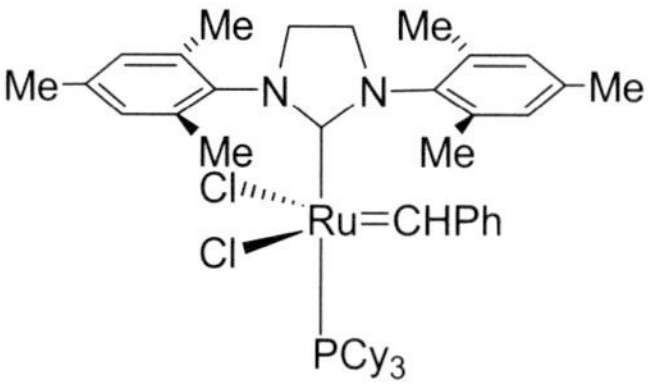

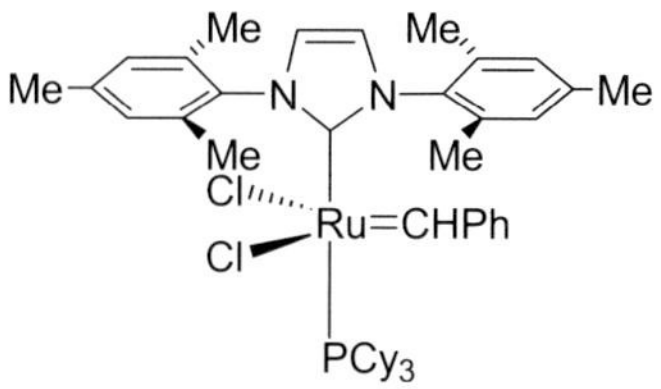

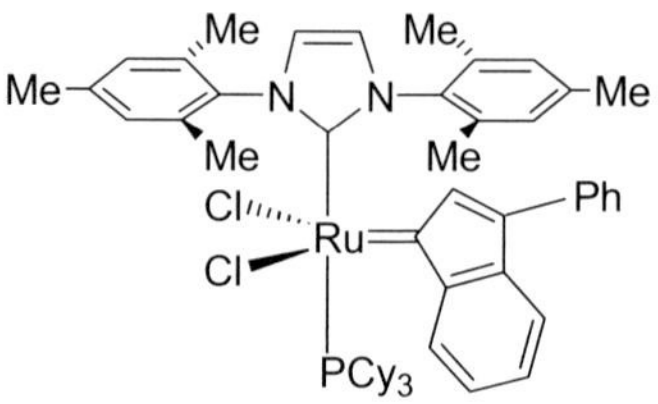

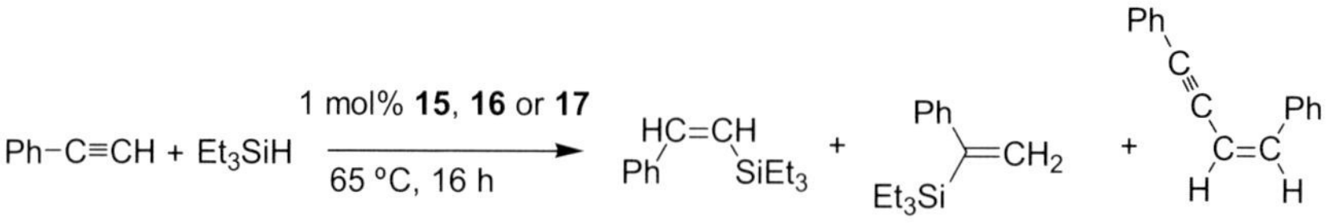

Scheme 2.2 Hydrosilylation of phenylethyne by **15–17**.

A combination of [Ru(PPh$_3$)$_3$Cl$_2$] and the bis-paracyclophane-derived imidazolium salts **21–23**, along with added AgOTf, gives an *in situ* prepared catalyst system that is effective for the addition of Ph$_2$SiH$_2$ across the C=O bond of aryl methyl ketones, as well as more hindered substrates (Scheme 2.3). Following acid work-up, the corresponding alcohols are isolated in good yields with high enantioselectivity. A range of representative results are given in Table 2.3, which suggests that the optimal ratio of Ru:NHC ligand in the hydrosilylation of acetophenone is 1:2.4 [15].

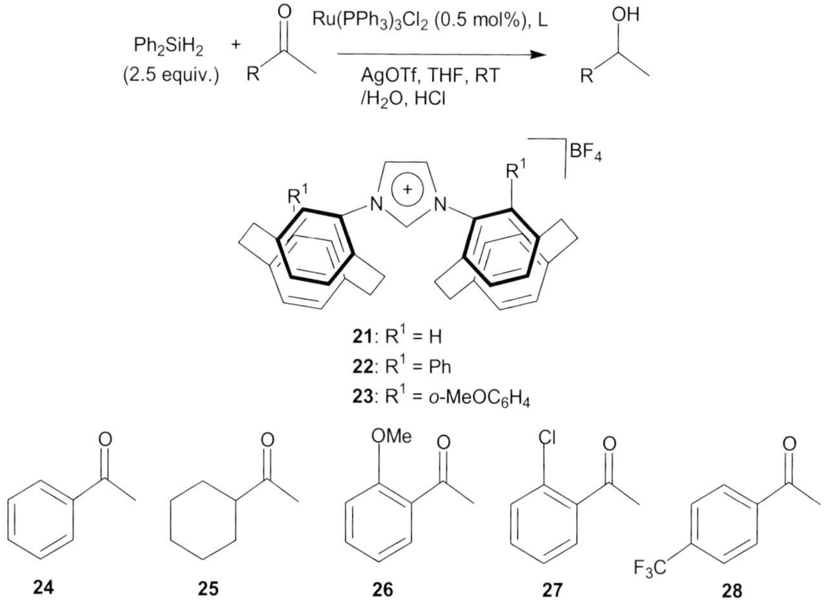

2.2 Hydrogenation and Hydrosilylation Reactions

Scheme 2.3 Ketone hydrosilylation with Ru paracyclophane NHC complexes.

Table 2.3 Yields and e.e.s of product alcohols following hydrosilylation of ketones by [Ru(PPh$_3$)$_3$Cl$_2$]: L (L = **21–23**) and Ph$_2$SiH$_2$.

Substrate	L	Ru:L	Time (h)	Yield (%)	e.e. (%)
24	21	1:2.4	16	98	90
24	22	1:2.4	16	98	97
24	23	1:2.4	16	98	97
24	23	1:1	16	96	83
25	23	1:2.4	36	93	58
26	23	1:2.4	15	92	96
27	23	1:2.4	20	90	97
28	23	1:2.4	48	81	77

2.3
Isomerization

Dinger and Mol [5] have reported the preparation of the hydride complexes [Ru-(SIMes)(PCy$_3$)(CO)HCl] (**29**), [Ru(PCy$_3$)$_2$(CO)HCl] (**1**) and [Ru(SIMes)(PCy$_3$)-(CO)(Ph)Cl] (**6**) upon reaction of **15** with MeOH, EtOH, ^{n}PrOH, C$_9$H$_{19}$OH and benzyl alcohol in the presence of base (**1** is also formed by an analogous reaction from the bis-phosphine alkylidene compound [Ru(PCy$_3$)$_2$(=CHPh)Cl$_2$] [16]). The phenyl complex **6** was shown to be active for the isomerization of oct-1-ene to oct-2-ene at 100 °C (**29** could not be isolated in sufficient purity to test for comparison) and to compare favorably with the activity found for **1**; notably, [Ru(PCy$_3$)$_2$-(CO)(Ph)Cl] **5** is a poor catalyst, implying that the detrimental effect of exchanging hydride for a phenyl group is offset in **6** by replacing PCy$_3$ for the SIMes ligand. An analogue of **29**, the isopropyl system [Ru(SIPr)(PCy$_3$)(CO)HCl] (**7**), has been isolated and shown to be moderately active for isomerization [6]. With 250 000 equivalents of oct-1-ene at 100 °C, isomerized products were formed in 67% yield, of which 75% was oct-2-ene (33:67 cis:trans). During studies with the dihydride species (**8**), Burling et al. have found that the dehydrogenation of phenethyl alcohol with hex-1-ene is limited by competitive isomerization of the alkene [13].

The unsaturated NHC complexes **16**, [Ru(IPr)(PCy$_3$)(=CHPh)Cl$_2$] (**30**) and [Ru(ICy)(PCy$_3$)(=CHPh)Cl$_2$] (**31**) all bring about isomerization of but-2-ene-1,4-diol (**32**, 95+% cis) at room temperature in 15 min to the trans isomer **33** in ca. 70% yield, while [Ru(SIMes)(PCp$_3$)(=CHCH=CMe$_2$)Cl$_2$] (**34**, Cp = cyclopentyl) gives a 95% yield in the same time period (Scheme 2.4) [17]. The attempted metathesis of allyl alcohol (**35**) with **15** in CH$_2$Cl$_2$ at slightly elevated temperature gave the expected product (but-2-ene-1,4-diol) along with propionaldehyde (**36**), as a result of isomerization (Scheme 2.4).

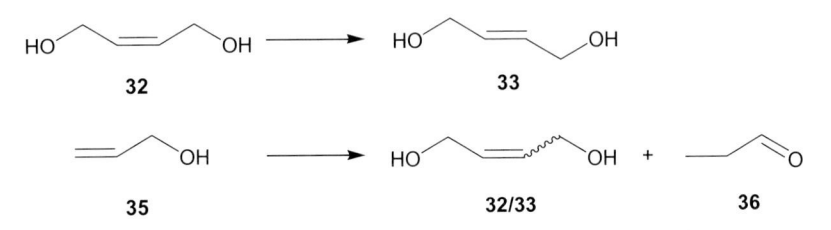

Scheme 2.4 Isomerization products of but-2-ene-1,4-diol and allyl alcohol.

The effect of increasing the catalyst to substrate ratio (from 1:100 to 1:10) is to increase the allyl alcohol conversion and to increase selectivity for both the diol and aldehyde (Table 2.4). Changing to oxygen donor containing solvents results in a decrease of the isomerization pathway (entries 5, 7, 8, Table 2.4), although this can be manipulated by running the reactions at higher temperature (entry 6, Table 2.4) [18].

Extension to the allylic ether and aniline derivatives **37** and **38**, respectively, under conditions that would be expected to result in ADMET (acyclic diene

2.3 Isomerization

Table 2.4 Competitive metathesis and isomerization of H$_2$C=CHCH$_2$OH (**35**) by **15**.

Entry	S:C (molar ratio)	Conditions	Conversion (%)	Metathesis product (%)	Isomerization product (%)
1	110:1	40 °C, 12 h, CH$_2$Cl$_2$	57	40	20
2	40:1	40 °C, 12 h, CH$_2$Cl$_2$	68.5	46	25
3	10:1	40 °C, 12 h, CH$_2$Cl$_2$	86.1	58	28
4	100:1	23 °C, 12 h, CH$_2$Cl$_2$	71.3	42	57
5	110:1	23 °C, 12 h, THF	66.6	65	5
6	110:1	60 °C, 12 h, THF	79.6	40	40
7	100:1	23 °C, 12 h, acetone	64.8	54	9
8	100:1	23 °C, 12 h, DME	66.1	52	13

metathesis) products (i.e., neat substrate, vacuum), led surprisingly only to the isomerization products **39** and **40** (Scheme 2.5). A Ru–H species, formed either *in situ* or present as a trace impurity, is the postulated catalytically active species for the isomerization; circumstantial support for this is the observation that if the reaction is run in CHCl$_3$ the product distribution is 10% metathesis product and 41% isomerization product, compared to a 98% yield of the latter in neat substrate. Cadot et al. have described the application of **15** in the isomerization of O- and N-allyl and O-homoallyl compounds to afford the corresponding enol ethers and enamines (Scheme 2.6) [19]. Thus, for example, treatment of allyl ethers **41** and **42** or the allyl amine **43** with **15** (3 mol%) gave the corresponding alcohols **44** and **45** (in ≥90% yield following acid work-up) and enamine **46**. Substitution of

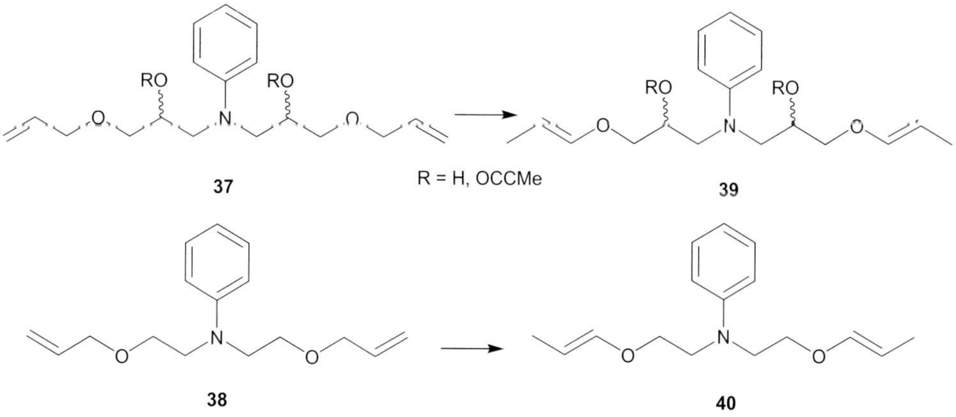

Scheme 2.5 Isomerization products using [Ru(SIMes)(PCy$_3$)(=CHPh)Cl$_2$] (**15**).

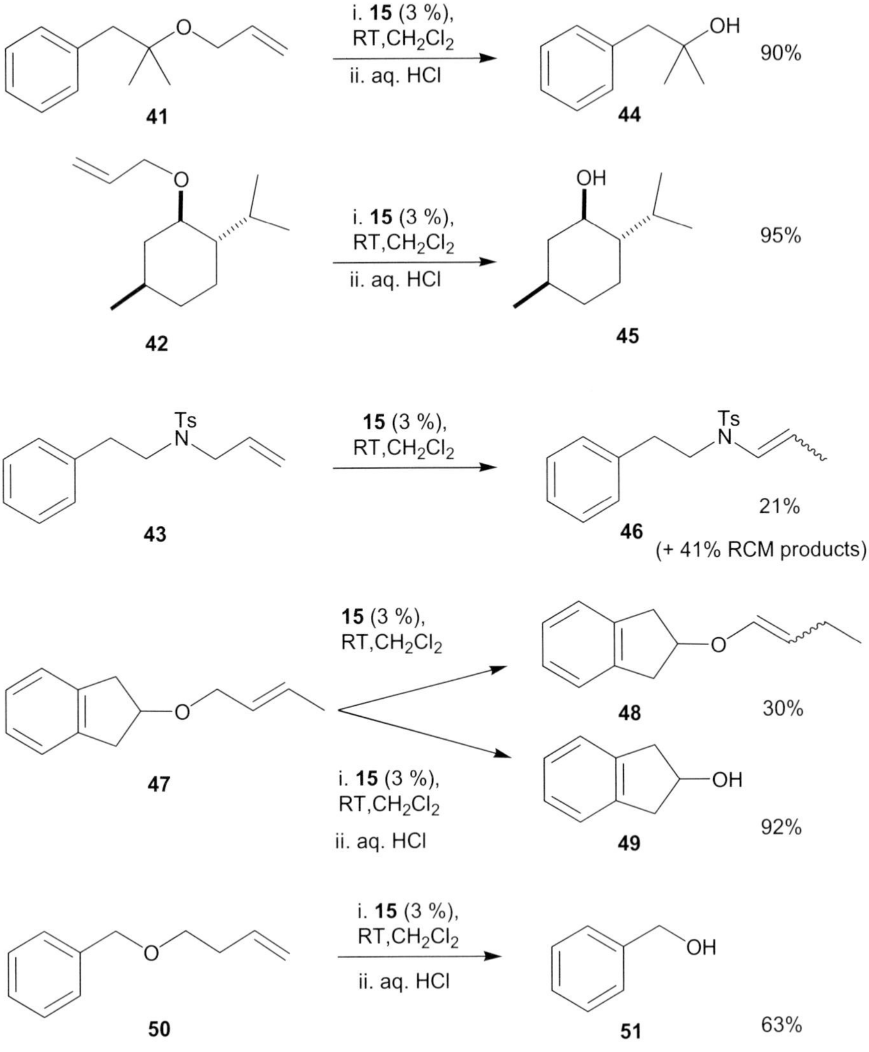

Scheme 2.6 Isomerization reactions catalyzed by [Ru(SIMes)(PCy₃)(=CHPh)Cl₂] (**15**).

the allyl chain (e.g., in **47**) allowed isolation of the stabilized enol product (**48**), or alternatively good conversion into the corresponding indanol (**49**) after work-up. Extension of this chemistry beyond allylic substrates is shown through the efficient formation of benzyl alcohol (**51**) from **50**.

Efforts have been made to probe the extent and mechanism of alkene isomerization under metathesis conditions. Exposure of oct-1-ene to **15** under typical ADMET (acyclic diene metathesis) conditions (neat substrate, 60 °C) gave a mixture of linear C_7–C_{16} alkenes. A profile showing the composition of the reaction mixture with time indicates that the metathesis occurs concurrently with isomerization. Changing the substrate from the terminal alkene to oct-2-ene similarly afforded a mixture of isomerization and metathesis products, thus proving that the methylidene complex [Ru(SIMes)(PCy$_3$)(=CH$_2$)Cl$_2$], which is formed during the reaction, is not the only mediator of the isomerization chemistry [20].

2.4
Other Reactivity

There have been several reports of ruthenium NHC catalyzed reactions covering a range of other processes. Çetinkaya and Dixneuf have described the use of arene Ru(II) NHC complexes for the synthesis of 2,3-dimethylfuran (**53**) via the intramolecular cyclization of (Z)-3-methylpent-2-en-4-yn-1-ol (**52**, Scheme 2.7). The 1,3-dialkylbenzimidazolidine-2-ylidene complexes **54–56** all show good activity (1 mol% catalyst loading) for the formation of 2,3-dimethylfuran at 60 °C in neat substrate, with the *p*-cymene methyl derivative **54** proving to be the most efficient [21].

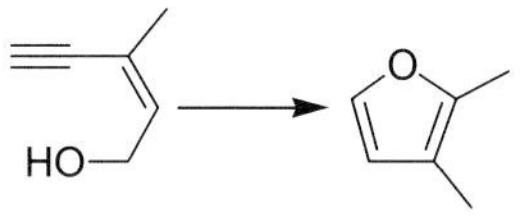

Scheme 2.7 Furan formation by ring cyclization of (Z)-3-methylpent-2-en-4-yn-1-ol.

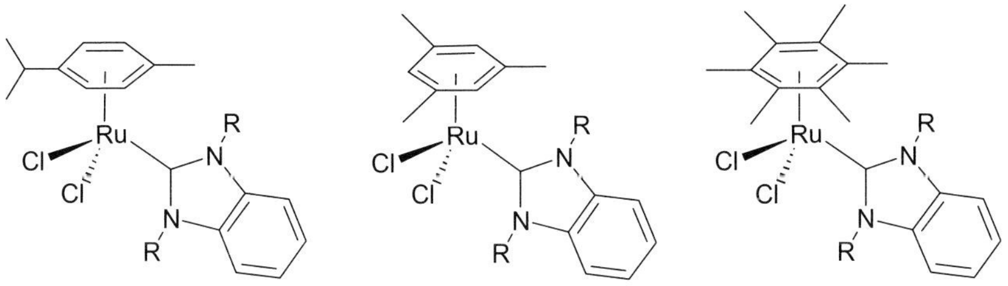

R = Me, Et

Subsequent studies have probed the effects of varying the substituents on the arene and NHC (Scheme 2.8) on catalytic efficiency; Table 2.5 summarizes some of these results. Elevated temperatures are required for catalysis with the mononuclear systems **57–63**. At 60 °C, the more electron-donating C$_6$Me$_6$ ligand affords a higher yield of product (entry 4; cf. entry 1), although increasing the temperature equalizes the efficiency of the two catalysts (entries 3 and 5). No particular trends are apparent upon varying the substituents of the N-heterocyclic carbene. Interestingly, changing to the dinuclear systems **64** and **65** allows the cyclization reaction to be performed at room temperature [22].

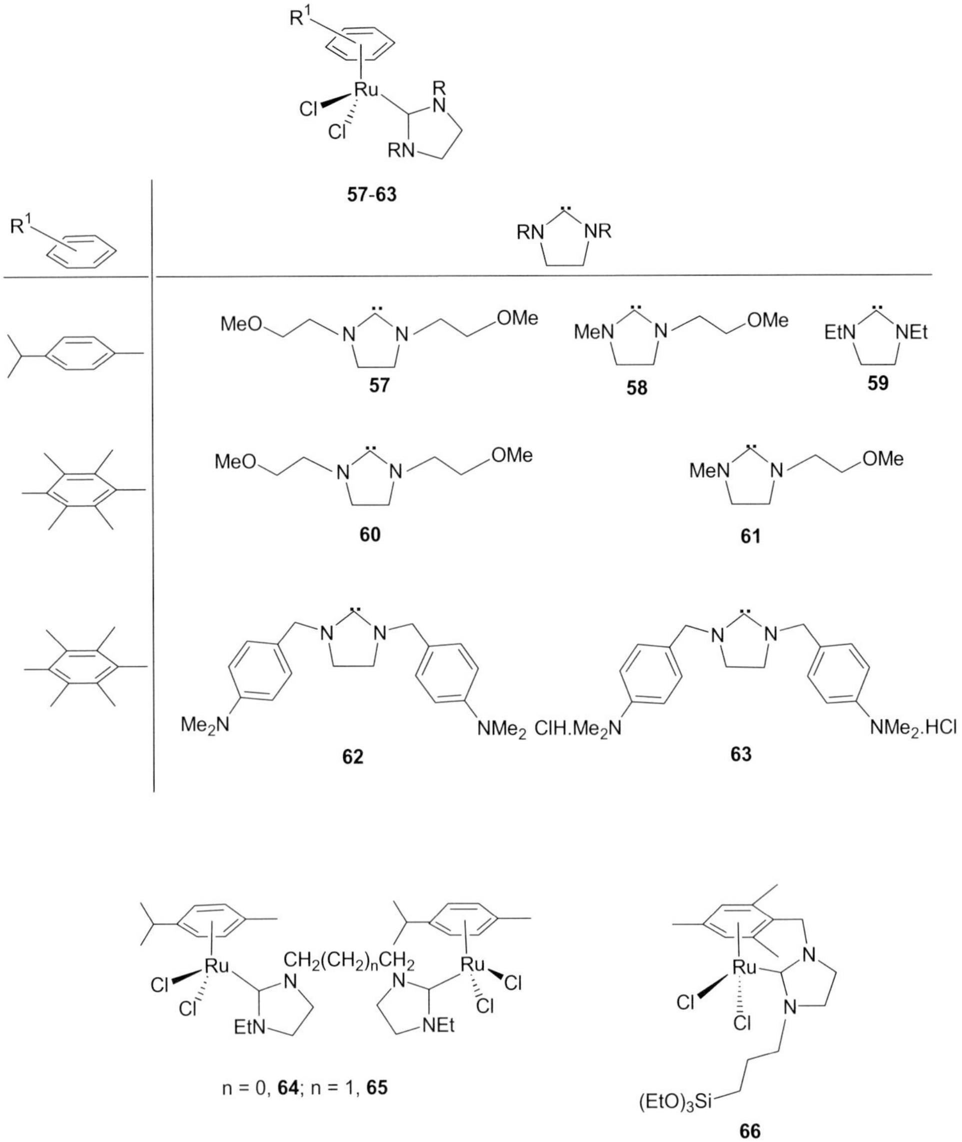

Scheme 2.8 Arene/NHC combinations used in catalytic furan synthesis.

In attempts to prepare derivatives that can be recycled through use of biphasic conditions, complexes **62** and **63** have been reported bearing p-NMe$_2$ substituted benzyl groups on the imidazole N atoms [23]. The quaternary salt **63** affords higher conversion than the parent complex **62** (1 h, 80 °C; **62**: 40% yield; **63**: 94% yield) and in addition is easily recyclable if the catalysis is run in a biphasic mixture of methylpent-2-en-4-yn-1-ol:water. A range of related pre-catalysts have been pre-

Table 2.5 Synthesis of 2,3-dimethylfuran by [(arene)Ru(NHC)Cl$_2$].

Entry	Catalyst	T (°C)	Time (h)	Yield (%)
1	57	60	1.5	44
2	57	60	16.5	87
3	57	80	1	79
4	60	60	1	60
5	60	80	1	67
6	58	80	1	87
7	61	80	1	80
8	59	80	1	74
9	62	80	1	40
10	63	80	1	94
11	64	22	1	49
12	65	22	1	97
13	66	22	0.5	61

pared with a Si(OR)$_3$ capped alkyl chain as one of the N-substituents of the carbene. Furan formation proved most successful with **66** (61% yield of 2,3-dimethylfuran, 30 min, ambient temperature). The silyl group allows the complex to be heterogenized through attachment to a silica surface, although activity is reduced compared to homogeneous conditions, with only a 39% yield of the furan after 1 h at room temperature [24]. The same workers have reported the application of **57**, **58**, **60** and **61** for the cyclopropanation of styrene with the diazomethane derivatives N$_2$CHCO$_2$Et, N$_2$CHC$_6$H$_5$ and N$_2$CHSiMe$_3$ (Scheme 2.9, Table 2.6) [25].

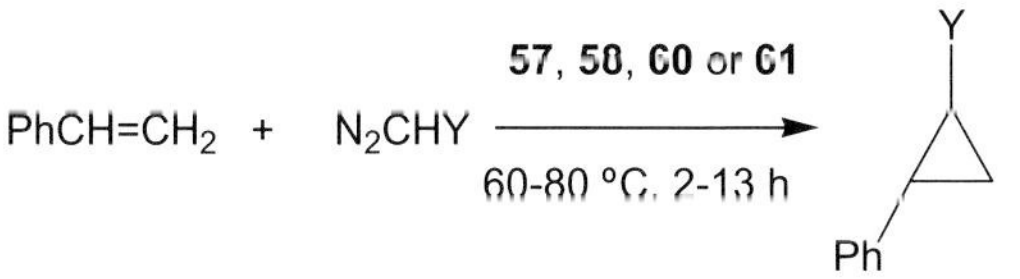

$$\text{PhCH=CH}_2 + \text{N}_2\text{CHY} \xrightarrow[\text{60-80 °C, 2-13 h}]{\textbf{57, 58, 60 or 61}} \underset{\text{Ph}}{\triangle}\text{Y}$$

Scheme 2.9 Cyclopropanation of styrene.

A comparison of the data reveals (a) that only N$_2$CHCO$_2$Et is a particularly effective substrate, (b) the efficiency of the reaction is temperature dependent (entries 1 and 2) and (c) that the presence of two bulky substituents and/or a more electron-donating arene affords higher catalytic activity at 80 °C (entries 2/3 and 3/8 respectively).

Table 2.6 Cyclopropanation of styrene by [(arene)Ru(NHC)Cl$_2$].[a]

Entry	Catalyst	Y[a]	T (°C)	Time (h)	Yield (%) (cis/trans)
1	57	CO$_2$Et	60	4	38 (18/82)
2	57	CO$_2$Et	80	4	52 (18/82)
3	58	CO$_2$Et	80	4	26 (24/76)
4	60	CO$_2$Et	60	4	34 (33/67)
5	60	CO$_2$Et	60	13	44 (27/73)
6	60	CO$_2$Et	80	4	54 (25/75)
7	60	CO$_2$Et	100	2	58 (22/78)
8	61	CO$_2$Et	80	4	44 (24/76)
9	60	Ph	80	2	39 (51/49)
10	60	SiMe$_3$	80	2	6

a) Conditions: catalyst (0.009 mmol), styrene (20 mmol), N$_2$CHY (1 mmol diluted in 1 mL styrene).

Demonceau and coworkers have utilized analogous [(arene)Ru(NHC)Cl$_2$] complexes for atom transfer radical polymerization (ATRP). Thus the mesityl-based complexes **67** and **68** prove effective for the ethyl 2-bromo-2-methylpropionate initiated ATRP of methyl methacrylate (MMA), more so than the ICy derivative **69**.

However, the latter proved to be the most efficient catalyst upon changing the substrate from MMA to styrene. In both cases, the reactions were characterized by good control of the number-average molecular weight and relatively low polydispersivities (ca. 1.3) [26]. ATRP of MMA and styrene can also be catalyzed by the benzylidene complexes [Ru(NHC)(L)(=CHPh)Cl$_2$] (NHC = ICy, L = PCy$_3$, **70**; NHC = (S)-CH(Me)(Ph), L = PCy$_3$, **71**; NHC = L = ICy, **72**; NHC = L = IPr, **73**) [27]. The bis-NHC complexes proved to be more active than **70**, although in comparison to the catalytic behavior of [Ru(PCy$_3$)$_2$(=CHPh)Cl$_2$] the presence of one or more NHC ligands leads to a broader molecular weight distribution of polymer.

The 16-electron complexes [(η^5-C$_5$Me$_5$)Ru(NHC)Cl] (NHC = ICy **74**, IMes **75**) prove to have very high activity for the dimerization of alkynes to give the coupled products **76–78** (Scheme 2.10) [28]. Thus, at room temperature, **74** (1 mol%) reacts to give quantitative conversion of PhC≡CH into a 76:16:8 mixture of **76–78** within 5 min, in a highly exothermic reaction. Changing the NHC from ICy to IMes results in a slower catalytic activity (95% conversion of PhC≡CH after 2 h) and a reversal of selectivity, with mainly the head-to-tail product **78** being formed (0:10:90 for **76:77:78**). This poorer activity and mixed selectivity also applies to reactions of the ICy complex with other alkynes (R = SiMe$_3$, ^{t}Bu, p-tolyl).

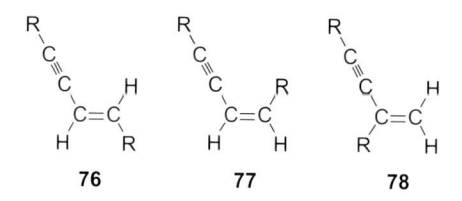

Scheme 2.10 Products of alkyne dimerization catalyzed by [(η^5-C$_5$Me$_5$)Ru(NHC)Cl] (NHC = ICy **74**, IMes **75**).

Peris and coworkers have shown that **12** oxidizes alkenes to aldehydes in the presence of sodium iodate at room temperature [10]. As shown in Table 2.7, aromatic alkenes are less reactive than aliphatic systems; it has been proposed that this enhancement of reactivity for more electron-rich alkenes also helps to explain the higher yield from 1-methylcyclohexene compared with cyclohexene. The regioselectivity for reduction of the double bond situated furthest from the OH functionality in 2-geraniol (shown in entry 5) remains to be understood.

Table 2.7 Oxidation of alkenes to aldehydes catalyzed by **12** in the presence of NaIO$_4$.[a]

Entry	Substrate	Time (h)	Yield (%)
1	Cyclohexene	2	6
2	Cyclohexene	24	72
3	1-Methylcyclohexene	24	>99
4	Styrene	24	23
5	2-Geraniol[b]	24	>99

a) Conditions: **12** (1 mol%), NaIO$_4$ (1.25 mmol), 1 mL CDCl$_3$/0.1 mL H$_2$O, room temp.
b) 2-Geraniol = Me$_2$C=CHCH$_2$CH$_2$CMe$_2$CH=CH$_2$OH.

2.5
Tandem Reactions [29]

2.5.1
Metathesis and Hydrogenation

Several examples have been reported in which [Ru(SIMes)(PCy$_3$)(=CHPh)Cl$_2$] (**15**) initially brings about a metathesis reaction and is then subsequently used as a hydrogenation catalyst upon addition of H$_2$, alcohol and/or base [30]. Louie and Grubbs have shown that in the presence of **15** (3 mol%) and ca. 6.5 atm H$_2$ at 70 °C, 4-phenylbut-3-en-2-one (**79**), formed by initial cross-metathesis of styrene and methyl vinyl ketone at 40 °C, is converted quantitatively into 4-phenylbutan-2-one (**80**). In contrast, with 1 atm H$_2$, ^{i}PrOH, NaOH (5 equivalents) and 1.1 equivalents of ethylenediamine (en), chemoselective reduction of the ketone functionality occurs (Scheme 2.11) [31].

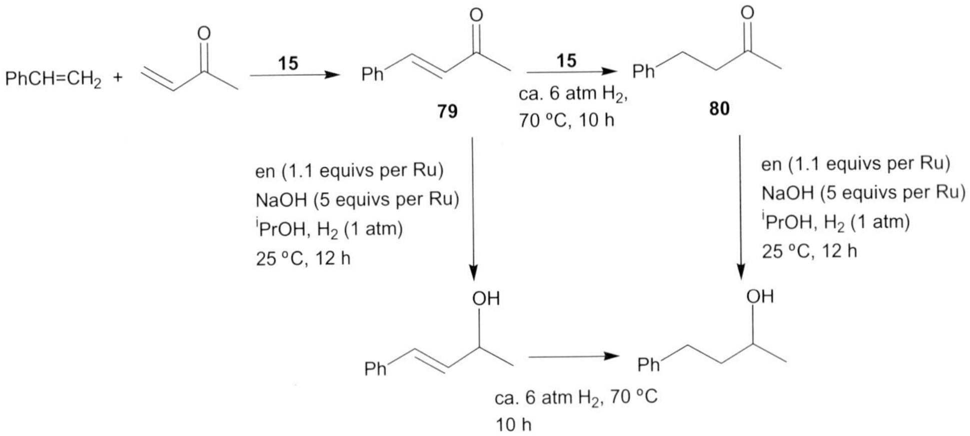

Scheme 2.11 Tandem metathesis/hydrogenation reactivity of **15**.

Independent synthesis has shown that in the presence of the diamine, H$_2$ and base, **15** is converted into [Ru(SIMes)(PCy$_3$)(en)HCl] (**81**). Transfer rather than direct hydrogenation provides an alternative methodology; RCM (ring-closing metathesis) of nona-1,8-dien-5-one followed by reflux in the presence of ^{i}PrOH/K$_2$CO$_3$ generated cyclohept-4-enol in 60% yield. The strength of this combined methodology is illustrated by the three-step, one-pot synthesis of (*R*)-(−)-muscone (**82**) (Scheme 2.12) [31]. Similarly, this tandem RCM/hydrogenation approach has been used to prepare phosphortriester-based cyclic dinucleotides (Scheme 2.13) [32].

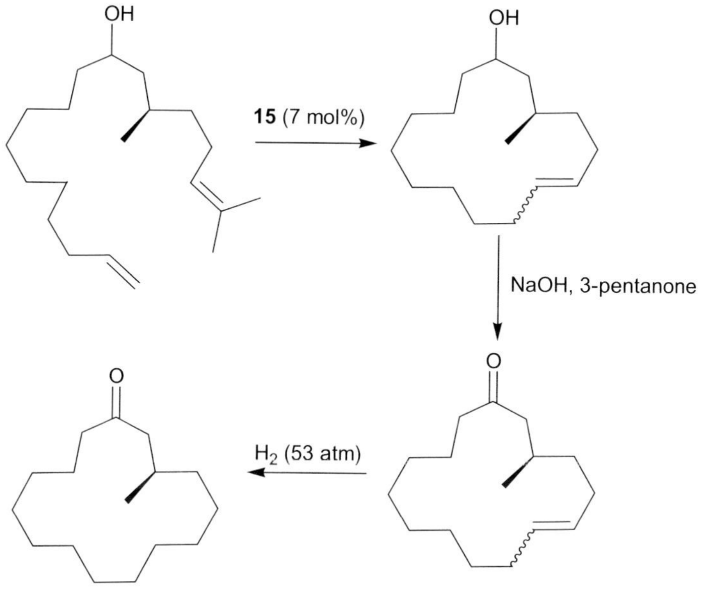

Scheme 2.12 Tandem RCM hydrogenation pathway to the synthesis of (R)-(−)-muscone (**82**).

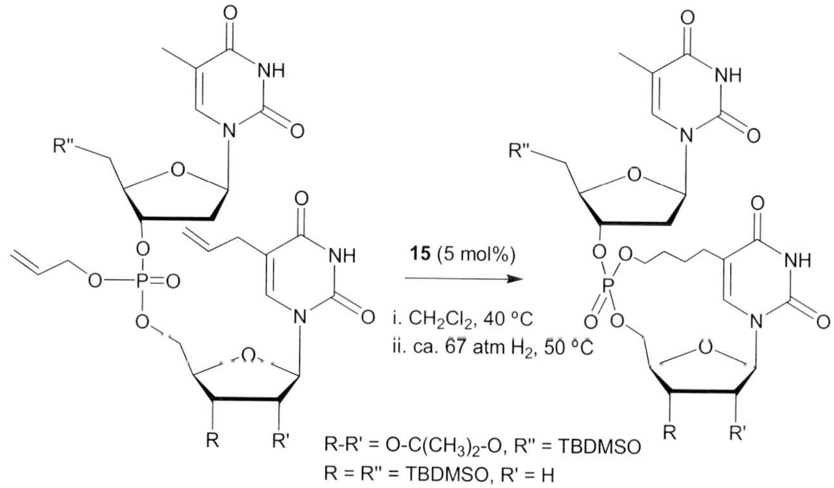

Scheme 2.13 Tandem catalysis route to phosphortriester-based cyclic dinucleotides.

In contrast to the direct hydrogenation reactions above, **15** fails to catalyze the hydrogenation of the ring-closed product 4-*tert*-butylhepta-1,6-dien-4-ol with 1 atm H_2 in toluene at ambient temperature. Addition of NaH or $LiAlH_4$ (13–14 mol%), however, brings about good conversions [33]. If excess hydride additive is employed along with water, H_2 is generated *in situ* without the need for external pressure. Thus, >70% conversion of **83** and **84** occurs upon reaction with 2 mol% **15** and 4 equivalents of NaH in the presence of 30 equivalents of water (Scheme 2.14).

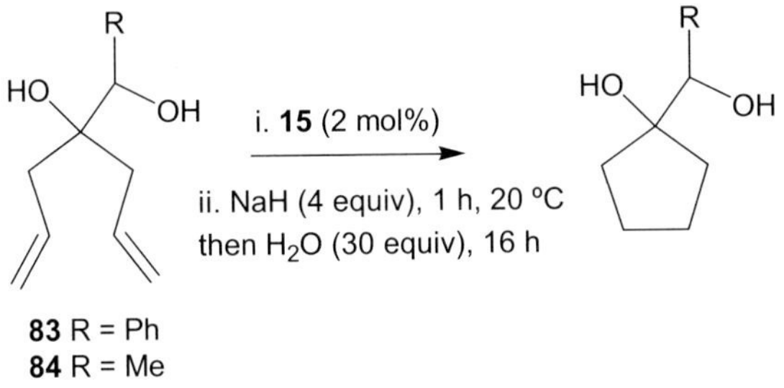

Scheme 2.14 Hydride enhanced RCM/hydrogenation catalysis.

2.5.2
Metathesis and Isomerization

Nolan and coworkers have described how $[Ru(IMes)(PCy_3)(=CHPh)Cl_2]$ (**16**) catalyzes competitive ring-closing metathesis and isomerization reactions depending upon solvent [34]. Under conditions of reflux, diene **85** can be converted into a 90:10 mixture of products **86** and **87** in dichloroethane, but a 20:80 mixture in toluene (Scheme 2.15).

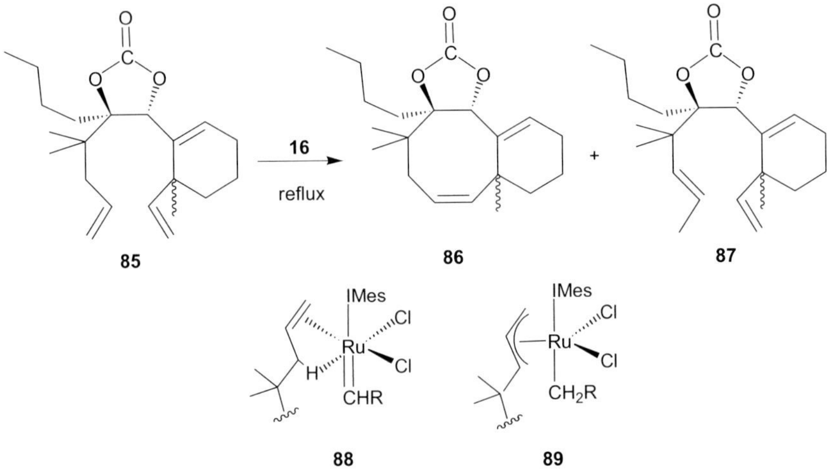

Scheme 2.15 Products resulting from competitive metathesis and isomerization reactivity of $[Ru(IMes)(PCy_3)(=CHPh)Cl_2]$ (**16**).

2.5 Tandem Reactions

A similar pattern of results was found with benzene and DME (Table 2.8), indicating that it is the coordinating ability of the solvent that is crucial in determining the reaction pathway. The authors have proposed that more coordinating solvents favor a lower coordinate ruthenium species (intermediate **88**), which precedes the ring-closing reaction, rather than the allyl complex intermediate **89**, which provides a pathway to isomerization.

Table 2.8 Solvent effects on the RCM and isomerization activity of **16** with diene **85**.

Solvent	% 86	% 87
DME	0	100
Toluene	20	80
Benzene	50–70	50–30
DCE	90	10

The solvent also plays a fundamental role in the attempted ring closure of diene **90**. Upon heating with **16** in toluene, CH_2Cl_2 or $ClCH_2CH_2Cl$, high yields of the 21-membered lactone from ring closure are formed, although in toluene 10–12% of the 20-membered ring product, resulting from isomerization and metathesis, is also produced [35].

90

The saturated carbene complex **15** catalyzes the isomerization of terminal alkenes in the presence of silylated ethers [36]. Treatment of $RCH_2CH=CH_2$ (R = Ph, $PhCH_2$, p-$MeOC_6H_4$, $HO(CH_2)_3$, BnO) with 5 mol% **15** at 50 °C in the presence of 10 equivalents of CH_2=CHOTMS affords good yields of the corresponding internal alkenes. Activity for isomerization can also be combined with RCM to provide a novel synthetic route to indoles. In the presence of **15** and 1 equivalent of CH_2=CHOTMS, **91** is isomerized to **92**, which can subsequently be ring closed at higher temperature in the presence of **15** alone (Scheme 2.16).

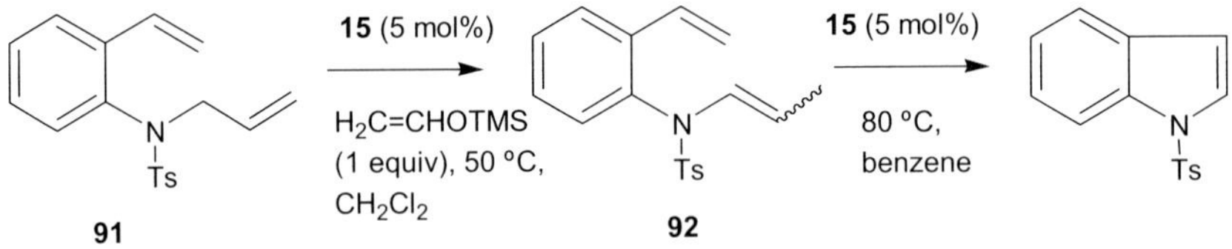

Scheme 2.16 RCM/isomerization pathway to indoles.

This methodology has been extended by the same research group to the combined cycloisomerization/RCM of α,ω-dienes [37]. Treatment of N,N-diallyl-p-toluenesulfonamide (**93**) with 5 mol% **15** and an equivalent of CH$_2$=CHOTMS gives a mixture of two isomerization products (**94** and **96**) and the ring-closed derivative **95** (Scheme 2.17, Table 2.9).

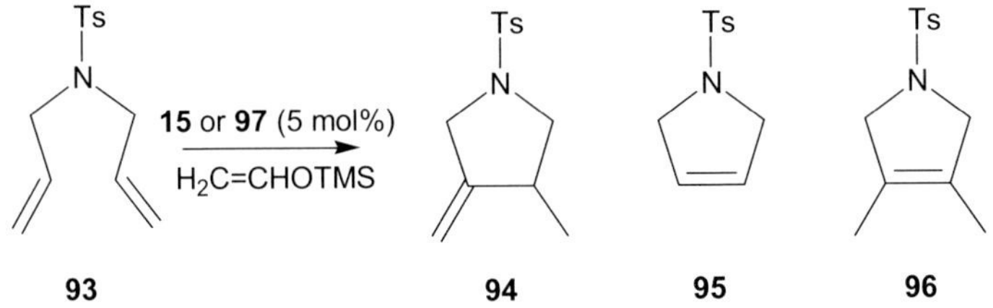

Scheme 2.17 RCM and isomerization products arising from N,N-diallyl-p-toluenesulfonamide.

Table 2.9 Cycloisomerization and RCM products of N,N-diallyl-p-toluenesulfonamide (**93**).

Catalyst	Solvent	T (°C)	% 94	% 95	% 96
15	CH$_2$Cl$_2$	22	65	21	0
	CH$_2$Cl$_2$	40	86	0	14
	Toluene	110	22	0	78
97	CH$_2$Cl$_2$	40	71	0	24

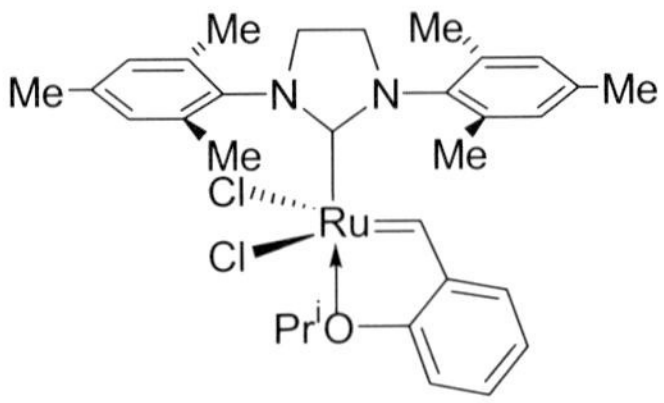

In the absence of the silyl ether, the RCM product **95** is the sole product; increased amounts of the isomerized compounds **94** and **96** are detected at higher temperature when silyl ether is present. Catalysis is maintained if **15** is replaced by the Hoveyda complex **97**.

The competition between cycloisomerization and isomerization has been probed further (Scheme 2.18). Treatment of **91** with 10 mol% **15** in refluxing xylene produces the optimum yield (81%) of the cycloisomerized indolidene product **98**. A series of related 3-methylene-2,3-dihydroindoles and 3-methylene-2,3-dihydrobenzofurans can be formed in good yield (Scheme 2.18) using this methodology.

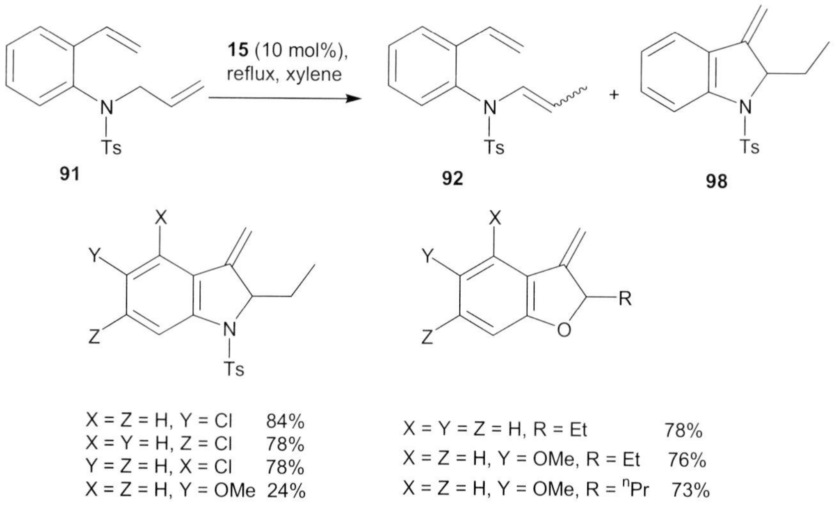

Scheme 2.18 Isomerization and cycloisomerization products catalyzed by **15**.

In the absence of silyl ether, a combination of [Ru(p-cymene)Cl$_2$]$_2$ (2.5 mol%), 1,3-bis(mesityl)imidazolinium chloride (5 mol%) and Cs$_2$CO$_3$ (10 mol%) results in the cycloisomerization of **93** into **94** in 100% yield over 1 h at 80 °C [38]. Mixed results were found upon changing the nature of the NHC salt; with 1,3-bis(2,6-diisopropylphenyl)imidazolinium chloride, a 2.3:1 ratio of **94** to **99** (Scheme 2.19) was produced, while the less electron donating, unsaturated salt 1,3-bis(mesityl)imidazolium chloride produced even less selectivity and required longer reaction times.

Replacement of the argon or nitrogen atmosphere by an atmosphere of ethyne (HC≡CH) completely reverses the selectivity to give exclusively **95**. The presence of an alkyne plays a key role in dictating the pathway of product formation; thus, with HC≡C(CH$_2$)$_3$CH$_3$ present, **95** is formed in 100% yield. The bulkier alkyne HC≡C^{t}Bu affords lower overall conversion and leads to more isomerization products (4:1, **94:99**), while HC≡CSiMe$_3$ produced greater conversion, but showed even less selectivity (**94:95:99** in a ratio of 3:12:5). While the effect of different alkynes is not readily explained, it has been postulated that the presence of an alkyne aids formation of a Ru-vinylidene intermediate, which is likely to be capable of performing metathesis. In the absence of alkyne, a coordinatively unsaturated Ru(NHC)Cl$_2$ entity could be formed, which would bring about isomerization of a diene via the favorable oxidative addition to the electron-rich Ru(II) center.

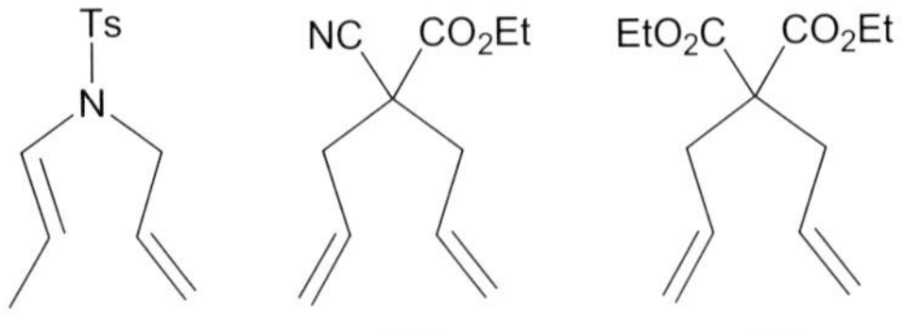

Scheme 2.19 Cycloisomerization chemistry of Ru(arene)(NHC) adducts.

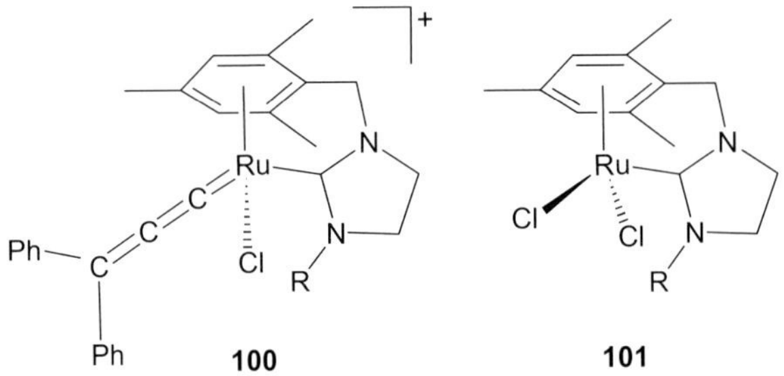

a R = CH$_2$CH$_2$OMe; b R = CH$_2$(2,4,6-C$_6$Me$_3$H$_2$)

Subsequent work on a related system revealed that the cationic allenylidene complex **100**, which can be isolated or alternatively prepared *in situ*, upon treatment of the dichloride precursor **101a,b** with propargyl alcohol in the presence of AgOTf, is active for cycloisomerization. With 2.5 mol% of **100a** present, near quantitative conversion to **94** can be achieved at 80 °C in chlorobenzene, although much lower conversion takes place in toluene. This solvent dependence is altered dramatically if the catalyst precursor is changed to **100b**, containing a bulkier substituent on the NHC, with good conversion to **94** recorded in both C$_6$H$_5$Cl (84%) and C$_6$H$_5$CH$_3$ (100%) [39]. In addition, changing the diene from a bis-allylamide to the 1,6-carbodienes **102** and **103** shuts down the cycloisomerization activity of **100a**, giving only metathesis products (Scheme 2.19) [40].

Tandem RCM/isomerization has allowed the synthesis of five-, six- and seven-membered ring cyclic enol ethers using **15** in the presence of a 95:5 mixture of N$_2$ and H$_2$ (Scheme 2.20) [41]. In the absence of any H$_2$, only ring-closed products are detectable, suggesting that a Ru hydride species is active for the isomerization process [42]. Although mechanistic details on the nature of the catalytically active species remain to be fully elucidated, isomerization is clearly subject to high levels of steric control. van Otterlo and coworkers have utilized the same

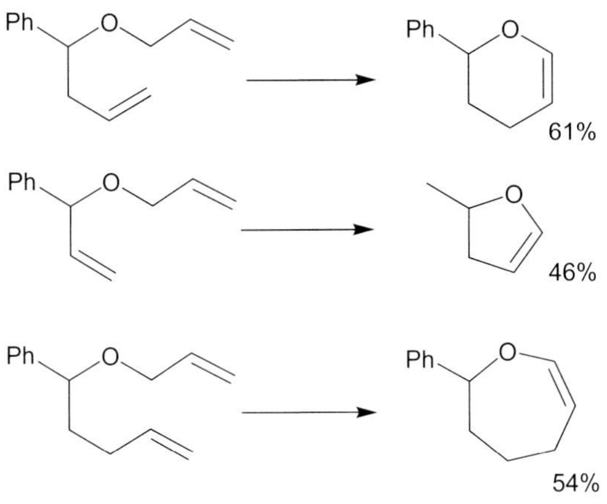

General conditions: **15** (10 mol%), 45-70 °C, CH_2Cl_2, 95:5 N_2:H_2

Scheme 2.20 Tandem RCM/isomerization route to cyclic enol ethers.

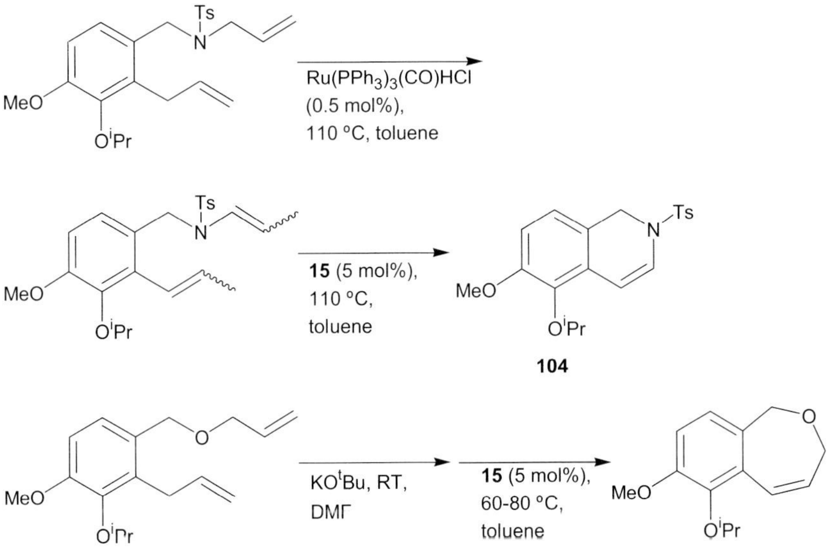

104

105

Scheme 2.21 Tandem RCM/isomerization to yield heterocycles.

isomerization/RCM capability of **15** for the synthesis of several six-, seven- and eight-membered benzo- and pyrido-fused heterocycles, such as **104** and **105** (Scheme 2.21) [43, 44].

Isomerization followed by ring-closure has allowed the synthesis of six-membered enamides in the presence of **16** [45].

2.5.3
Tandem Reactions not Involving Metathesis

The reversible conversion of **8** and **9** described in Section 2.2 has been utilized in a tandem dehydrogenation–hydrogenation process that allows an indirect Wittig reaction of a temporarily oxidized alcohol to be performed, ultimately leading to new C–C bond formation (Scheme 2.22). Fundamental to this cycle is the crossover reaction between alcohol and alkene (Scheme 2.23), which occurs with 100% conversion with 5 mol% **8**.

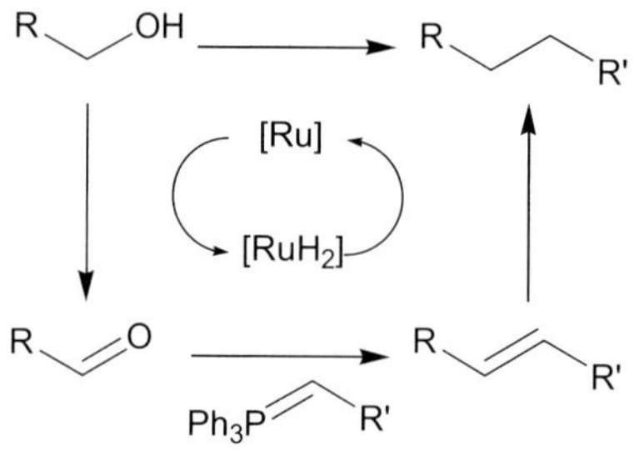

Scheme 2.22 Indirect Wittig reaction of alcohols.

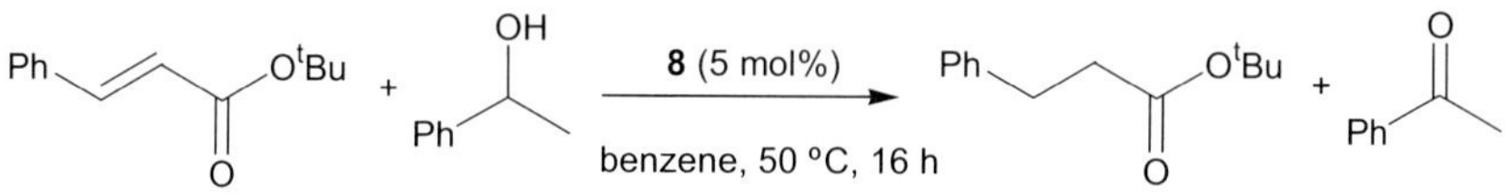

Scheme 2.23 Crossover reaction of *tert*-butyl cinnamate with phenethyl alcohol catalyzed by **8**.

Benzyl alcohol undergoes conversion into $PhCH_2CH_2CO_2Bn$ in 90% yield in the presence of the ester ylide $Ph_3P=C(CO_2Bn)H$ and 5 mol% **8** (80 °C, 24 h). The reactivity also proved applicable with a range of other substituted ylides and aromatic alcohols under optimized conditions with the catalyst loading reduced to 1 mol% (Scheme 2.24) [8].

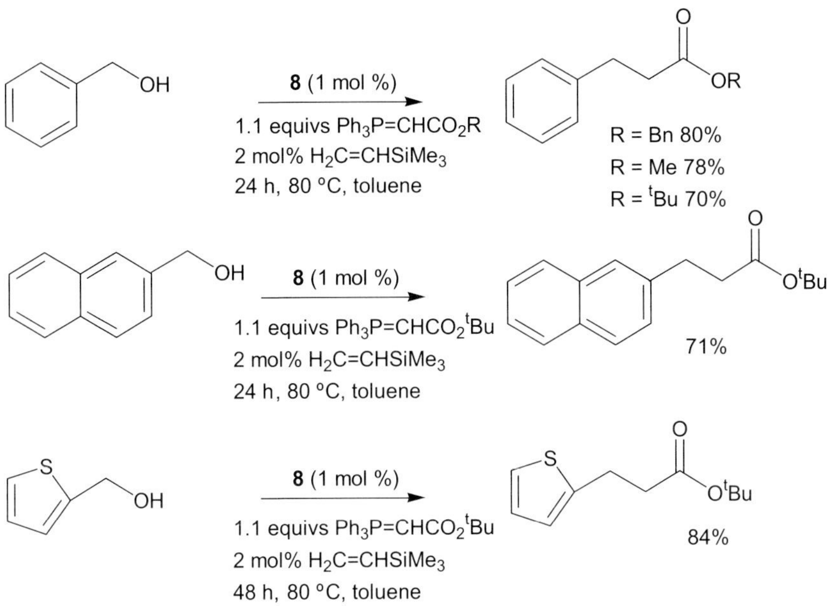

Scheme 2.24 Functionalized alkane products catalyzed by **8**.

2.6
Conclusions

The above sections reveal that despite there being only a relatively small number of reported studies on the catalytic applications of Ru NHC complexes in organic synthesis, quite a diverse range of reactions have been investigated. Without doubt, Ru NHC complexes show considerable promise, particularly in hydrogenation and isomerization reactions, but it is apparent that, in most cases, there is very little mechanistic understanding underpinning the catalysis. Only through such work will the influence of the NHC and its substituents (saturated vs. unsaturated, aryl vs. alkyl substituents, bulky vs. less bulky substituents) and the role of other ancillary ligands on ruthenium, as well as the importance of physical variables, including solvent and temperature, become more transparent and allow existing reactions to be made more efficient and new catalysis to be developed.

References

1 Trnka, T. M., Grubbs, R. H. *Acc. Chem. Res.* **2001**, *34*, 18–29; Grubbs, R. H. *Tetrahedron* **2004**, *60*, 7117–7140.
2 Lee, H. M., Smith Jr., D. C., He, Z., Stevens, E. D., Yi, C. S., Nolan, S. P. *Organometallics* **2001**, *20*, 794–797.
3 Diggle, R. A., Macgregor, S. A., Whittlesey, M. K. *Organometallics* **2004**, *23*, 1857–1865.
4 Dharmasena, U. L., Foucault, H. M., dos Santos, E. N., Fogg, D. E., Nolan, S. P. *Organometallics* **2005**, *24*, 1056–1058.
5 Dinger, M. B., Mol, J. C. *Eur. J. Inorg. Chem.* **2003**, 2827–2833.
6 Banti, D., Mol, J. C. *J. Organomet. Chem.* **2004**, *689*, 3113–3116.
7 Jazzar, R. F. R., Macgregor, S. A., Mahon, M. F., Richards, S. P., Whittlesey, M. K. *J. Am. Chem. Soc.* **2002**, *124*, 4944–4945.
8 Edwards, M. G., Jazzar, R. F. R., Paine, B. M., Shermer, D. J., Whittlesey, M. K., Williams, J. M. J., Edney, D. D. *Chem. Commun.* **2004**, 90–91.
9 Csabai, P., Joó, F. *Organometallics* **2004**, *23*, 5640–5643.
10 Poyatos, M., Mata, J. A., Falomir, E., Crabtree, R. H., Peris, E. *Organometallics* **2003**, *22*, 1110–1114.
11 Danopoulos, A. A., Winston, S., Motherwell, W. B. *Chem. Commun.* **2002**, 1376–1377.
12 Chiu, P. L., Lee, H. M. *Organometallics* **2005**, *24*, 1692–1702.
13 Burling, S., Whittlesey, M. K., Williams, J. M. J. *Adv. Synth. Catal.* **2005**, *347*, 591–594.
14 Maifield, S. V., Tran, M. N., Lee, D. *Tetrahedron Lett.* **2005**, *46*, 105–108.
15 Song, C., Ma, C., Ma, Y., Feng, W., Ma, S., Chai, Q., Andrus, M. B. *Tetrahedron Lett.* **2005**, *46*, 3241–3244.
16 Dinger, M. B., Mol, J. C. *Organometallics* **2003**, *22*, 1089–1095.
17 Bielawski, C. W., Scherman, O. A., Grubbs, R. H. *Polymer* **2001**, *42*, 4939–4945.
18 Sworen, J. C., Pawlow, J. H., Case, W., Lever, J., Wagener, K. B. *J. Mol. Catal. A: Chem.* **2003**, *194*, 69–78.
19 Cadot, C., Dalko, P. I., Cossy, J. *Tetrahedron Lett.* **2002**, *43*, 1839–1841.
20 Lehman Jr., S. E., Schwendeman, J. E., O'Donnell, P. M., Wagener, K. B. *Inorg. Chim. Acta* **2003**, *345*, 190–198.
21 Kücükbay, H., Çetinkaya, B., Guesmi, S., Dixneuf, P. H. *Organometallics* **1996**, *15*, 2434–2439.
22 Çetinkaya, B., Özdemir, I., Bruneau, C., Dixneuf, P. H. *J. Mol. Catal. A: Chem.* **1997**, *118*, L1–L4.
23 Özdemir, I., Yiğit, B., Çetinkaya, B., Ülkü, D., Tahir, M. N., Arici, C. *J. Organomet. Chem.* **2001**, *633*, 27–32.
24 Çetinkaya, B., Gürbüz, N., Seçkin, T., Özdemir, I. *J. Mol. Catal. A: Chem.* **2002**, *184*, 31–38.
25 Çetinkaya, B., Özdemir, I., Dixneuf, P. H. *J. Organomet. Chem.* **1997**, *534*, 153–158.
26 Delaude, L., Delfosse, S., Richel, A., Demonceau, A., Noels, A. F. *Chem. Commun.* **2003**, 1526–1527.
27 Simal, F., Delfosse, S., Demonceau, A., Noels, A. F., Denk, K., Kohl, F. J., Weskamp, T., Herrmann, W. A. *Chem. Eur. J.* **2002**, *8*, 3047–3052.
28 Baratta, W., Herrmann, W. A., Rigo, P., Schwarz, J. *J. Organomet. Chem.* **2000**, *594*, 489–493.
29 For a recent overview of tandem reactions, see: Fogg, D. E., dos Santos, E. N. *Coord. Chem. Rev.* **2004**, *248*, 2365–2379.
30 Schmidt, B. *Eur. J. Org. Chem.* **2004**, 1865–1880.
31 Louie, J., Bielawski, C. W., Grubbs, R. H. *J. Am. Chem. Soc.* **2001**, *123*, 11312–11313.
32 Børsting, P., Nielsen, P. *Chem. Commun.* **2002**, 2140–2141.
33 Schmidt, B., Pohler, M. *Org. Biomol. Chem.* **2003**, *1*, 2512–2517.
34 Bourgeois, D., Pancrazi, A., Nolan, S. P., Prunet, J. *J. Organomet. Chem.* **2002**, *643–644*, 247–252.

35 Fürstner, A., Thiel, O. R., Ackermann, L., Schanz, H.-J., Nolan, S. P. *J. Org. Chem.* **2000**, *65*, 2204–2207.
36 Arisawa, M., Terada, Y., Nakagawa, M., Nishida, A. *Angew. Chem. Int. Ed.* **2002**, *41*, 4732–4734.
37 Terada, Y., Arisawa, M., Nishida, A. *Angew. Chem. Int. Ed.* **2004**, *43*, 4063–4067.
38 Sémeril, D., Bruneau, C., Dixneuf, P. H. *Helv. Chim. Acta* **2001**, *84*, 3335–3341.
39 Çetinkaya, B., Demir, S., Özdemir, I., Toupet, L., Sémeril, D., Bruneau, C., Dixneuf, P. H. *New J. Chem.* **2001**, *25*, 519–521.
40 Çetinkaya, B., Demir, S., Özdemir, I., Toupet, L., Sémeril, D., Bruneau, C., Dixneuf, P. H. *Chem. Eur. J.* **2003**, *9*, 2323–2330.
41 Sutton, A. E., Seigal, B. A., Finnegan, D. F., Snapper, M. L. *J. Am. Chem. Soc.* **2002**, *124*, 13 390–13 391.
42 Recently, Hong et al. have shown that decomposition of **15** affords a hydride-containing dinuclear Ru complex with a bridging carbide ligand. This will catalyze the isomerization of allylbenzene to 1-phenyl-1-propene in 76% yield at 40 °C. Hong, S. Y., Day, M. W., Grubbs, R. H. *J. Am. Chem. Soc.* **2004**, *126*, 7414–7415.
43 van Otterlo, W. A. L., Pathak, R., de Koning, C. B. *Synlett* **2003**, 1859–1861.
44 The same group have used a combination of $Ru(PPh_3)_3(CO)HCl$ and **15** for isomerization/RCM to afford dioxenes (van Otterlo, W. A. L., Ngidi, E. L., de Koning, C. B. *Tetrahedron Lett.* **2003**, *44*, 6483–6486) and 6-membered N,O- and 7-membered N,S-fused systems (van Otterlo, W. A. L., Morgans, G. L., Khanye, S. D., Aderibigbe, B. A. A., Michael, J. P., Billing, D. G. *Tetrahedron Lett.* **2004**, *45*, 9171–9175).
45 Kinderman, S. S., van Maarseveen, J. H., Schoemaker, H. E., Hiemstra, H., Rutjes, F. P. J. T. *Org. Lett.* **2001**, *3*, 2045–2048.

3
Cross-coupling Reactions Catalyzed by Palladium N-Heterocyclic Carbene Complexes

Natalie M. Scott and Steven P. Nolan

3.1
Introduction

Palladium-catalyzed reactions have become increasingly important in the field of organic synthesis [1]. In particular, palladium cross-coupling reactions are now widely used to form carbon–carbon and carbon–nitrogen bonds. Such processes include the Suzuki–Miyaura [2], Heck [3–7], telomerization of alkenes [8–10] arylamination [1c, d] and α-arylation of ketones [11] reactions, often involving coupling of an aryl halide with a nucleophilic partner. Although a wide variety of palladium catalysts efficiently promote such couplings (usually combinations of palladium salts or complexes with phosphorus ligands, such as phosphines [12], phosphates [13], and phosphine oxides [14]), these catalysts are often very sensitive to air and moisture, suffer from significant P–C bond degradation at elevated temperatures, and require air-free handling to minimize ligand oxidation. The recent emergence of N-heterocyclic carbene ligands has produced more stable, yet highly reactive, metal catalysts where phosphine analogues were less effective or ineffective [8, 16–23].

In general, palladium-catalyzed cross-coupling reactions are thought to proceed via a mechanism that involves a Pd(0)/Pd(II) couple in a distinctive three-step process (Scheme 3.1) [1a]. The first step involves an oxidative addition reaction of an aryl halide with a palladium(0) center. Then, a transmetallation reaction yields a palladium(II) species that contains the two moieties to be coupled. The final step releases the product through a reductive elimination process that simultaneously regenerates the active palladium(0) catalyst (Scheme 3.1). In this mechanism, the nature of the aryl halide substrate is very important in determining the rate-limiting step. For example, for aryl chlorides and unactivated aryl bromides, the oxidative addition appears to be the rate-limiting step. As a result, aryl chlorides usually undergo coupling reactions more slowly than aryl bromides or iodides (C–Cl > C–Br > C–I); however, their use as chemical feedstock would economically benefit several industrial processes [24].

N-Heterocyclic Carbenes in Synthesis. Edited by Steven P. Nolan
Copyright © 2006 WILEY-VCH Verlag GmbH & Co. KGaA, Weinheim
ISBN: 3-527-31400-8

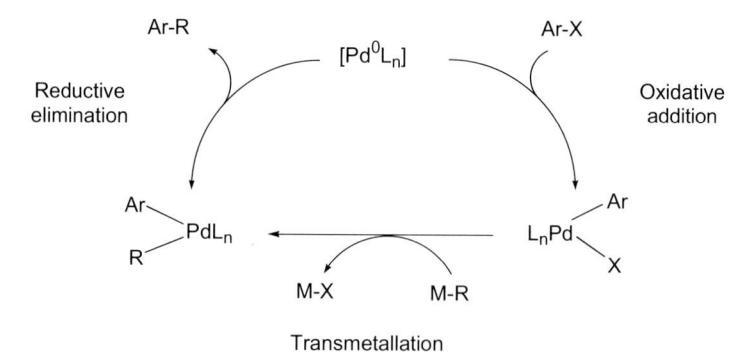

Scheme 3.1 General mechanism for palladium-catalyzed cross-coupling reactions.

This chapter overviews recent NHC palladium complexes that are catalytically active in various cross-coupling reactions. It is divided into three main sections. While the first focuses on the synthesis of (NHC)Pd(0) complexes, the second concentrates on the preparation of recent (NHC)Pd(II) complexes. Finally, catalytic activity of various palladium/NHC complexes, which have been shown to catalyze various cross-coupling reactions, will be discussed.

3.2
Palladium(0) NHC Complexes

Palladium(0) complexes enter into catalytic cycles more readily than their Pd(II) counterparts since they are primed for oxidative addition. In contrast to the plethora of Pd(II) complexes stabilized by NHC ligands (see below), examples of (NHC)Pd(0) are relatively scarce [5, 6, 9, 25–27]. Cloke and coworkers were the first to isolate a homoleptic (NHC)$_2$Pd complex [25]. While a synthetic route involving the co-condensation of palladium vapor with N,N'-di-*tert*-butylimidazol-2-ylidene (I^{t}Bu) led to the complex (I^{t}Bu)$_2$Pd, it was low yielding and strictly limited to easy to sublime NHC ligands (Scheme 3.2).

Scheme 3.2 Synthesis of bis-NHC palladium complexes by the metal vapor method.

Solution-phase synthetic procedures leading to (NHC)$_2$Pd complexes center around ligand exchange reactions involving palladium centers bearing weakly coordinating ligands such as phosphines or olefins. For example, the 14-electron species, Pd{P(*o*-tolyl)$_3$}$_2$, reacts with free NHC to afford the first mixed [(NHC)Pd(PR$_3$)] complex and then complete displacement of all phosphine li-

3.2 Palladium(0) NHC Complexes

gands to afford bis-(NHC)$_2$Pd. Alternatively, [(NHC)Pd(PR$_3$)] can also be synthesized by mixing the [(NHC)$_2$Pd] complex with free phosphine (P(o-tolyl)$_3$, PCy$_3$), suggesting that some reversibility is associated with the reaction (Scheme 3.3) [28, 29].

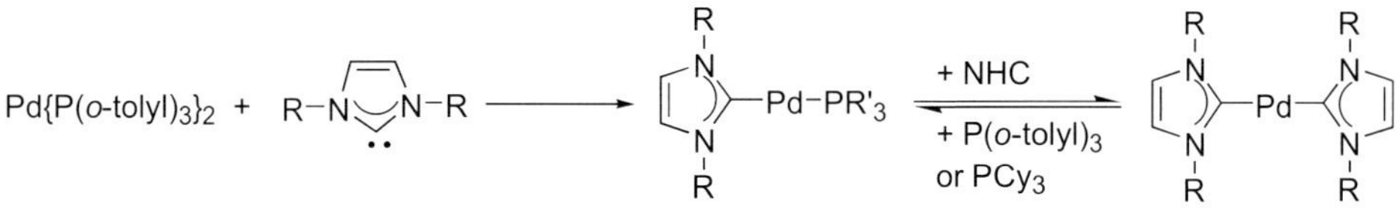

Scheme 3.3 Synthesis of bis-NHC palladium complexes.

One alternative approach to (NHC)$_2$Pd complexes (Scheme 3.4) is the *in situ* reduction of Pd(II) species in the presence of NHC. Although the electrochemical reduction of palladium(II) complexes was investigated and reported, no complexes have been isolated and characterized via this technique so far [30].

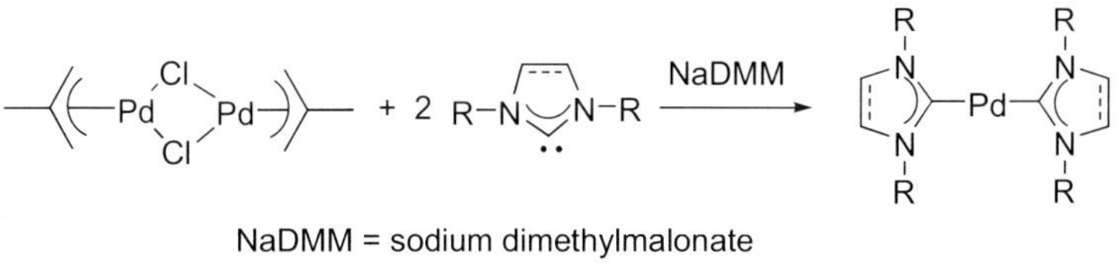

NaDMM = sodium dimethylmalonate

Scheme 3.4 Synthesis of Pd(NHC)$_2$ complexes in the presence of reducing agents.

Derivatives of (NHC)$_2$Pd complexes can be readily prepared from the reaction of Pd(COD)(alkene) (COD = cyclooctadiene) with two equivalents of free NHC [NHC = 1,3,4,5-tetramethylimidazol-2-ylidene (IMe)] to afford the complexes Pd-(IMe)$_2$(alkene) [alkene = maleic anhydride (MAH), tetracyanoethylene (TCNE)] (Fig. 3.1) [5]. Spectroscopic studies showed significant back-bonding of the olefin onto the Pd center, suggesting that the NHC ligands are strongly donating. Although the olefin complexes are envisaged as approaching Pd(II), these complexes readily undergo oxidation reactions typical of Pd(0).

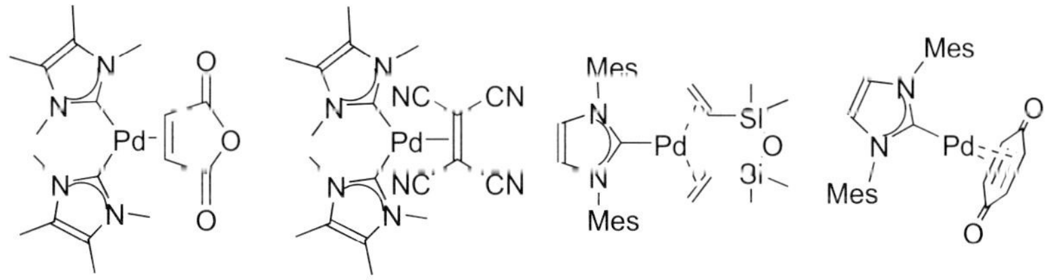

Fig. 3.1 Examples of (NHC)Pd(0) complexes.

Reports of monocarbenepalladium(0) complexes without additional phosphine donors are scarce. Beller et al. have synthesized several monocarbenepalladium(0)-diolefin complexes [6, 7]. The addition of IMes [1,3-bis(2,4,6-trimethylphenyl)imidazol-2-ylidene] to either a palladium(0)diallyl ether complex [Pd$_2$(dae)$_3$] (dae = diallylether) in 1,1,3,3-tetramethyl-1,3-divinyl-disiloxane or (COD)Pd(quinone) [quinone = *p*-benzoquinone (BQ), 1,4-naphthoquinone (NQ)] at low temperatures results in the complexes (IMes)Pd(dae), (COD)Pd(BQ) and (COD)Pd(NQ), respectively (Fig. 3.1). Notably, these olefin complexes showed remarkable productivities and selectivity in several telomerization and Heck reactions.

3.3
Palladium(II) N-Heterocyclic Carbene Complexes

Since the first catalytic applications of palladium/NHC complexes were realized in 1995 [4], numerous complexes have been synthesized using various methods. The reaction of Pd(OAc)$_2$ and imidazolium salts was one of the first methods reported to afford palladium(II) NHC complexes. The imidazolium salts are deprotonated *in situ* through the acetate base that is incorporated into the palladium salt precursor. In most cases, the reaction requires two equivalents of the ligand per metal center and leads to the formation of (NHC)$_2$PdCl$_2$ (Scheme 3.5).

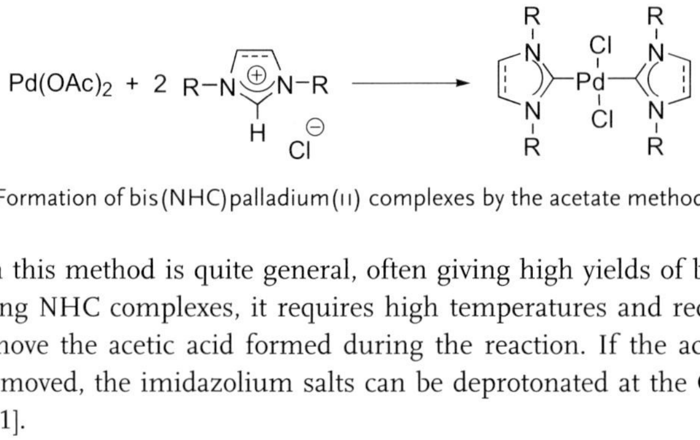

Scheme 3.5 Formation of bis(NHC)palladium(II) complexes by the acetate method.

Although this method is quite general, often giving high yields of both simple and chelating NHC complexes, it requires high temperatures and reduced pressure to remove the acetic acid formed during the reaction. If the acetic acid is not fully removed, the imidazolium salts can be deprotonated at the C4 position (Fig. 3.2) [31].

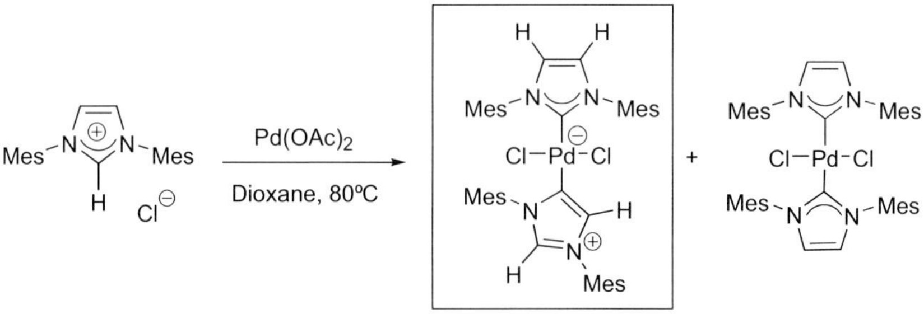

Fig. 3.2 Abnormal binding of NHC to a palladium center.

A second common method for the synthesis of Pd(II) complexes is the reaction of palladium salts with isolated or *in situ* generated NHC ligands. While the isolation of free NHC or a reliable means of deprotonation is required, this method is very versatile and has afforded many complexes bearing monocarbene, bis-carbene, chelating carbene, and mixed chelating ligands that consist of at least one carbene [32].

3.4
Palladium/NHC Complexes as Catalysts

Generally, it is accepted that the presence of only one NHC ligand per palladium center results in a very active cross-coupling catalyst in comparison to bis-carbene palladium(II) complexes. Consequently, significant efforts have concentrated on preparing monomeric NHC-bearing palladium(II) complexes to study their catalytic activity. Highlighted below are the most successful and promising NHC/palladium complexes that display the highest catalytic activity towards various cross-coupling processes.

3.4.1
C–N Bond-forming Reactions: the Hartwig–Buchwald Reaction

Initially, our efforts focused on the reaction between one equivalent of NHC with $PdCl_2(S)_2$ (S = MeCN, PhCN). The nitrile ligands were readily displaced by the carbene, affording dimeric [(NHC)PdCl$_2$]$_2$ complexes in high yields (Scheme 3.6) [33].

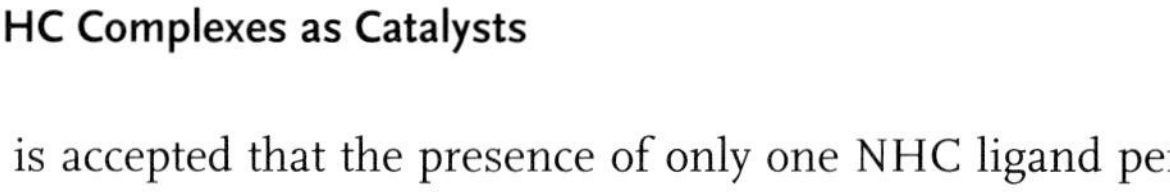

Scheme 3.6 Synthesis of dimeric monocarbene dichloropalladium(II) complexes.

The well-defined [(NHC)PdCl$_2$]$_2$ complexes have been used in the aryl amination of more economical and more difficult to activate aryl chlorides and amines/anilines, using DME as solvent and potassium *tert*-amylate (KO^{t}Am) as base. When activated aryl bromides and chlorides were used, the products were obtained in just a few minutes (Scheme 3.7). As evidence of the robust nature of the catalyst, the aminations can be performed in reagent grade solvent (without the need to exclude water) in air with little deleterious effect on either the yield of the product or reaction time.

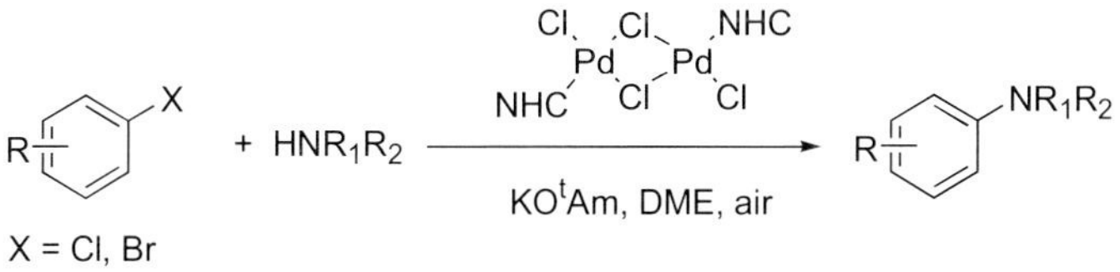

Scheme 3.7 Aryl amination mediated by [(NHC)PdCl$_2$]$_2$.

A key step in the design of catalysts for aryl amination is to eliminate the activation step [palladium(II) to palladium(0) before oxidative addition]. Obviously, well-defined palladium(0) complexes of (NHC)Pd(0) or mixed phosphine/NHC of the type (NHC)Pd(R$_3$P) would be best suited. These complexes have been examined and were shown to be efficient catalysts for the C–N bond-forming reaction at mild temperature. A second approach is to sidestep two required activation pathways in the catalytic cycle and eliminate the need for the preactivation and oxidative processes by using well-defined catalysts that are "oxidative addition" adducts, such as NHC-stabilized palladacycles (Scheme 3.8) [34].

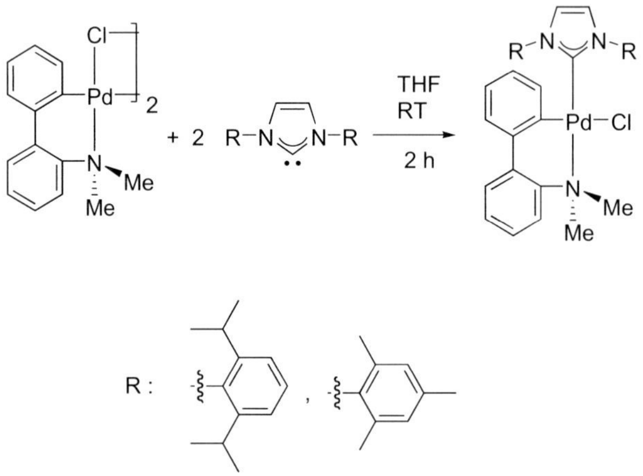

Scheme 3.8 Synthesis of NHC-palladacycles.

These complexes showed remarkable activity, despite using low catalyst loadings (0.5 mol%), in short reaction times for the aryl amination of aryl chlorides, triflates, and bromides. In addition, both aryl and alkyl primary and secondary amines were well tolerated.

The catalytic activity of the NHC-stabilized palladacycles mirrors that obtained employing the recently reported (NHC)Pd(R-allyl)Cl (R = H, Me, Me$_2$, Ph) system [35, 36]. This series of complexes are air and moisture stable and can be prepared on a multigram scale in high yields (Scheme 3.9). (IPr)Pd(R-allyl)Cl (IPr = 1,3-bis(2,6-diisopropylphenyl)imidazol-2-ylidene; R = H, Ph) compounds are now also commercially available from Strem Chemicals Inc. and Umicore AG.

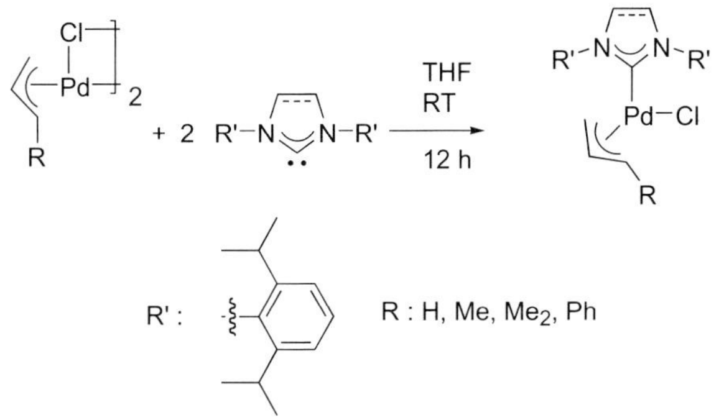

Scheme 3.9 Synthesis of [(NHC)Pd(R-allyl)Cl] complexes.

The catalytic behavior of the NHC allyl palladium series allowed excellent conversions into amine products at temperatures as low as room temperature. Furthermore, the alkoxide base lacking β-hydrogen atoms has a dual action in this system: (a) it acts as an efficient base for the catalytic process and (b) it is involved in the activation of the catalyst through nucleophilic attack on the palladium allyl moiety. Simple modification of the allyl moiety surrounding the palladium center resulted in dramatic changes in the catalytic performance of the system (Scheme 3.10). For example, by increasing the steric bulk at the terminal allyl moiety, a more rapid generation of the active catalytic species was achieved. As a result, amination reactions could be performed at room temperature with extremely low catalyst loadings (10 ppm), leading to the coupling of an unprecedented range of aryl bromides, triflates and chlorides with amines at room temperature in extremely short reaction times.

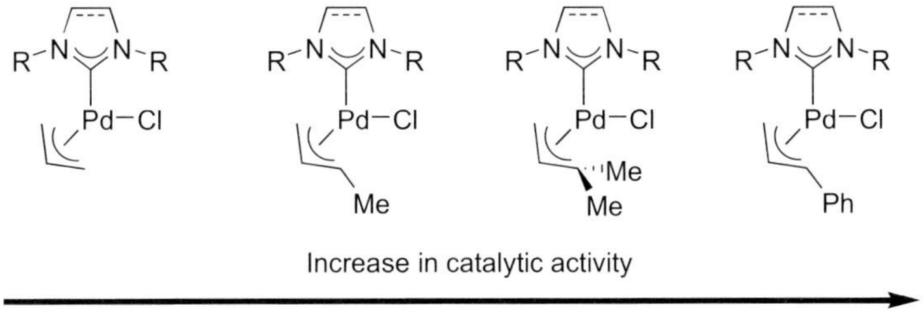

Scheme 3.10 Substitution effects on the (NHC)Pd(R-allyl)Cl system.

Another monocarbene palladium(II) series that allows easy modification and fine tuning of the ancillary ligands is the [(IPr)Pd(R$_2$-acac)Cl] series (Fig. 3.3) [37, 38]. These complexes are straightforward to synthesize and, like the above allyl series, are air and moisture stable precatalysts. Refluxing a mixture of Pd(R$_2$-acac)$_2$ (acac = acetylacetonate) and IPr·HCl (both commercially available) in dioxane affords complexes of the type [(IPr)Pd(R$_2$-acac)Cl] in high yields.

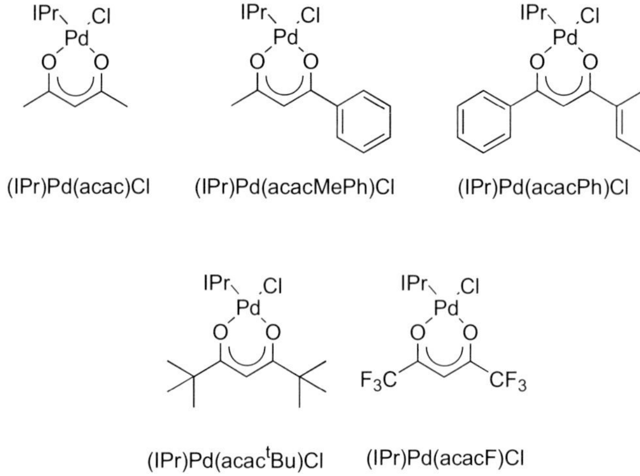

Fig. 3.3 [(IPr)Pd(R₂-acac)Cl] series of complexes.

Structures shown: (IPr)Pd(acac)Cl, (IPr)Pd(acacMePh)Cl, (IPr)Pd(acacPh)Cl, (IPr)Pd(acac^{t}Bu)Cl, (IPr)Pd(acacF)Cl.

Studies of the effect of substitution on the acac ligand in the Buchwald–Hartwig amination of dibutylamine and 4-chlorotoluene at 50 °C showed no conversion after 20 h with the (IPr)Pd(acacF)Cl complex. Interestingly, even at 80 °C no precipitation of palladium black was observed, indicating that the pathway leading to the catalytically active Pd(0) species was completely shut down by the presence at the terminal position of the strongly electron-withdrawing trifluoromethane groups. However, switching to electron-donating and strongly sterically demanding *tert*-butyl groups at the terminal position of the acac scaffold provided an extremely active but short-lived catalyst. Moreover, the activity of the (IPr)Pd(acacMePh)Cl complex, in terms of reaction rate and longevity of the catalyst, was an exact average between the results obtained for the regular (IPr)Pd(acac)Cl and (IPr)Pd(acacPh)Cl complexes.

Overall, the (IPr)Pd(R₂-acac)Cl complexes, with the exception of R = CF₃, were found to be efficient catalysts for the Buchwald–Hartwig amination. A wide range of aryl chlorides and bromides could be coupled with alkyl- or arylamines under mild conditions and short reaction times. This catalytic system appears to be particularly active for the formation of sterically hindered products such as tri- or tetrasubstituted diarylamines. Furthermore, this catalytic system does not bind with halopyridines and, therefore, can produce a large array of 2- and 3-aminopyridines in high yields (Scheme 3.11).

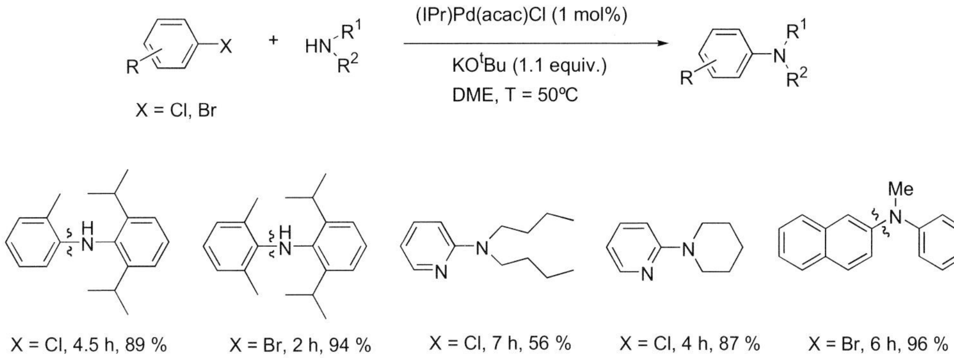

3.4 Palladium/NHC Complexes as Catalysts

Scheme 3.11 Aryl amination mediated by (IPr)Pd(acac)Cl.

3.4.2
C–C Bond-forming Reactions: α-Arylation of Ketones

The α-arylation of ketones is a powerful method of C–C bond formation. Since the discovery of direct coupling between simple ketones and aryl halides without the use of tin or silicon intermediates, the synthesis of α-aryl ketones has received renewed attention [11] as it constitutes the key step in the synthesis of a wide range of complex systems [1f, 39]. The enolate form of the ketone can be generated efficiently *in situ* and acts as transmetalating agent in the catalytic cycle. Extremely active NHC palladium systems have been reported for the α-arylation of esters at room temperature and for the arylation of amides [40]. Since ketones that possess α-protons can be deprotonated in the presence of strong bases (causing condensation of two ketone molecules), this side reaction can be minimized if a rapid oxidative addition of aryl halide and a fast reductive elimination to the desired product are at play.

α-Arylation of ketones can be performed in the presence of a catalytic amount of the thermally robust (NHC)Pd(allyl)Cl as precatalyst and NaO^{t}Bu as base (Scheme 3.12) [41]. Optimization of the reaction conditions found that IPr and SIPr (SIPr = *N,N'*-bis(2,6-diisopropylphenyl)dihydroimidazol-2-ylidene) resulted in higher yields of products at mild temperature and short reaction times. Furthermore, lowering the catalyst loading of (SIPr)Pd(allyl)Cl to 1 mol% [Na(O^{t}Bu), THF, 60 °C] in this system enabled the coupling of simple ketones and aryl chlorides, bromides, or triflates to proceed in high yield at mild temperature [35, 42, 43].

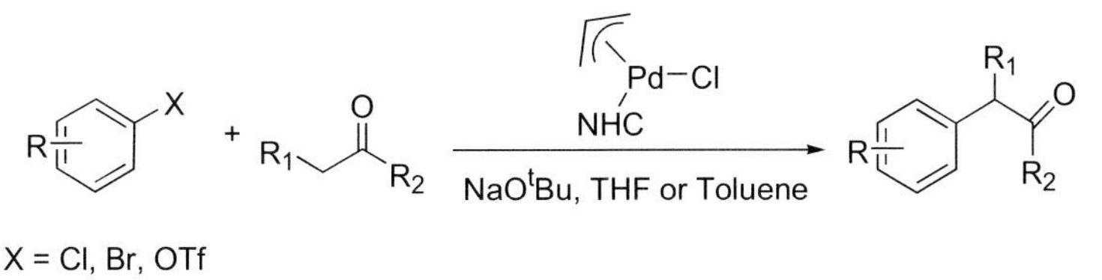

Scheme 3.12 α-Arylation of ketones catalyzed by (NHC)Pd(allyl)Cl complexes.

The reactivity of the (IPr)Pd(acac)Cl was also investigated as a precatalyst for the α-arylation of ketones [38]. Under mild reaction conditions (1 mol% of (IPr)Pd(acac)Cl, Na(O^{t}Bu), toluene, 60 °C), a large array of aryl chlorides and bromides could be efficiently coupled with aryl ketones. As previously observed in N-aryl amination, sterically demanding substrates were afforded in high yields and short reaction times. Additionally, ketones could be α-arylated with polyphenyl halides, providing a new type of substrate that has been previously inaccessible via other palladium-catalyzed cross-coupling reactions (Fig. 3.4).

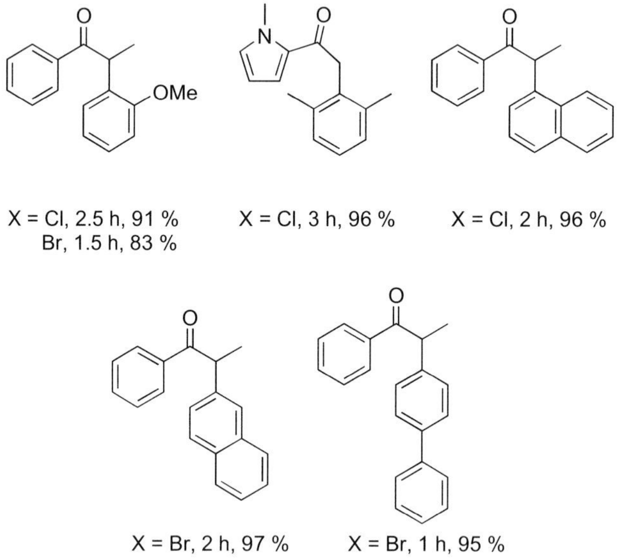

Fig. 3.4 New α-arylated ketones produced by (IPr)Pd(acac)Cl mediated coupling of aryl ketones with aryl chlorides and bromides (RX).

3.4.3
Suzuki–Miyaura Cross-coupling of Aryl Chlorides with Arylboronic Acids

An extremely versatile reaction, which has found extensive use in natural product and pharmaceutical intermediate synthesis, is the Suzuki–Miyaura reaction [44]. This reaction commonly makes use of arylboronic acids, ArB(OH)$_2$, as substrates together with aryl halides or triflates (Ar'X, X = halogen or triflate). Initial research centered on an *in situ* formed zero-valent (NHC)Pd(0) catalytic species. This was achieved using Pd$_2$(dba)$_3$, as the palladium source, the carbene IMes, and Cs$_2$CO$_3$ as base in dioxane [45]. Although this reagent combination afforded a moderate yield in the coupling of 4-chlorotoluene with phenylboronic acid, the catalytic protocol could be simplified using air-stable IMes·HCl that is deprotonated *in situ* with the base [46]. While a 30 min activation period was required to insure the reduction of Pd(II) to Pd(0) and the deprotonation of IMes·HCl, isolated yields for the Suzuki–Miyaura coupling of both aryl chlorides and triflates with phenylboronic acids reached an improved >95% (Scheme 3.13). In this study, NHC ligands bearing bulky ortho-substituted aryl groups on the nitrogen side arm afforded the highest yields, and clearly indicated the importance of steric factors on the effectiveness of this catalytic system. Furthermore, steric factors associated with ortho-substituted reagents decreased the yields and reaction rates (Scheme 3.13).

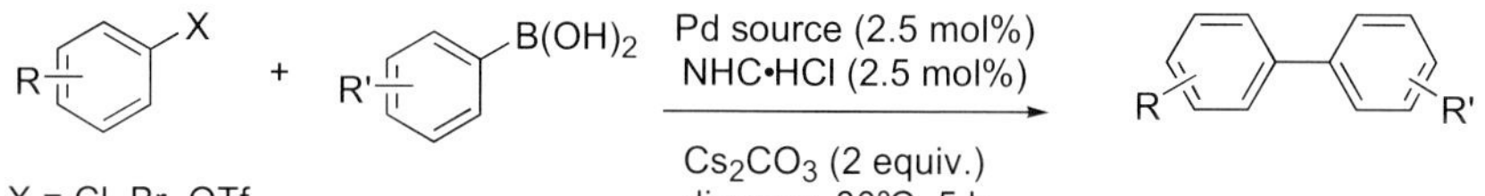

$$R\text{–}Ar\text{–}X + R'\text{–}Ar\text{–}B(OH)_2 \xrightarrow[\text{Cs}_2\text{CO}_3\text{ (2 equiv.)}, \text{ dioxane, 80°C, 5 h}]{\text{Pd source (2.5 mol\%)}, \text{ NHC·HCl (2.5 mol\%)}} R\text{–}Ar\text{–}Ar\text{–}R'$$

X = Cl, Br, OTf

Scheme 3.13 Cross-coupling of aryl halides and phenylboronic acids mediated by an *in situ* generated (NHC)Pd system.

The Suzuki–Miyaura reaction was subsequently extended to air-stable Pd(II) precatalysts, using the previously described (NHC)Pd(palladacycle)Cl [17] and (NHC)Pd(R-allyl)Cl complexes [36]. The former complex proved to be one of the most efficient catalytic systems for the synthesis of di- and tri-ortho-substituted biaryls when used in conjunction with NaO^{t}Bu as base, and isopropanol as solvent. Remarkably, this protocol coupled sterically hindered aryl chlorides with sterically hindered boronic acids at room temperature in minutes. The improvement in activity, compared with the above reactions in dioxane, which required longer reaction times and a slight excess of acid, suggested that the solvent was not innocent after all. The use of additives in the form of alcohol or water dramatically enhances the performance of the (IPr)Pd(palladacycle)Cl system due to the increasing solubility of the boronic acid [47] and by forming an unstable palladium hydride upon attack of the ^{i}PrO group followed by β-hydride elimination (Scheme 3.14).

Notably, the mechanism incorporating the catalytically active species (NHC)Pd, and not a (NHC)$_2$Pd species, has been the subject of much speculation. While (NHC)$_2$Pd complexes may very well act as precatalysts in these reports, recent work by Hollis et al. on the reaction of phenyl halides and phenylboronic acid mediated by a series of chelated and monocarbene complexes of palladium showed that the monocarbene system gave consistently higher yields of the desired product than did the chelating derivatives [48].

Economically and industrially, the conditions for the NHC palladacycle Pd complex are very appealing, especially regarding the use of an inexpensive and environmentally friendly solvent without pre-drying or purification. However, (NHC)Pd(palladacycle)Cl has two industrial disadvantages: (a) synthesis of the complex requires harsh reaction conditions and is low yielding and (b) slow addition of the aryl chloride is required when performing the coupling to avoid dehalogenation as a side reaction. Accordingly, we have examined the readily prepared (NHC)Pd(allyl)Cl as a precatalyst using the same simple catalytic conditions to confirm if a generality of the effect observed for the palladacycle complex existed.

Scheme 3.14 Activation of (NHC)Pd(palladacycle)Cl in the presence of alkoxide base.

We have reported recently the reaction profiles of various Pd complexes bearing N-heterocyclic carbene (NHC) or phosphine ligands (Fig. 3.5) and compared their activity to the (IPr)Pd(palladacycle)Cl complex [42].

The simple reaction conditions allowed, in most cases, the cross-coupling of aryl chlorides with arylboronic acids, yielding tri-ortho-substituted biaryls in high yields, in short times and mild temperature conditions. As a general trend, (IPr)Pd complexes perform better than (IMes)Pd complexes. At 50 °C, all IPr-bearing complexes performed identically in terms of yield and reaction time. Unfortunately, these catalysts failed to mediate the reaction at room temperature,

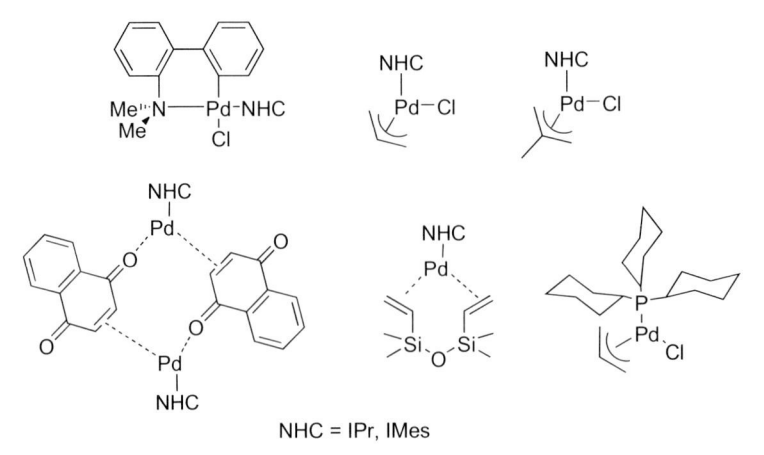

Fig. 3.5 Complexes tested in the Suzuki–Miyaura reaction.

with the exception of (IPr)Pd(palladacycle)Cl, which allowed for the coupling in that range of time. This suggested that generation of the active (NHC)Pd species depends not only on temperature but also on the steric environment around the palladium center. As also seen for aryl amination and α-arylation of ketone coupling reactions, a simple modification at the terminal allyl moiety surrounding the palladium center dramatically influences the catalytic performance of the (IPr)Pd(R-allyl)Cl system (Scheme 3.15) [36].

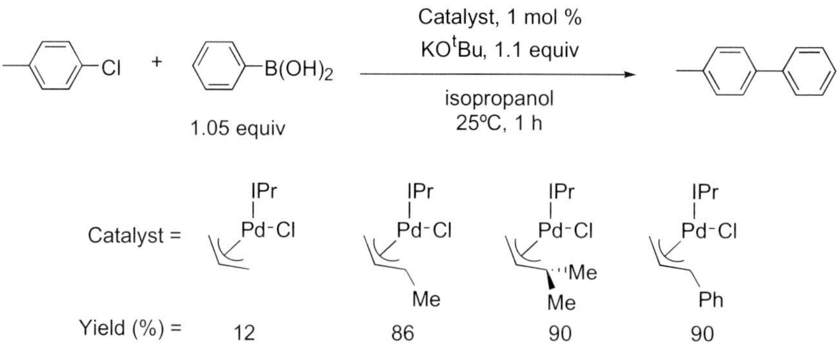

Scheme 3.15 Substitution effect at the allyl moiety of (IPr)Pd(R-allyl)Cl on the catalytic activity for the Suzuki–Miyaura cross-coupling reaction.

As a result, increasing the steric congestion around the palladium center enables the coupling of a large array of aryl bromides, triflates and chlorides with boronic acids at room temperature in extremely short reaction times. Remarkably, the catalyst loadings could be decreased to as little as 50 ppm despite using aryl chlorides as substrates, without sacrificing the yield of coupling product. While the unsubstituted (IPr)Pd(allyl)Cl complex displays no activity at room temperature, it can withstand high thermal conditions. This stability was shown in several cross-coupling microwave-assisted reactions that resulted in similar conversions to that obtained with conventional heating, with no formation of Pd black – a feature that was not successful using (IPr)Pd(palladacycle)Cl as precatalyst.

3.4.4
C–H Bond-forming Reactions: Dehalogenation of Aryl Halides

Dehalogenation of aryl halides, especially aryl chlorides, is an important transformation in organic synthesis and environmental remediation. The high thermal stability imparted by the NHC on palladium permits dehalogenation reactions at high or microwave-assisted temperatures. These aryl dehalogenation reactions, in the absence of coupling partners and operating with adequate bases, can be performed in the presence of electron-rich NHC/Pd complexes, either generated *in situ* or with well-defined (NHC)Pd(allyl)Cl and (NHC)Pd(palladacycle)Cl complexes [49, 50]. *In situ* catalytic systems were generated from imidazolium salts in the presence of a base. Metal alkoxides with β-hydrogen atoms, particularly methoxides, gave the best efficiency as base. Presumably, the reaction takes place by an alkoxide attack on the aryl palladium intermediate followed by β-hydrogen transfer to the palladium center. The aryl palladium hydride reductively eliminates the dehalogenated product and regenerates the Pd(0) species (Scheme 3.16).

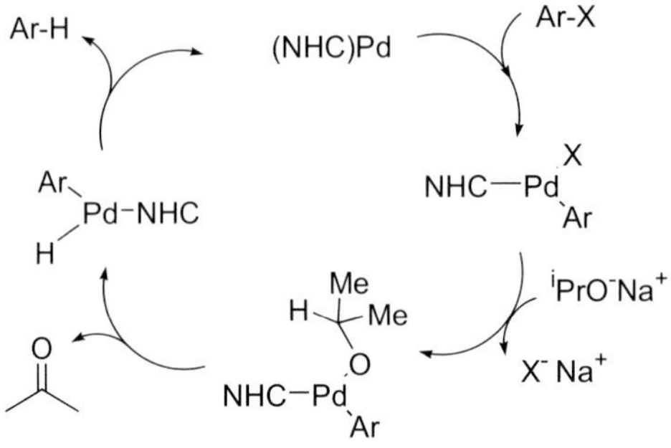

Scheme 3.16 Postulated mechanism for the catalytic dehalogenation of aryl halides.

While the *in situ* prepared palladium systems show high tolerance to functional groups present on the aryl moiety, the reaction temperature is relatively high (80 °C). Using the well-defined air- and moisture-stable (NHC)Pd(allyl)Cl complexes, aryl chlorides can be dehalogenated either under mild heating conditions (60 °C) or through microwave heating, where the reaction time is reduced to 2 min and the amount of catalyst to as low as 0.025 mol% (Scheme 3.17).

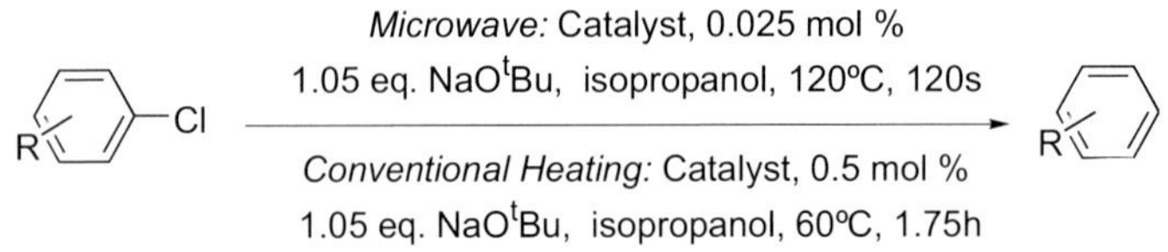

Scheme 3.17 Dehalogenation reactions mediated by (NHC)Pd(allyl)Cl.

As mentioned above, a disadvantage of (NHC)Pd(palladacycle)Cl in Suzuki–Miyaura cross-coupling was that slow addition of the aryl halide was required to avoid dehalogenation. However, in the absence of a boronic acid and the use of a stronger base, KO^{t}Bu, formation of dehalogenated products in quantitative yields occurred at room temperature in short reaction times. This noteworthy reaction was the first example of a homogeneous dehalogenation performed at room temperature. Furthermore, the catalyst loading could be reduced to only 1 mol% using technical grade isopropanol as solvent (Scheme 3.18) [51].

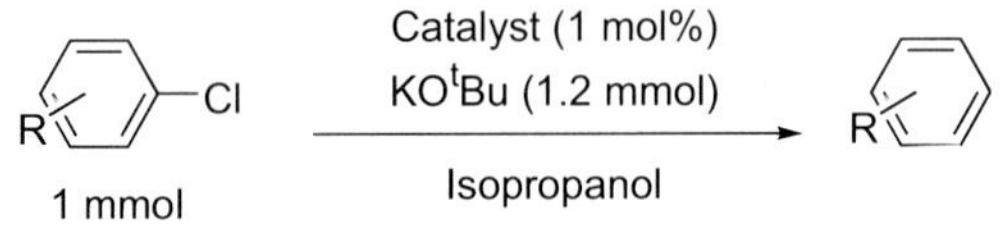

Scheme 3.18 Catalytic dehalogenation of aryl chlorides mediated by (NHC)Pd(palladacycle)Cl at room temperature.

3.4.5
C–C Bond-forming Reactions: Hydroarylation of Alkynes

A synthetic method that allows C–C bond formation of inexpensive starting materials (arenes) without prefunctionalization is the hydroarylation of alkynes. This reaction proceeds through transformation of C–H bonds. We have recently reported the synthesis of (NHC)Pd(OAc)$_2$ complexes in which the presence of the NHC was thought to stabilize this unusual series (Scheme 3.19) [52, 53].

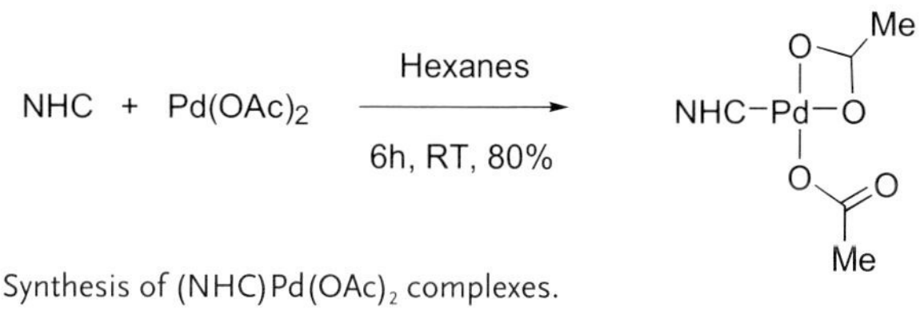

Scheme 3.19 Synthesis of (NHC)Pd(OAc)$_2$ complexes.

These NHC-stabilized palladium acetate complexes were found to be unaffected by harsh acidic conditions and show complex coordination behavior and versatile chemistry. This behavior contrasts with that of PR$_3$/Pd/acetate complexes, which lack stability due to the propensity of the supporting ligand to oxidize via an intramolecular process [54]. The complexes (IPr)Pd(OAc)$_2$ and (IPr)Pd(OOCCF$_3$)$_2$(H$_2$O) (Fig. 3.6) were investigated as precatalysts in the hydroarylation of alkynes and were found to active catalysts, allowing several arenes to react with various alkynes to form stilbenes in moderate to high yields (Scheme 3.20) [52]. While the insertion of a second alkyne molecule can also occur, when a sterically demanding NHC ligand is employed, such as IPr, this process is inhibited.

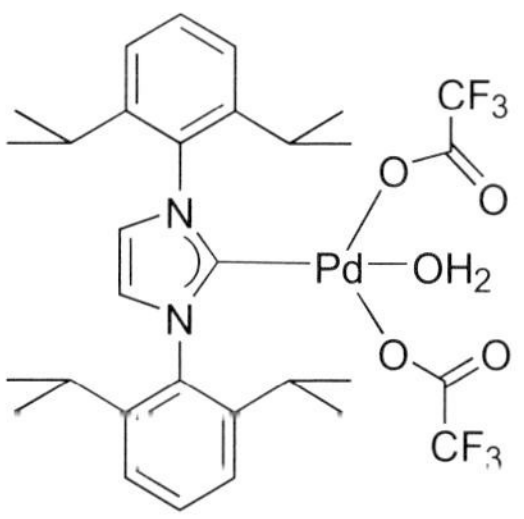

Fig. 3.6 Structure of (IPr)Pd(OOCCF$_3$)$_2$(H$_2$O).

The versatility of the (NHC)Pd(OAc)$_2$ (NHC = IMes and IPr) as a precatalyst is also realized in Suzuki–Miyaura cross-coupling reactions. This precatalyst efficiently mediates the Suzuki–Miyaura coupling of sterically hindered di-ortho and tri-ortho substituted diaryls under mild reaction conditions. In addition, this protocol was successfully employed in coupling sterically hindered aryl and activated alkyl chlorides bearing β-hydrogen atoms (Scheme 3.21) [53].

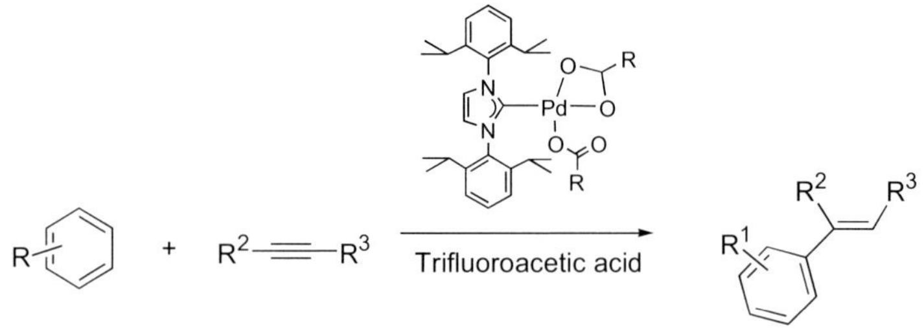

Scheme 3.20 Hydroarylation of alkynes mediated by (IPr)Pd(OAc)₂ complexes.

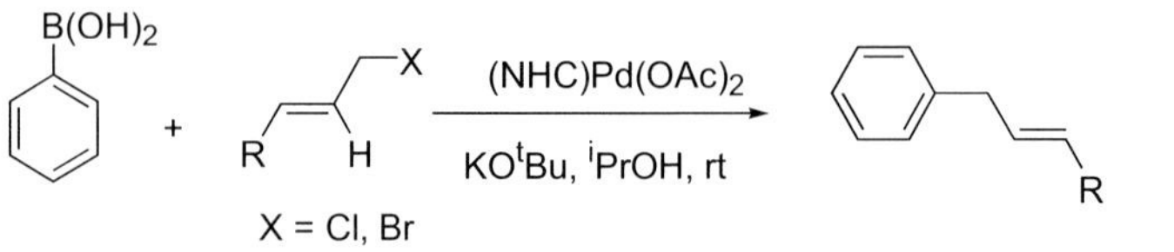

Scheme 3.21 Suzuki–Miyaura cross coupling of activated C(sp³) chlorides and bromides.

3.5 Conclusion

The use of N-heterocyclic carbenes as ligands for metals, especially Pd, in cross-coupling reactions has been very successful. Development of various well-defined air- and moisture-stable monocarbenepalladium(II) precatalyst systems has resulted in highly efficient coupling for C–C and C–N bond-forming reactions under mild conditions at room temperature. By increasing the steric hindrance of the co-ligands surrounding the palladium center, the 12-electron active (NHC)Pd(0) species has been generated more rapidly. This insight allowed highly efficient NHC/Pd-catalyzed cross-coupling reactions to be performed at room temperature in short reaction times.

References

1 (a) Christmann, U., Vilar, R. *Angew. Chem. Int. Ed.* **2005**, *44*, 366–374.
(b) A. F. Littke, G. C. Fu, *Angew. Chem. Int. Ed.* **2002**, *41*, 4176–4211.
(c) Wolfe, J. P., Wagaw, S., Marcoux, J. F., Buchwald, S. L. *Acc. Chem. Res.* **1998**, *31*, 805–818. (d) Hartwig, J. F. *Angew. Chem. Int. Ed.* **1998**, *31*, 852–860.
(e) Muci, A. R., Buchwald, S. L. *Top. Curr. Chem.* **2002**, *219*, 131–209.
(f) Hiyama, T., Shirawara, E. *Handbook of Organopalladium Chemistry for Organic Synthesis*, Wiley-VCH, New York, **2002**, vol. 1, pp. 285–309.

2 (a) For a review see: Miyaura, N., Suzuki, A. *Chem. Rev.* **1995**, *95*, 2457–2483. (b) Suzuki, A. *Metal-Catalyzed Cross-Coupling Reactions*, Wiley-VCH, Weinheim, **1998**, pp. 49–97 and cited references therein. (c) Hamann, B. C.,

Hartwig, J. F. *J. Am. Chem. Soc.* **1998**, *120*, 7369–7379. (d) Reez, M. T., Lohmer, G., Schwickardi, R. *Angew. Chem. Int. Ed.* **1998**, *37*, 481–483. (e) Littke, A. F., Fu, G. C. *J. Org. Chem.* **1999**, *64*, 10–11.

3 McGuiness, D. S., Green, M. J., Cavell, K. J., Skelton, B. W., White, A. H. *J. Organomet. Chem.* **1998**, 165–178.

4 Herrmann, W. A., Elison, M., Fischer, J., Köcher, C., Artus, G. R. J. *Angew. Chem. Int. Ed.* **1995**, *34*, 2371–2374.

5 McGuiness, D. S., Cavell, K. J., Skelton, B. W., White, A. H. *Organometallics*, **1999**, *18*, 1596–1605.

6 Selvakumar, K., Zapf, A., Spannerberg, A., Beller, M. *Chem. Eur. J.* **2002**, *8*, 3901–3906.

7 Selvakumar, K., Zapf, A., Beller, M. *Org. Lett*, **2002**, *4*, 3033–3034.

8 Vicui, M. S., Zinn, F. K., Stevens, E. D., Nolan, S. P. *Organometallics*, **2003**, *22*, 3175–3177.

9 Jackstell, R., Andreu, M. G., Frisch, A., Selvakumar, K., Zapf, A., Klein, H., Spannenberg, A., Röttger, D., Briel, O., Karch, R., Beller, M. *Angew. Chem. Int. Ed.* **2002**, *41*, 986–989.

10 Jackstell, R., Frisch, A., Beller, M., Rottger, D., Malaun, M., Bildtein, B. *J. Mol. Catal. (A).* **2002**, *185*, 105–112.

11 (a) Hamann, B., Hartwig, J. F. *J. Am. Chem. Soc.* **1997**, *119*, 12 382–12 383. (b) Palucki, M., Buchwald, S. L. *J. Am. Chem. Soc.* **1997**, *119*, 11 108–11 109. (c) Satoh, T., Kawamura, Y., Mirua, M., Nomura, M. *Angew. Chem. Int. Ed.* **1997**, *36*, 1740–1742. (d) Terao, Y., Fukuoka, Y., Satoh, T., Miura, M., Nomura, M. *Tetrahedron Lett.* **2002**, *43*, 101–104. (e) Fox, J. M., Huang, X., Chieffi, A., Buchwald, S. L. *J. Am. Chem. Soc.* **2000**, *122*, 1360–1370. (f) Culkin, D. A., Hartwig, J. F. *Acc. Chem. Res.* **2003**, *36*, 234–245.

12 (a) Wolfe, J. P., Singer, R. A., Yang, B. H., Buchwald, S. L. *J. Am. Chem. Soc.* **1999**, *121*, 9550–9561. (b) Bei, X., Turner, H. W., Weinberg, W. H, Guram, A. S., Petersen, J. L. *J. Org. Chem.* **1999**, *64*, 6797–6803. (c) Zapf, A., Ehrentraut, A., Beller, M. *Angew. Chem. Int. Ed.* **2000**, *22*, 4153–4155.
(d) Littke, A. F., Dai, C., Fu, G. C. *J. Am. Chem. Soc.* **2000**, *122*, 4020–4028.
(e) Pickett, T. E., Richards, C. J. *Tetrahedron Lett.* **2001**, *42*, 3767–3769.
(f) Feuerstein, M., Laurenti, D., Bougeant, C., Doucet, H., Santelli, M. *Chem. Commun.* **2001**, 325–326. (g) Feuerstein, M., Doucet, H., Santelli, M. *Synlett* **2001**, 1458–1460. (h) Liu, S.-Y., Choi, M. J., Fu, G. C. *Chem. Commun.* **2001**, 2408–2409.

13 Zapf, A., Beller, M. *Chem. Eur. J.* **2000**, *6*, 1830–1833.

14 Li, G. Y. *Angew. Chem. Int. Ed.* **2001**, *40*, 1513–1516.

15 Markò, I. E., Stérin, S., Buisine, O., Mignani, G., Branlard, P., Tinant, B., Declerq, J. -P. *Science* **2002**, *298*, 204–207.

16 Huang, J., Stevens, E. D., Nolan, S. P., Petersen, J. L. *J. Am. Chem. Soc*. **1999**, *121*, 2674–2678.

17 Navarro, O., Kelly III, R. A., Nolan, S. P. *J. Am. Chem. Soc.* **2003**, *125*, 16 194–16 195.

18 Clyne, D. S., Jin, J., Genest, E., Gallucci, J. C., RajanBabu, T. V., *Org. Lett*, **2000**, *2*, 1125–1128.

19 Powell, M. T., Hou, D, R., Perry, M. C., Cui, X. H., Burgess, K. *J. Am. Chem. Soc.* **2001**, *123*, 8878–8879.

20 Albrecht, M., Miecznikowski, J. R., Samuel, A., Faller, J. W., Crabtree, R. H. *Organometallics* **2002**, *21*, 3596–3604.

21 Mehlhofer, I. E., Stressner, T., Herrmann, W. A. *Angew. Chem. Int. Ed.* **2002**, *41*, 1745–1747.

22 Loch, J. A., Albrecht, M., Peris, E., Mata, J., Faller, J. W., Crabtree, R. H. *Organometallics* **2002**, *21*, 700–706.

23 Poyatos, M., Mas-Marza, E., Mata, J. A., Sanau, M., Poric, E. *Eur. J. Inorg. Chem.* **2003**, 1215–1221.

24 Cornils, B., Herrmann, W. A. *Applied Homogeneous Catalysis with Organometallic Compounds*, VCH, Weinheim, **1996**.

25 Arnold, P. A., Cloke, F. G. N., Geldbach, T., Hitchcock, P. B. *Organometallics* **1999**, *18*, 3228–3233.

26 Böhm, V. P. W., Gstöttmayr, C. W. K., Weskamp, T., Herrmann, W. A. *J. Organomet. Chem.* **2000**, *595*, 186–190.
27 Caddick, S. Cloke, F. G. N., Clentssmith, C. K. B., Hitchcock, P. B., McKerrecher, D., Titcomb, L. R., Williams, M. R. V. *J. Organomet. Chem.* **2001**, *617–618*, 635–639.
28 Titcomb, L. R., Caddick, S., Cloke, F. G. N., Wilson, D. J., McKerrecher, D. *Chem. Commun.* **2001**, 1388–1389.
29 Gstöttmayr, C. W. K., Böhm, V. P. W., Herdtweck, E., Groschem, M., Herrmann, W. A., *Angew. Chem. Int. Ed.* **2000**, *41*, 1363–1365.
30 Pytkowicz, J., Roland, S., Mangeney, P., Meyer, G., Jutand, A. *J. Organomet. Chem.* **2003**, *678*, 166–179.
31 Lebel, H., Janes, M. K., Charette, A. B., Nolan, S. P. *J. Am. Chem. Soc.* **2004**, *126*, 5046.
32 See references in Viciu, M. S., Nolan. S. P. *Top. Organomet. Chem.* **2005**, *14*, 241–278.
33 Viciu, M. S., Kissling, R. M., Stevens, E. D., Nolan, S. P. *Org. Lett.* **2002**, *4*, 2229–2231.
34 Viciu, M. S., Kelly, R. A. III, Stevens, E. D., Naud, F., Studer, M., Nolan, S. P. *Org. Lett.* **2003**, *5*, 1479–1482.
35 Viciu, M. S., Germaneau, R. F., Navarro, O., Stevens, E. D., Nolan. S. P. *Organometallics* **2002**, *21*, 5470–5472.
36 Marion, N., Navarro, O., Mei, J., Scott, N. M., Stevens, E. D., Nolan, S. P. *J. Am. Chem. Soc.* **2006**, *128*, 4101–4111.
37 Navarro, O., Marion, N., Scott, N. M., Stevens, E. D., Amoroso, D., Bell, A., Nolan, S. P. *Tetrahedron* **2005**, 9716–9722.
38 Marion, N., Ecarnot, E., Navarro, O., Stevens, E.D., Nolan, S.P. manuscript submitted.
39 (a) Abramovitch, R. A., Barton, D. H. R., Finet, J.-P. *Tetrahedron* **1988**, *44*, 3039–3071. (b) Muratake, H., Hayakawa, A., Natsume, M. *Chem. Pharm. Bull.* **2000**, *48*, 1558–1566.
40 Lee, S., Hartwig, J. F. *J. Org. Chem.* **2001**, *66*, 3402–3415.
41 Viciu, M. S., Germaneau, R. F., Nolan, S. P. *Org. Lett.* **2002**, *4*, 4053–4056.
42 Navarro, O., Oonishi, Y., Kelly, R. A., Stevens, E. D., Briel, O., Nolan, S. P. *J. Organomet. Chem.* **2004**, *689*, 3722–3727.
43 Weber, L., Strege, P. E., Fullerton, T. J., Dietsche, T. J., Trost, B. M. *J. Am. Chem. Soc.* **1978**, *11*, 3416–3426.
44 S. P. Stanforth, *Tetrahedron* **1998**, *54*, 263–303.
45 Zhang, C., Huang, J., Trudell, M. L., Nolan, S. P. *J. Org. Chem.* **1999**, *64*, 3804–3805.
46 Grasa, G. A., Viciu, M. S., Huang, J., Zhang, C., Trudell, M. L., Nolan. S. P. *Organometallics* **2002**, *21*, 2866–2873.
47 Ito, T., Molander, G. A. *Org. Lett.* **2001**, *3*, 393–396.
48 Vargas, V. C., Rubio, R. J., Hollis, T. K., Salcido, M. E. *Org. Lett.* **2003**, *5*, 4847–4849.
49 Viciu, M. S., Grasa, G. A., Nolan, S. P. *Organometallics* **2001**, *20*, 3607–3612.
50 Navarro, O., Kaur, H., Mahjoor, P., Nolan, S. P. *J. Org. Chem.* **2004**, *69*, 3173–3180.
51 Navarro, O., Marion, N., Oonishi, Y., Kelly, R. A. III, Nolan, S. P. *J. Org. Chem.* **2006**, *71*, 685–692.
52 Viciu, M. S., Stevens, E. D., Peterson, J. L., Nolan, S. P. *Organometallics* **2004**, *23*, 3752–3755.
53 Singh, R., Viciu, M. S., Kramareva, N., Navarro, O., Nolan, S. P. *Org. Lett.* **2005**, *7*, 1829–1832.
54 Tsuju, J. *Palladium Reagents and Catalysts – Innovations in Organic Synthesis*, Wiley-VCH, New York, **1998**, p. 2.

4
Pd-NHC Complexes as Catalysts in Telomerization and Aryl Amination Reactions

David J. Nielsen and Kingsley J. Cavell

4.1
Introduction

Pd-NHC catalyst systems have proven competitive with phosphane-ligated systems in many C–C bond-forming reactions such as the Heck reaction and the Suzuki, Kumada, Negishi, Sonogashira, etc. cross-coupling reactions. However, these catalytic processes fall outside the scope of this review and are addressed in the accompanying chapters of this book. The current chapter examines the impact of Pd-NHC catalyst systems on C–E coupling, as demonstrated by the telomerization and Buchwald–Hartwig aryl amination reactions.

The use of NHC ligands in catalysis continues to develop through the introduction of new NHC ligand architectures, and by the provision of additional donor groups on NHC ligands to generate chelating ligands. Recent advances have highlighted the importance of monoligated Pd(0) complexes as the active species in several catalytic cycles; thus the chemistry of these and related Pd(0)-NHC complexes is of some interest and will be discussed in detail in relation to the important and topical telomerization and aryl amination reactions. Additionally, we adress the nature and behavior of possible catalytic intermediates, particularly with respect to redox processes occurring at the metal center. In common with the mechanisms proposed for many palladium-catalyzed reactions, these processes involve the metal center undergoing oxidation and reduction during each circuit of the catalytic cycle. N-Heterocyclic carbenes appear to be very effective in promoting such reactions; indeed, a rich redox chemistry connecting "Pd(II)-NHC" and "Pd(0) + imidazolium" systems has been elucidated in recent years. These redox reactions also have relevance to the use of imidazolium-based ionic liquids as solvents in catalysis, a rapidly growing area given the attractiveness of ionic liquids as replacements for traditional organic solvent-based systems.

4.2
Telomerization

4.2.1
Definition and Background

Telomerization is an important reaction that builds functionalized long-chain dienes from 1,3-diene feedstocks in an atom efficient manner. The reaction entails the transition metal catalyzed coupling of two molecules of a conjugated 1,3-diene (the taxogen) with the addition of one molecule of a nucleophile HNu (the telogen), yielding mixtures of functionalized cis/trans dienes (telomers) that predominantly consist of "linear" (*n* or anti-Markovnikov, **1**) and "branched" (*iso*, **2**) isomers (Scheme 4.1). A transition metal catalyst, typically palladium, together with supporting ligand(s), is necessary to achieve worthwhile product yields. By-products of the reaction include trienes, **3**, resulting from linear 1,3-diene dimerization without nucleophile addition, and the product of the Diels–Alder addition of two 1,3-diene molecules, yielding 4-vinylcyclohexene (**4**) in the case of buta-1,3-diene. Several excellent reviews provide a good background to the telomerization reaction [1–3].

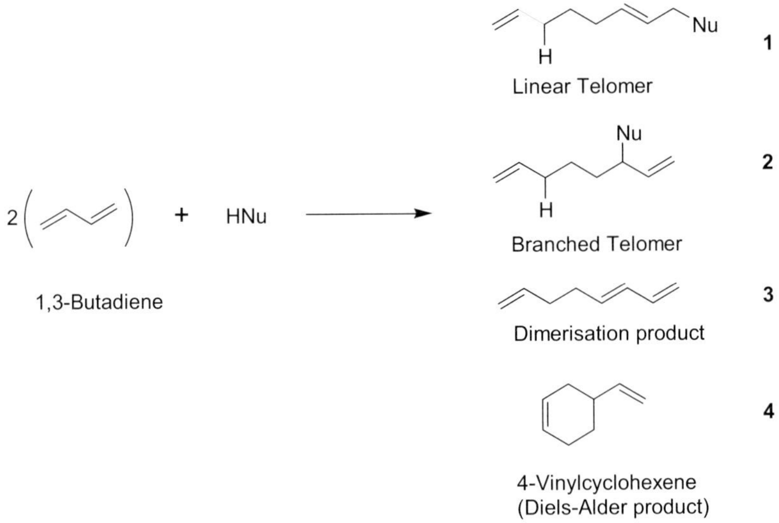

Scheme 4.1 Products and by-products of the telomerization of buta-1,3-diene with nucleophile HNu.

Buta-1,3-diene is the most frequently used 1,3-diene for telomerization, due to its low price and high reactivity [4], yielding telomers **5** and **6**. Isoprene (2-methylbuta-1,3-diene) is another attractive substrate that provides a potential route into terpenoid compounds [5, 6]. However, non-symmetrical 1,3-dienes increase the number of possible telomer products due to the additional complication of head-to-head, tail-to-tail, or mixed head/tail couplings, for which selectivity is often difficult to control. A range of nucleophiles have been investigated, including water, alcohols, aldehydes, amines, carboxylic acids, nitroalkanes, and compounds with active methylene groups [1, 7]. Telomerization with water (hydrodimerization) or ammonia is attractive as the products may be partially or completely hydrogenated to give long-chain alcohols or amines (which are used to build detergents). Various telomers are useful as plasticizers, surfactants, diesel fuel additives, and as intermediates in fine chemical and natural product syntheses [8–10]. Telomerization may be made enantioselective in the presence of suitable chiral supporting ligands or chiral auxiliaries [10].

5   **6**

In summary, the telomerization reaction is attractive due to its versatility, 100% atom efficiency (providing economic and ecological benefits that derive from the minimization of waste and simplification of the process chemistry), and its ability to lengthen the carbon chain while concurrently incorporating various functional groups and conserving unsaturation in the product. The value-adding potential, a consequence of employing cheap substrates (e.g., buta-1,3-diene, methanol, ethylene glycol, water, or ammonia), is an added bonus. The capacity to use a wide variety of telogens, including polyfunctional nucleophiles, means that a range of new and potentially valuable products, such as polyethers derived from carbohydrates [9, 11] (which are useful, potentially degradable, non-ionic surfactants), may be generated.

The telomerization of buta-1,3-diene with nucleophiles such as MeOH [12] and phenol [13] was first reported in 1967, catalyzed by palladium in the presence of PPh$_3$. Since then there have been many synthetic applications described, using a range of Pd-phosphane/phosphite catalyst systems, with further development resulting in generally modest improvements in catalyst activity and selectivity [4, 8]. A notable advance was the addition of NEt$_3$ as co-catalyst (ca. 2 mol%) in Pd(II)-PR$_3$ catalyst systems [4, 8], which was effective in promoting the reduction of Pd(II) to active Pd(0) species [14], thus increasing the observed productivity without affecting selectivity. However, despite advances in catalyst activity and selectivity, these rarely approached the levels necessary for commercialization, *vide infra*, and it is the recent application of NHC ligands in homogeneous catalysis [15, 16] that has led to renewed interest in the telomerization reaction.

4.2.1.1 Commercial Viability of the Telomerization Reaction

The nature of the catalyst, the nucleophile, and the 1,3-diene can have a significant impact on catalyst activity and product selectivity in the telomerization reaction, with both electronic and steric factors affecting the product distribution. As a consequence, each specific reaction must be optimized to yield the desired product with satisfactory selectivity. When low-cost feedstocks such as buta-1,3-diene, isoprene and MeOH are used, the high cost of the palladium catalyst (and, to a greater or lesser extent, the cost of the supporting ligands) becomes paramount, although this may be mitigated by low catalyst loadings or catalyst recycling. Before the introduction of Pd-NHC systems, commercialization of telomerization reactions was greatly hindered by the high catalyst loadings (with a few exceptions often in the order of 3 mol%) required to achieve acceptable conversions (turnover numbers, TONs) and/or reaction rates (turnover frequencies, TOFs = TON·h^{-1}), and by the relatively poor selectivity of the catalysts in generating the desired (usually linear) telomer [4, 9, 17]. Some catalyst systems also produced significant amounts of by-products **3** and **4** [8].

A consideration of catalyst cost versus substrate and product values requires TONs for palladium-catalyzed telomerization in the order of 10^5-10^6, which corresponds to maximum Pd loadings of only 0.001 mol% [4, 18]. High chemo- and regioselectivity for the desired linear telomer, typically >95% [17], should also be obtained to facilitate product isolation without expensive separation steps. Catalysts that can selectively telomerize the 1,3-dienes from mixed alkene feedstocks would, economically, be even more attractive.

The first substantial commercialization of telomerization was by the Kuraray company in Japan, where buta-1,3-diene was hydrodimerized with water in aqueous sulfolane using a Pd/phosphonium salt catalyst system [3]. Extraction with hexane allowed isolation of the octa-2,7-dienol thus produced, which was hydrogenated to yield octan-1-ol for use as feedstock for plasticizer production [3].

Telomerization reactions employing palladium catalysts immobilized in ionic liquids, specifically imidazolium salt ionic liquids, have also recently received attention and will be addressed in Section 4.2.5.

4.2.2
Catalyst Design: Ligand Selection

Until recently, homogeneous palladium/phosphane catalyst systems had yielded the greatest success in terms of activities and product selectivities for the telomerization of buta-1,3-diene and isoprene, although most employed relatively high catalyst loadings. Reports of telomerization reactions achieving reasonable results (TONs around 100 000) using low Pd loadings (in the order of 0.001 mol%) are limited to those of Beller et al., who studied the telomerization of buta-1,3-diene with methanol using various Pd(OAc)$_2$/PR$_3$ catalyst systems [4, 8]. Of the different phosphanes trialed, only PPh$_3$ and PCy$_3$ showed significant activity at low Pd loadings, with PPh$_3$ being superior [4]. Preformed Pd^0(diallylether)(triphenylphosphane) complexes (**7**) [19] with fixed Pd:PPh$_3$ ratios of 1:1 were also tested in the telomerization of MeOH and buta-1,3-diene but showed no advantage over the *in situ* generated catalysts with a similar Pd:PPh$_3$ ratio [8]. Study of low concentration Pd(0)/Pd(II)-PPh$_3$ catalyst systems showed that reaction rates and product selectivity were influenced by Pd:PPh$_3$ ratio, reaction temperature, and the diene:MeOH ratio. In the presence of large excesses of PPh₃, higher overall TON was observed at the expense of reaction rate (TOF) and regioselectivity for **5** [4]. Higher MeOH:butadiene ratios (2:1) improved the regioselectivity for **5**, especially when lower reaction temperatures were employed [8], whereas higher temperatures gave increased reaction rates and conversions, but reduced the chemoselectivity for telomers **5** and **6** with respect to **3** and **4**. The best regioselectivity for **5** was found with a Pd:PPh₃ ratio of 1:1 at reaction temperatures < 50 °C [4, 8]. Thus, by variation of the reaction conditions and the Pd:PPh₃ ratio these systems could be tuned to produce acceptable TONs, TOFs, chemoselectivities (the proportion of telomers **5** and **6** to by-products **3** and **4**), or regioselectivities (the proportion of linear telomer **5** to branched **6**). However, to date, no one Pd/phosphane system has been developed that can efficiently produce the desired **5** in good yield with high selectivity. Thus the recent development of the highly active and selective palladium/NHC catalyst systems [18, 20–22] described in Section 4.2.4 has, presently, largely overshadowed the results obtained with Pd/phosphane catalysts.

7

An area of some success for Pd/phosphane catalyst systems involves the use of water-soluble phosphanes, enabling good product yields to be obtained in aqueous biphasic systems where product separation is facilitated by simple solvent extraction, and catalyst recycling is often feasible [3, 23]. The use of biphasic systems can also prevent telomer products that are in themselves active nucleophiles from undergoing further reaction [24]. Palladium-NHC catalyst systems that can efficiently telomerize 1,3-dienes under aqueous conditions have not yet been reported.

4.2.3
Mechanism of the Pd-catalyzed Telomerization of Buta-1,3-diene with Methanol

Detailed mechanistic work on telomerization reactions catalyzed by Pd-NHC compounds has not been reported. However, the reaction mechanism and nature of the intermediates active in Pd-phosphane catalyzed telomerization have been extensively studied and the conclusions drawn for those catalysts are likely to be applicable to Pd-NHC systems given the parallels between these two ligand classes, especially when sterically demanding ligands are considered [25, 26]. Although it will be assumed that very similar mechanisms apply for both ligand types, actual catalytic performances may differ due to the high basicity and steric characteristics of NHCs.

The mechanism of the telomerization of buta-1,3-diene with methanol catalyzed by Pd-phosphane systems (particularly PPh$_3$) was elucidated in studies by Jolly and coworkers [7], with more recent complementary studies by Beller et al. to explain the observed regioselectivity of the reaction [8]. Thus, the following discussion focuses on the role of Pd-phosphane complexes and the nature of the intermediates in the telomerization reaction of buta-1,3-diene with MeOH. Studies specific to NHC ligands are limited to DFT calculations of the relative stabilities of various tricoordinate Pd^0(NHC)(ethene)$_2$ complexes as models for rate-determining telomerization intermediates [18]. Although the Pd/PR$_3$-catalyzed telomerization of buta-1,3-diene with methanol is the most thoroughly studied, mechanistic work on the telomerization of buta-1,3-diene in the presence of ammonia, yielding linear aminodienes [23], and the telomerization of isoprene with amines [5] has also been reported. Alternative telomerization mechanisms invoking bimetallic intermediates bridged by supporting ligands and/or η^3,η^3-octadiendiyl protoelomers (for which intermediates have also been isolated and characterized [27]) have been proposed [6]. However, bimetallic species are considered unlikely to form in the presence of bulky NHC spectator ligands; mechanisms that invoke such species are thus less relevant to the current discussion and will not be discussed further.

The generally accepted mechanism for the telomerization of buta-1,3-diene and MeOH (Scheme 4.2) invokes trigonal-planar Pd(0) and Pd(II) species containing only one spectator ligand, L, with the remaining coordination sites occupied by two molecules of buta-1,3-diene or the various coupling products derived therefrom [5, 8, 23]. The active catalyst is thus a Pd(0) complex bearing one strongly coordinated ligand L together with one or more weakly bound ligand(s) (e.g., solvent or substrate); formation of the L-Pd(0/II) fragment is thus critical to the reaction. Arguably, analogous processes would operate in the cases where L is a phosphane or an NHC; here the often-stated behavior of NHCs as phosphane-mimics allows the catalytic cycle to be drawn with some confidence, particularly when sterically demanding ligands are considered. A bulky spectator ligand L will assist in preventing agglomeration of low-valent Pd(0) species formed during the catalytic cycle. However, it will become apparent that the unique electronic properties of NHC ligands play an important role in the superior performance of Pd-NHC systems compared with Pd-phosphane catalyst systems.

If Pd(II) precursors are employed an initial reduction step is required to yield a Pd(0) precatalyst species that can enter the catalytic cycle, and this may lead to the observation of an induction period before the commencement of telomerization [4]. Under telomerization conditions, where there is a high concentration of buta-1,3-diene, any weakly bound ligand(s) on the Pd(0) precatalyst are displaced by two molecules of buta-1,3-diene, yielding entry into the catalytic cycle at **8**. The telomerization reaction then proceeds by oxidative coupling of the two coordinated buta-1,3-diene molecules to form the Pd^{II}(η^3,η^1-octadiendiyl)L complex (**9**). Electron-rich phosphanes facilitate this coupling [5] and it is likely that NHC ligands behave in a similar manner. Intermediate complexes of type **9** can be synthesized by the reaction of Pd^{II}(η^3-allyl)$_2$ with buta-1,3-diene (and other 1,3-

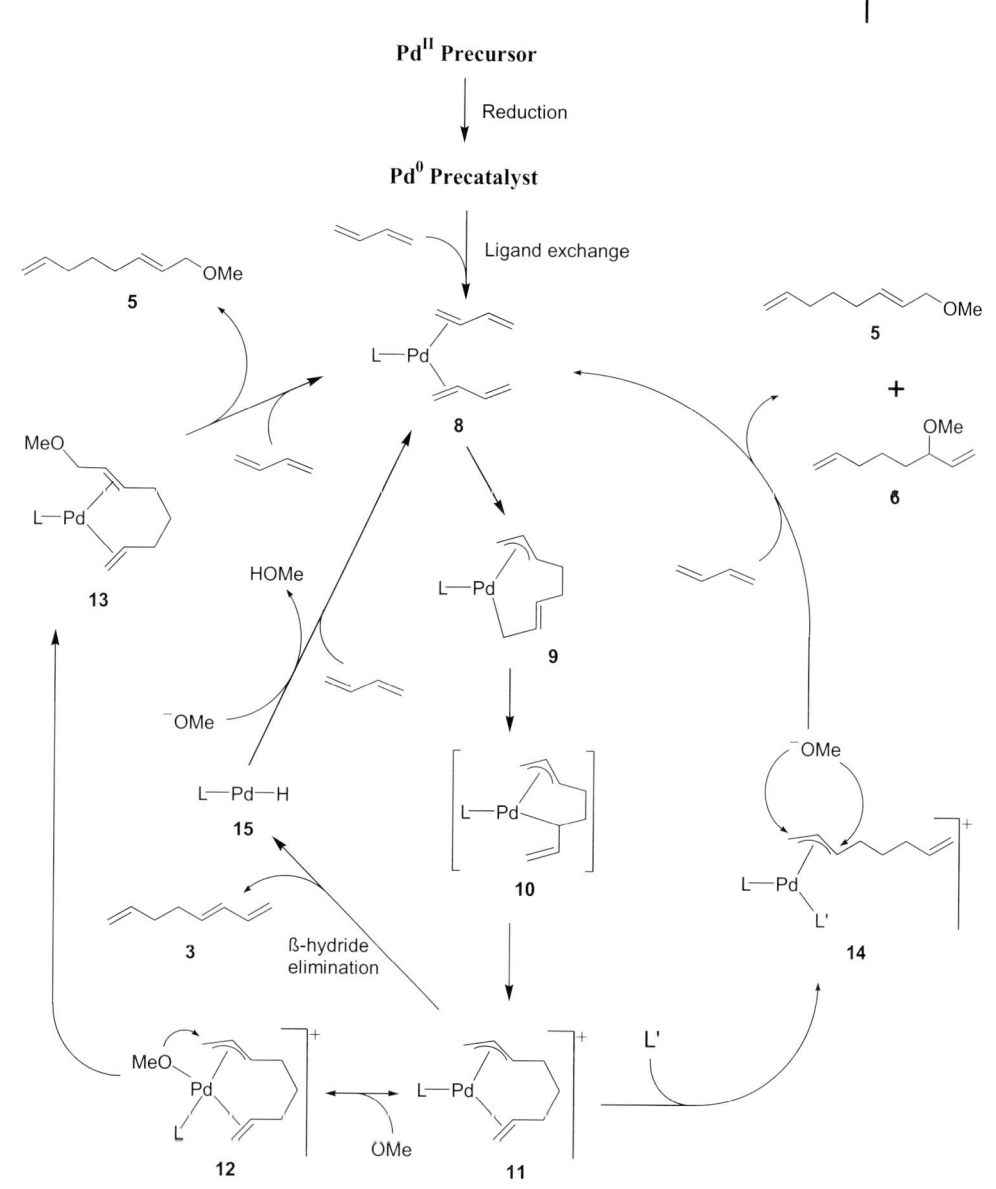

Scheme 4.2 Proposed mechanism for the telomerization of buta-1,3-diene with methanol, showing intermediates and pathways to various products and by-products.

dienes) in the presence of a phosphane ligand, L [7, 28, 29]. They can be isolated at low temperature and examples have been characterized by X-ray crystallography [28, 30, 31] (Fig. 4.1). The Pd(II) complex **9** (L = PPh$_3$) is stable towards rearrangement to the η^3,η^3-octadiendiyl complex [32]. Rearrangement of **9** to the Pd(6-η^1,η^3-

octadiendiyl)L intermediate **10**, bearing a pendant vinyl group, has been proposed to explain the exclusive protonation of the η^3,η^1-octadiendiyl chain at the C6 position (as shown by deuteration studies using MeOD or CH$_3$COOD) [7]; this protonation yields the cationic η^3,η^2-octadienyl complex **11** [7, 8, 23]. However, intermediates of type **10** have yet to be detected, and DFT calculations by Szabó indicate that the steric strain caused by adoption of the five-membered palladacycle in **10** is considerably greater than that in complex **9** [33]. Furthermore, coordination of the internal C6 carbon to palladium further destabilizes **10** with respect to **9** through electronic effects [33]. Electronic and steric factors contribute to the stability of the reactant, product, and transition structures involved in electrophilic attack on the η^3,η^1-octadiendiyl chain at C6, via a process that maximizes d–π type hyperconjugative interactions, resulting in the observed high regioselectivity for protonation at C6 rather than C8 [33].

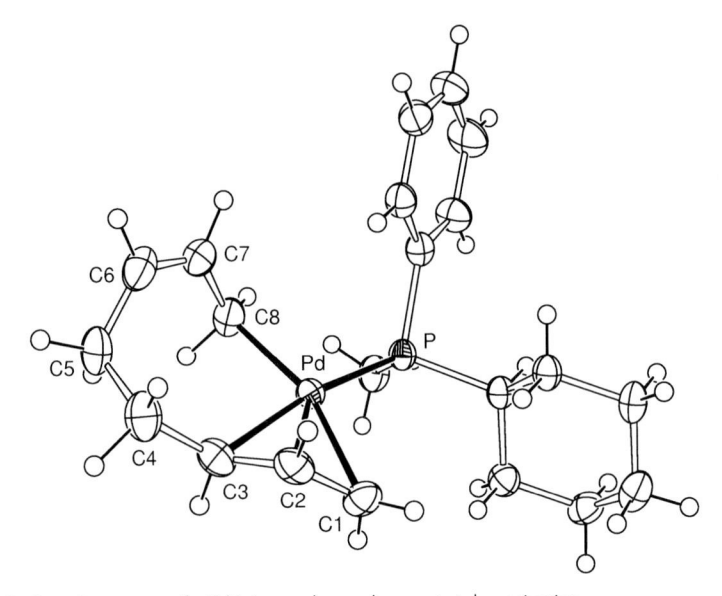

Fig. 4.1 Crystal structure of a Pd(II) complex analogous to telomerization intermediate **9**, showing the numbering scheme for the η^1,η^3-octadiendiyl chain. (From Storzer et al. [30].)

Coordination of methoxide to the Pd(II) center of **11** to give square planar **12** (showing arbitrary ligand configuration about Pd in Scheme 4.2) has been proposed by Jolly [7] on the basis of ^{13}C NMR evidence [29]. In reactions promoted by ammonia, Prinz et al. propose external (i.e., intermolecular) attack by the soft nucleophile ammonia on complex **11** [23]. Importantly for the regioselectivity of the telomerization reaction, the π-accepting η^2-alkene moiety of the η^3,η^2-octadienyl ligand in **11** or **12** will act to direct nucleophilic attack to the ligand trans to itself, and thus to C1 rather than C3 of the η^3-allyl moiety [23], although if L is also a π-acceptor this influence may be small. The combination of a rigid conformation engendered by the η^3,η^2-octadienyl ligand and a sterically demanding L should also give steric preference to attack at C1. Beller et al. suggest that this is the critical driving force for regioselectivity in the telomerization reaction [8]. Nucleophilic attack by methoxide at C1 of the η^3,η^2-octadienyl ligand of **11** or **12** yields complex **13**, which contains the linear telomer **5** as a neutral η^2,η^2-diene ligand. Jolly has shown that **5** can be displaced and **8** (L = PMe$_3$) regenerated by reaction of **13** with excess buta-1,3-diene [7]. Moreover, in the presence of the extremely bulky NHC, Me$_2$IPr [Me$_2$IPr = 1,3-bis(di-isopropylphenyl)-4,5-dimethylimidazolin-2-ylidene, **16**], transformation of the coordinated telomer diene in complex **13** to the linear triene **3** via subsequent elimination of alcohol has been ruled out [34], although conversion of **5** into **3** has been reported under certain conditions [13].

Alternatively, displacement of the η^2-alkene moiety of the η^3,η^2-octadienyl chain of **11** or **12** by some other ligand L′ (e.g., additional equivalents of phosphane, excess buta-1,3-diene) gives complex **14** (showing arbitrary L, L′ orientation in Scheme 4.2), which bears an η^3-octadienyl ligand where the alkyl chain conformation is no longer fixed. Methoxide attack on the η^3-allyl moiety of **14** is no longer directed to C1 through steric or electronic effects as L′ is likely to be a π-acceptor ligand under telomerization conditions; attack at C3 produces the branched telomer **6**, which is liberated in the presence of excess buta-1,3-diene to regenerate **8**, and poor regioselectivity is observed.

The linear dimerization product octa-1,3,7-triene (**3**) is generated by β-hydrogen elimination of C4-H on the η^3-octadienyl ligand of **11**, **12**, or **14**, giving **3** and a Pd^{II}(hydrido)(L) complex (**15**) that undergoes Pd-H abstraction by methoxide, in the presence of buta-1,3-diene, to regenerate complex **8**. Notably, however, the coordinated η^3,η^2-octadienyl ligand of **11** or **12** imposes a conformation that is likely to preclude C4-H interaction with Pd. Alternatively, to explain octa-1,3,7-triene formation, Beller et al. propose hydrogen abstraction by methoxide from C4 of the η^3,η^2-octadienyl chain on **11** or **12**, giving **3** by β-hydrogen elimination [34].

Complex **11** is considered by Jolly [7] to be the key telomerization/dimerization "switch", and by Beller et al. [8] to be critical in selecting for linear telomers, i.e., it is this complex that allows both chemo- and regioselective control of the reaction in the presence of bulky supporting ligands, L. In support of these suggestions, both linear and branched telomers, as well as linear 1,3-diene dimerization products, can be generated from complex **11** depending on the outcome (or absence) of directed attack of the nucleophile on this species. The chemo- and regioselective control (or loss thereof) of the telomerization reaction thus stems from the formation of key intermediates of type **11**, with the latter (and perhaps also the former) largely dependent on the geometry enforced by the η^3,η^2-octadienyl ligand; the overall rate of reaction is determined by the stability of the same intermediate. The role of the coordinated η^2-alkene moiety of complexes **11** or **12** in affecting chemo- and regioselectivity is likely to be three-fold. Specifically, the resulting imposed η^3,η^2-octadienyl ligand conformation inhibits β-hydride elimination, the η^3,η^2-octadienyl ligand is positioned for nucleophilic attack at C1 (which is further enforced by a bulky spectator ligand) and, finally, the π-acceptor properties of the η^2-alkene moiety electronically directs nucleophilic attack to C1.

The latter factor may be especially significant when NHC ligands are employed rather than phosphanes, as the former are not effective π-acceptors compared to the η^2-alkene moiety of the η^3,η^2-octadienyl ligand. Furthermore, NHC ligands exhibit poor π-acceptor properties compared to other available ligands (most likely 1,3-dienes in the absence of phosphanes) that may displace the η^2-alkene moiety of the η^3,η^2-octadienyl ligand under telomerization conditions. Thus, having a π-acceptor alkene ligand trans to C1, in combination with an NHC ligand (**11**, L = NHC), will always result in strongly electronically directed nucleophilic attack at C1 and the predominant formation of linear telomers. These effects account for the proclivity of the nucleophile to attack the coordinated η^3,η^2-octadienyl chain almost exclusively at C1 in the presence of NHC ligands, and provide insight into the impressive regioselectivity observed when NHC ligands are used in telomerization. However, if the η^2-alkene moiety of the η^3,η^2-octadienyl ligand in **11** is displaced, yielding **14** (L = NHC), isomerization, either by L/L' transposition or through flipping of the η^3-allyl ligand, to place the NHC trans to C1 is possible. A mixture of linear and branched telomers will then result through diminished regioselective control of the nucleophilic attack. The loss of regioselective control and influence on activity observed in the presence of high concentrations of π-acceptor ligands [4, 8] (i.e., when excess phosphanes or high buta-1,3-diene:-MeOH ratios are employed) are thus explained by the formation of **14** from **11**. Thermodynamic considerations also favor the production of linear telomers **1** over branched **2** due to the internal double bond in the former; however, Beller et al. found that the reaction is not under thermodynamic control [8].

Although regioselectivity is likely to be primarily under electronic control in the presence of NHC ligands, steric effects also play a role in both the chemo- and regioselectivity of the telomerization reaction. Additionally, when non-symmetrical 1,3-dienes such as isoprene are used, different telomers are possible, depending on the relative orientations of the ligands in the precursor PdL(1,3-diene)$_2$ complex; these orientations may be a consequence of the steric constraints imposed by bulky ligands, L. Thus, in the presence of sterically demanding tris(2,4,6-trimethoxyphenyl)phosphane, the telomerization of isoprene with amines was found to yield predominantly head-to-head coupled telomers [5]. In respect of regioselectivity, steric effects related to the bulk of the supporting ligand L on complexes of type **11** can play a part in directing nucleophilic attack to C1, as this position is the least substituted and therefore most accessible part of the η^3-allyl moiety. Conversely, inductive effects favor nucleophilic attack at the more substituted C3 carbon. However, nucleophiles of only moderate steric bulk have been shown to have difficulty entering at the internal carbon, especially when it is further substituted [35].

The steric influence on the chemoselectivity of the telomerization reaction is highlighted by the almost complete switch from telomers to linear dimerization product octa-1,3,7-triene (**3**) (as a mixture of cis/trans isomers) observed when a combination of the sterically demanding NHC ligand **16** and nucleophile of moderate steric bulk (secondary alkoxide) were employed [34]. Use of the less bulky IMes [IMes = 1,3-bis(2,4,6-trimethylphenyl)imidazolin-2-ylidene, **17**] ligand and/

or a less bulky nucleophile restored the selectivity to telomers with minimal formation of **3**. Notably, the use of tertiary alcohols in the presence of **16** resulted in very low levels of buta-1,3-diene conversion [34], suggesting that interaction of the nucleophile with intermediate **11** is crucial to both the formation of telomers and linear dimerization products. The production of **3** increases at higher temperature; however, the regioselectivity of the telomerization reaction for linear or branched telomers shows relatively little temperature influence [8].

Other by-products of the telomerization process arise from reactions peripheral to the catalytic cycle shown in Scheme 4.2. Most significant of these is 4-vinylcyclohexene (**4**), formed by the Diels–Alder coupling of two buta-1,3-diene molecules. Formation of such self-coupling by-products becomes increasingly significant at higher reaction temperatures (particularly above 100 °C) and at high 1,3-diene concentrations [8].

4.2.4
Pd-(NHC) Complexes as Telomerization Catalysts

In recent years phosphane ligands have been supplemented or supplanted in many catalytic reactions by nucleophilic carbenes (NHCs). Benefits such as ease of handling and better control of metal/ligand stoichiometry are possible with NHCs as they are less prone to dissociate from the metal center, they do not suffer from P–C bond cleavage at elevated temperatures, and are not as sensitive to trace oxygen [15]. However, NHCs may participate in reductive coupling reactions, liberating imidazolium salts in the presence of certain other ligands (specifically, hydrocarbyl and hydrido ligands) on the metal center [36], although there is no report to date of this behavior under telomerization conditions.

NHCs possess somewhat greater σ-donor (although trialkylphosphanes such as P^{t}Bu$_3$ and PCy$_3$ may be comparable) and much poorer π-acceptor properties than phosphanes [25]. Additionally, the steric regime of NHCs has been likened to that of a fence, rather than the cone of typical trialkyl/aryl phosphanes; a comparison of steric and electronic parameters of selected phosphane and NHC ligands is available [25, 26]. In ruthenium-NHC complexes the large N-substituents on IMes (**17**) and IPr (**18**) can prevent these ligands from closely approaching the metal center and in turn reduce the M–NHC bond strength [26].

Applications of Pd-NHC catalyst systems to telomerization are limited to reports describing the reaction of buta-1,3-diene with alcohols (including phenols) [17, 18, 20, 21, 37, 38] and amines [22]. However, the reported activities and chemo- and regioselectivities far exceed those of phosphane ligated systems, especially with respect to the telomerization of buta-1,3-diene with methanol [18]. Palladium-NHC systems have recently been applied on pilot plant scale for the highly selective production of more than 15 000 kg of telomers from buta-1,3-diene and MeOH [39]. These in turn have been selectively converted into oct-1-ene [37, 38] for use in oct-1-ene/ethylene copolymers (these so-called "high performance polyethylenes" are the fastest growing class of polyolefin).

Me₂IPr, 16   **IMes, 17**   **IPr, 18**

Me₂IMes, 19   **Cl₂IMes, 20**   **SIPr, 21**

SIMes, 22

23   **24**   **25**: *p*-benzoquinone; NHC = SIMes
26: naphthoquinone; NHC = IMes, IPr   **27**

The active catalysts in all reported Pd-NHC systems are thought to be Pd(0) complexes bearing a single NHC ligand, with more sterically demanding ligands such as IMes (**17**) showing the greatest activity. The first reported example of the application of NHCs as spectator ligands in telomerization used Pd^{II}ClMe(tmiy)$_2$ (tmiy = 1,3,4,5-tetramethylimidazolin-2-ylidene) as catalyst precursor in the presence of 1 mol% NaOH [40]. The resulting system showed modest activity (TOF ca. 50 h^{-1}) but very high selectivity for the linear telomer in the telomerization of buta-1,3-diene with MeOH. Since that time various precatalyst systems have been employed to generate the active catalytic species, including Pd^0(NHC)(dvds) (**23**, dvds = 1,3-divinyl-1,1,3,3-tetramethyldisiloxane or similar complexes bearing one NHC ligand and a weakly bound diene such as diallyl ether or hepta-1,6-diene) [17, 18, 20, 37], Pd^0(NHC)(olefin) complexes (olefin = dimethyl fumarate, **24**; *p*-benzoquinone, **25**; naphthoquinone, **26**) [41], Pd^0(NHC)$_2$ complexes [17, 41], and Pd^{II}Cl(η^3-allyl)(NHC) complexes (**27**) [18, 22], together with *in situ* systems derived from imidazolium salts and Pd(0/II) precursors [18, 21, 37, 41]. The Pd⁰(NHC)(dvds) systems (**23**) are analogues of some of the more successful Pd-phosphane systems, such as **7**, described in Section 4.2.2.

The Pd⁰(NHC)(dvds) complexes **28–32** are readily prepared by several routes [20]. A convenient preparation is by the reaction of a commercially-obtained Pd(0)-dvds solution with the free NHC, followed by low-temperature recrystallization from pentane [18, 34]; Pd⁰(IMes/IPr)(dvds) complexes **28** and **31** are commercially available through Umicore, as are [Pd⁰(IMes/IPr)(naphthoquinone)]$_2$ complexes **26**. The dvds complexes **23** are significantly more stable than the diallyl ether or hepta-1,6-diene analogues and may be handled even with exposure to atmospheric oxygen [18].

28: Ar = Mesityl, R = H
29: Ar = Mesityl, R = Me
30: Ar = Mesityl, R = Cl
31: Ar = 2,6-di-(isopropyl)phenyl, R = H
32: Ar = 2,6-di-(isopropyl)phenyl, R = Me

Complex **28** is highly active for the telomerization of buta-1,3-diene with methanol at 70–90 °C, and TONs of 96 000 may be achieved while maintaining almost complete conversion and selectivity for the linear telomer **5**, Table 4.1 (entry 4) [18, 20]. Lower loadings of **28** in the presence of an 80-fold excess of IMes.HCl (i.e., the imidazolium salt precursor to IMes) lead to exceptionally high TONs of 1.54×10^6, at the expense of reduced conversion, but without detrimental effects on the extremely high selectivity for **5**, with negligible amounts of **3** and **4** produced, Table 4.1 (entry 5) [18]. The excess imidazolium salt is perhaps acting as a reservoir of IMes in case of decomposition, but the retention of regioselectivity implies that the catalytically active species contains only one NHC ligand (i.e., resembles complex **11**) [18]. Preformed and *in situ* prepared catalysts, *vide infra*, performed similarly (Table 4.1, entries 2 to 4) [18, 34]. At lower temperatures using **28**, the regioselectivity is even more pronounced; at 50 °C **5** is nearly the sole product (<1% impurities), although this is achieved at the expense of conversion (57%, with TON = 57 000 and TOF = 3560 h^{-1}) [20]. The results listed in Table 4.1 remain the best reported to date for the telomerization of buta-1,3-diene with methanol.

Subsequent variation of the NHC ligand on **28** led to a range of complexes, **29–32**, bearing IMes and IPr ligands modified to allow investigation of the steric and electronic influences of the ligand [18]. Notably, N-aryl substituted imidazolin-2-ylidenes are in themselves considerably less basic than their N-alkyl counterparts, by at least five pK_a units, due to the electron-withdrawing inductive effect of the

Table 4.1 Comparative representative telomerization reactions of buta-1,3-diene with methanol catalyzed by Pd-(NHC) systems.

Entry	Catalyst	Ligand	Pd (mol%)	Conditions[a]	Yield (%)[b]	Regio. (%)[c]	Chem. (%)[d]	TON[e]	TOF[f]	Ref.
1	Pd(OAc)$_2$/3(PPh$_3$)[g]	PPh$_3$	0.001	A	26	96	87	26 000	1625	[18]
2	Pd(OAc)$_2$/4(IMes.HCl)	IMes	0.001	A	94	98	>99	94 000	5875	[18]
3	Pd(dba)$_2$/2(IMes)	IMes	0.001	B	92	97	98	92 000	5750	[21]
4	28	IMes	0.001	A	96	98	>99	96 000	6000	[18]
5	28	IMes	0.00005	B*	77	98	99	1 540 000	96 250	[18]
6	29	Me$_2$IMes	0.001	A	93	98	99	93 000	5813	[18]
7	30	Cl$_2$IMes	0.001	A	96	98	>99	96 000	6000	[18]
8	31	IPr	0.001	A	90	92	97	90 000	5625	[18]
9	32	Me$_2$IPr	0.001	A	2	91	–	2000	125	[18]
10	42	IMes	0.001	A	94	98	99	94 000	5875	[18]
11	43	IPr	0.001	A	46	92	96	46 000	2875	[18]
12	42	IMes	0.0001	B**	89	98	98	890 000	55 625	[18]
13	24	IMes	0.0049	C	84	98	–	8574	8574	[41]
14	25	SIMes	0.0034	C	89	97	–	13 215	13 215	[41]
15	Pd(SIPr)$_2$	SIPr	0.0042	C*	85	88	–	9994	9994	[41]
16	Pd(dba)$_2$/2(34)	34[h]	0.001	B	73	98	96	73 000	4560	[21]
17	Pd(dba)$_2$/2(35)	35[h]	0.001	B	8	99	75	8000	500	[21]
18	Pd(dba)$_2$/2(36)	36[h]	0.001	B	88	97	97	88 000	5500	[21]
19	Pd(dba)$_2$/2(37)	37[h]	0.001	B	82	97	97	82 000	5125	[21]

a) A: 16 h, 70 °C, 1.0 mol% NaOMe, MeOH:buta-1,3-diene = 2:1. B: 16 h, 90 °C, 1.0 mol% NaOMe, MeOH:buta-1,3-diene = 2:1; B*: 80 equivalents IMes.HCl relative to Pd; B**: 40 equivalents IMes.HCl relative to Pd. C: 1 h, 90 °C, 3.0 mol% NaOMe, MeOH:buta-1,3-diene = 2:1; C*: 20 min reaction time.
b) Total yield of telomers **5** and **6**.
c) Regioselectivity for linear telomer **5** = (Yield **5**)/(Yield **5** + **6**) × 100.
d) Chemoselectivity for telomers = (Yield **5** + **6**)/(Yield **5** + **6** + **3** + **4**) × 100.
e) TON = (mol product) (mol Pd)$^{-1}$ for telomers **5** and **6**.
f) TOF = (mol product) (mol Pd)$^{-1}$ hour^{-1} for telomers **5** and **6**.
g) Comparative system, showing best Pd-PPh$_3$ selectivity.
h) Ligand = NHC derived from the corresponding imidazolium/benzimidazolium salt.

aryl moieties [42]. Additionally, the steric effects of large N-substituents can dominate the full expression of electronic effects by preventing close M–L approach, e.g., in a series of Ru(NHC) complexes it was found that SIPr (**21**) and SIMes (**22**) ligands (in this nomenclature "S" implies "saturated backbone", i.e., imidazolidin-2-ylidenes or related imidazolinium salts) were only slightly more basic than their unsaturated analogues [26]. Thus, the incorporation of methyl (Me$_2$IMes, **19**) and chloro (Cl$_2$IMes, **20**) substituents at the C4,5 "backbone" positions of the imidazolin-2-ylidene ring of **17** led to ligands possessing somewhat greater and lesser basicity, respectively, than the parent IMes [42]. Steric influences on selectivity and activity in the telomerization reaction were probed using complexes **31** and **32**, bearing the 2,6-di-(isopropyl)phenyl analogues [IPr (**18**) and Me$_2$IPr (**16**), respectively] of IMes and Me$_2$IMes [18]. The observed variation in activity due to electronic effects was minor but followed trends predicted (DFT calculations) by the stability of proposed intermediates of type [Pd(η^3,η^2-octadienyl)(NHC)]$^+$ (**11**), with the more basic Me$_2$IMes showing slightly lower activity than IMes, and the less basic Cl$_2$IMes facilitating a slightly faster reaction (Table 4.1, entries 6 and 7) [18]. In these cases excellent regio- and chemoselectivity were retained; however, steric factors had a negative impact on both activity and selectivity when the Pd^0(IPr/Me$_2$IPr)(dvds) complexes **31** and **32** were employed as precatalysts. Using **31** a slight reduction in conversion and chemoselectivity were observed, with a considerable drop in the regioselectivity (Table 4.1, entry 8) [18]. A much greater effect is apparent with the Me$_2$IPr ligand (**32**), where a catastrophic loss of activity is observed, resulting in only 2% conversion, although regioselectivity remains similar to that obtained with **31** (Table 4.1, entry 9) [18]. This dramatic change in activity results from a seemingly relatively minor variation of NHC ligands (i.e., from IMes to IPr to Me$_2$IPr), although notably the addition of Me groups in the latter case has a significant effect on the orientation of the N-aryl substituents, obvious in the X-ray structures of **31** and **32** [18]. Nevertheless, it emphasizes that the steric demands of ligands such as IPr and Me$_2$IPr are significant and that the environment around Pd in key telomerization intermediates is crowded. Furthermore, the combination of **32** with bulkier nucleophiles (such as secondary alcohols) was sufficient to push the product distribution exclusively from telomers to the linear dimerization product **3** [34]. These effects may be rationalized entirely by steric factors given the catalytic performance of the electronically similar Me$_2$IPr and Me$_2$IMes ligated systems **32** and **29** (Table 4.1, entries 9 and 6). Concurrent with the report of Beller et al. [20], a preliminary series of telomerization experiments were undertaken by the Cavell group [41] using monoligated "Pd(NHC)" catalysts derived from Pd^0(IMes)(dimethyl fumarate)$_2$ (**24**), [Pd^0(SIMes)(*p*-benzoquinone)]$_2$ (**25**), or Pd^0(SIPr)$_2$ complexes. All precursors gave active buta-1,3-diene/methanol telomerization systems (Table 4.1, entries 13–15), although the regioselectivity of the Pd^0(SIPr)$_2$-catalyzed system was significantly reduced compared to the IMes complexes (Table 4.1, entry 15). Steric effects of the SIPr ligand are the likely cause of this reduction, rather than the result of a series of bis-NHC complexes persisting throughout the catalytic cycle, as the dissociation of one NHC ligand from electron-rich Pd⁰(NHC)₂ complexes is facile in the presence of π-acceptor ligands [43].

Complexes **28–31** have also been applied to catalyze the telomerization of buta-1,3-diene with a wider range of alcohols and phenols. Here good results for primary and benzylic alcohols (similar to those reported for methanol) were found, with secondary alcohols (specifically *iso*-propanol) retaining high selectivity for the linear telomer, but with significantly reduced chemoselectivity and conversion [18]. Notably, *in situ* generated Pd-imidazolium catalysts are less active than defined complexes for telomerization with higher alcohols [21]. Telomerization with phenols showed lower, but still reasonable, conversions and high chemoselectivity, but the regioselectivity suffered markedly when sterically hindered phenols (e.g., 2,4,6-trimethylphenol) were used [18]. The telomerization of tertiary alcohols catalyzed by these systems has not been reported.

In general, results comparable to those obtained using preformed complexes **28–32** may be achieved from *in situ* generated catalyst systems, using either the free NHC or the imidazolium salt precursor, in combination with appropriate Pd(0) or Pd(II) sources (Table 4.1, entries 2 and 3) [18, 21]. Typically, two [21] or four [18, 21, 41] equivalents of imidazolium salt relative to Pd have been used. These *in situ* catalyst systems offer significant advantages over the use of preformed complexes in that very simple reaction protocols may be employed. A series of *in situ* generated Pd catalysts derived from mono- and bis-Fc-substituted (Fc = ferrocenyl) imidazolium and benzimidazolium salts (**33–37**) showed interesting telomerization activities ascribed to the steric bulk of the Fc substituents [21]. Unsymmetrical salts **34** and **35** bearing N-Fc and N-Me substituents showed variable conversions but very high regioselectivities (Table 4.1, entries 16 and 17); the conversion increased significantly when methylene or ethylene groups separated the Fc and benzimidazolium centers in salts **36** and **37** (Table 4.1, entries 18 and 19) [21]. No activity was observed using 1,3-bis(ferrocenyl)imidazolium tetraphenylborate (**33**), is most likely due to the extreme steric bulk of the corresponding carbene [21]. The *in situ* testing of potentially chelating ligands derived from bis-imidazolium salts **38** and **39** [20], and the picolyl-functionalized imidazolium salts **40a** and **40b** [41], showed no telomerization activity, as would be expected from a mechanism that relies on monoligated Pd (Section 4.2.3). The chiral mono-imidazolium salt **41** was also tested under *in situ* conditions (Table 4.1, entry 2) and showed good regioselectivity (>98% **5**), but poor overall yield and chemoselectivity (5% and 71%, respectively) with a TON of 7000 [18].

38 2BF$_4^{\ominus}$

39 2OTs$^{\ominus}$

40a I$^{\ominus}$ Dipp

40b Br$^{\ominus}$ Mes

41 BF$_4^{\ominus}$

The Pd^{II}Cl(η^3-allyl)(NHC) complexes **27** of Nolan et al. [44] in the presence of alkoxide bases yield monoligated Pd^0(NHC) species [45] that may also be expected to form effective telomerization catalysts. Correspondingly, PdCl(η^3-allyl)(IMes) **42** and the IPr analogue **43** (commercially available from Strem) showed comparable activities, under identical conditions, to their Pd(NHC)(dvds) counterparts for the telomerization of buta-1,3-diene with MeOH (Table 4.1, entries 10–12).

42; Ar = Mesityl
43; Ar = 2,6-diisopropylphenyl

However, **43** was inactive as catalyst for the telomerization of buta-1,3-diene with amines, and only poorly active in the presence of KO^{t}Bu [22]. Abstraction of the chloride ion from **43** with AgPF$_6$ or AgBF$_4$ in MeCN to yield cationic solvent systems **44** and **45** was required to give highly active telomerization catalysts (Scheme 4.3) [22]. The strongly bound chloride ligand probably occupies a site on the metal center required for buta-1,3-diene coordination. However, cationic complexes generated *in situ* from **43** and sodium salts of non-coordinating anions showed significantly higher activities than isolated **44** and **45** for the telomerization of buta-1,3-diene with amines [22]. Furthermore, the nature of the non-coordinating anion was found to affect the activity of the *in situ* generated catalyst. The best results were achieved using a PF$_6$ salt of **43** – this showed high selectivity and activity for the telomerization of buta-1,3-diene with morpholine, achieving complete conversion in only 15 min at a Pd loading of 0.2 mol% (TOF = 2000 h^{-1}) [22]. Regarding the telomerization of buta-1,3-diene with amines, direct comparisons between Pd-phosphane and Pd-NHC catalyzed systems are difficult due to the use of different substrates under different conditions.

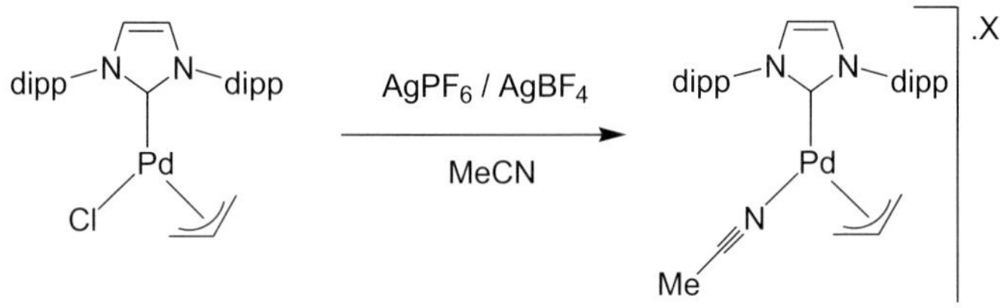

Scheme 4.3 Preparation of active telomerization precatalysts by halide abstraction from Pd^{II}Cl(η^3-allyl)(IPr).

4.2.5
Telomerization in Imidazolium-based Ionic Liquids

Unsymmetrically substituted 1,3-dialkylimidazolium salts **46** have found wide application as low-melting non-aqueous ionic liquids (ILs) that have favorable properties as reaction media for various catalytic reactions [46]. In comparison to reactions undertaken in traditional molecular solvents, processes performed in ILs offer specific and often significant economic and environmental benefits in terms of catalyst immobilization and ease of product separation, where simple phase separation is often effective; both factors facilitate catalyst recycling that can raise the overall TON of a reaction. Finally, the intimate relationship between imidazolium salts and NHC ligands leads to possible behavior that is specific to imidazolium-based ILs [36].

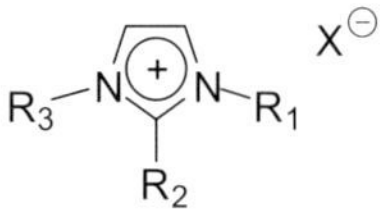

Imidazolium ILs have been applied as solvents in a wide variety of Pd-catalyzed homogeneous reactions [46]. However, caution must be used when identifying the actual catalytic species involved when Pd is used in such ionic liquids, especially when metal precursors are used in combination with bases capable of deprotonating the C2-H (or even the C4,5-H [47]) position of the imidazolium ring. For example, catalytically active PdBr$_2$(bmiy)$_2$ complexes (bmiy = 1-n-butyl-3-methylimidazolin-2-ylidene) were isolated from [bmim]Br (bmim = 1-n-butyl-3-methylimidazolium) under Heck reaction conditions using Pd(OAc)$_2$ as precatalyst; however, Pd-(bmiy) complexes were not formed in [bmim]BF$_4$. Correspondingly, the catalytic activity of the Pd-[bmim]BF$_4$ system was much reduced [48]. Suzuki coupling reactions performed in [bmim]BF$_4$ using Pd(PPh$_3$)$_4$ and a halide source gave PdX(bmiy)(PPh$_3$)$_2$ (X = Cl, Br), which were found to be highly active catalysts [49]. Furthermore, EXAFS studies have confirmed that coordinating chloride ions are required to give catalytically active PdCl$_2$(NHC)$_2$ complexes from the combination of Pd(OAc)$_2$ and imidazolium ILs in the absence of phosphanes [50]. Non-coordinating anions such as hexafluorophosphate and tetrafluoroborate lead to deposition of Pd metal [50]. More unforeseen was the often facile oxidative addition of imidazolium C2–H bonds to Group 10 M(0) complexes to yield [M^{II}(hydrido)(NHC)] (M = Ni [51], Pd [51] and Pt [52]) complexes. Similar reports now exist on the oxidative addition of imidazolium C4,5-H protons to Pt(0) [53]; related Pd(0) and Ni(0) chemistry is also known [54]. The activation of imidazolium C4,5–H bonds is significant as C2-Me blocked imidazolium ILs were thought to be resistant to redox processes in the presence of low-valent metal species.

The unintentional generation of Pd-(NHC) complexes may or may not be of benefit to the catalytic reaction in question. Although more active catalytic species may be formed, there remains the possibility that the interaction results in inactive and/or less selective catalysts, or acts to incorporate the IL into the product via NHC–hydrocarbyl reductive coupling [36]. An example of this non-innocent IL behavior [55] was observed in the telomerization of buta-1,3-diene with methanol using Pd-phosphane catalyst systems in [bmim]$^+$ and [emim]$^+$ (emim = 1-ethyl-3-methylimidazolium) salts of non-coordinating anions [56]. Unexpectedly, very low activities were observed. The addition of several equivalents of 1,3-dialkylimidazolium salts was sufficient to poison the Pd-phosphane catalyst [56], with the likely formation of catalytically inactive L$_2$Pd-(bmiy/emiy) (emiy = 1-ethyl-3-methylimidazolin-2-ylidene) complexes, in which coordination sites required for catalysis were blocked. However, running the reactions in 1-n-butyl-2,3-dimethylimidazolium ILs, where the imidazolium C2 position was blocked with a methyl group, showed activity comparable to the Pd-phosphane reference systems run in MeOH, with somewhat improved chemoselectivity (ca. 80%) and regioselectivity [56]. Notably, the steric demands (and basicity) of the bmiy/emiy and IMes ligands vary greatly, and so could be expected to show different reactivities under catalytic conditions.

Reports of telomerization reactions conducted in imidazolium ILs are limited and, interestingly, no report of the application of highly active Pd-NHC systems such as Pd(IMes)(dvds) **(28)** to the telomerization reaction in imidazolium ILs is available. However, there are several reports where "conventional" Pd-phosphane systems (as discussed in Section 4.2.2) have been transferred from molecular solvents to ILs. Aside from the influence of imidazolium C2-blocking as described previously, the telomerization of buta-1,3-diene with methanol in ILs (typically [bmim]BF$_4$ or similar) yields results [56, 57] that are satisfactory for phosphane ligated systems (Section 4.2.2) but markedly inferior to Pd-IMes systems. The palladium-catalyzed telomerization of buta-1,3-diene with HNEt$_2$ shows more robustness towards 1,3-dialkylimidazolium salts, and Pd-TPPTS-[bmim]BF$_4$ (TPPTS = sodium triphenylphosphane trisulfonate) systems yield linear telomers with high selectivity at reaction temperatures below 40 °C, at the expense of conversion (ca. 70%) [58]. Again, these results are inferior to Pd-NHC catalyzed systems in molecular solvents [22]. An additional report on the hydrodimerization of buta-1,3-diene catalyzed by a PdCl$_2$(methylimidazole)$_2$ complex derived from [bmim]Cl decomposition in the presence of [PdCl₄]⁻ and water highlights another facet of the non-innocence of imidazolium ILs [59]. Using this catalyst system in [bmim](BF$_4$/PF$_6$) almost exclusive regioselectivity for the linear telomer was observed, although overall conversions remained below 50% [59]. As expected, the use of IL solvents was found to greatly facilitate product separation and catalyst recycling in the aforementioned telomerization processes [56, 58, 59].

4.3
Buchwald–Hartwig Amination Reactions Catalyzed by Pd(NHC) Complexes

4.3.1
Introduction

The Pd-catalyzed reaction of amines with aryl halides, known as the Buchwald–Hartwig amination reaction, is an effective method for the formation of C(sp^2)–N(sp^3) bonds [60] and may also be used to form C(sp^2)–N(sp^2) bonds when applied to indole and imine substrates [61]. The reaction in its present form was first reported in 1995 [62], and is based upon a less useful Pd-catalyzed cross-coupling of tin amides and aryl halides developed in the early 1980s [63]. Pd-catalyzed aryl amination as a means of generating C_{aryl}–N bonds has significant advantages over classical organic methodologies in terms of generality, milder conditions, tolerance of a wide variety of functional groups, and reduced synthetic complexity. The reaction takes place between an aryl halide (or pseudo-halide, i.e., aryl-triflate, -tosylate or -diazonium) and an amine in the presence of base (usually a strong alkoxide base) and catalyst to produce aryl-substituted amines (Scheme 4.4); several excellent reviews on the topic are available [60, 61, 64].

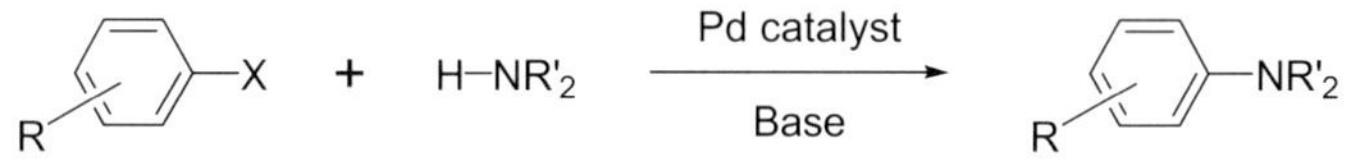

Scheme 4.4 Aryl amination coupling of aryl halides and amines.

Aryl amination is another process where the application of Pd-NHC systems in various guises, as either preformed Pd(NHC) complexes or as simple Pd(0/II) precursors in combination with imidazolium salts or free NHC ligands, have been applied to an area previously dominated by Pd-phosphane catalysts. The amination reaction shows similarities to telomerization, in that both reactions are thought to be catalyzed by Pd(0)-L species where the Pd center shifts between the Pd(0) and Pd(II) oxidation states during the catalytic cycle, and both result in the formation of a C–E (E = non-carbon element) bond. The development of new NHC-based aryl amination catalysts has complemented that of phosphane-based catalyst systems, and much of the work conducted with Pd-NHC catalysts has focused on the coupling of industrially relevant aryl chloride substrates. Despite the lower reactivity of Ar-Cl substrates, their advantages in terms of price and availability make them attractive substrates; several catalyst systems can perform the amination of aryl chlorides at room temperature, *vide infra*.

4.3.2
Mechanism of Aryl Amination

As with the telomerization reaction, monoligated 12-electron [Pd⁰(NHC)] complexes **47** have been implicated in the catalytic cycle of the aryl amination reaction [65] (Scheme 4.5). Rate studies by Hartwig et al. on the amination of aryl chlorides in the presence of Pd and a sterically demanding basic phosphane (P^{t}Bu$_3$) ruled out rate-limiting oxidative addition of Ar-Cl to Pd⁰(P^{t}Bu$_3$)$_2$ [66]. Kinetic studies on stoichiometric systems using Pd⁰(I^{t}Bu)$_2$ (I^{t}Bu = 1,3-di-*tert*-butylimidazolin-2-ylidene) showed a rate-determining dissociative mechanism to operate for the oxidative addition of Ar-Cl (i.e., initial dissociation of I^{t}Bu followed by oxidative addition of Ar-Cl to the [Pd⁰(I^{t}Bu)] fragment), producing a complex of type **48** [65]. The observation of reversible, thermally-induced dissociation of an I^{t}Bu ligand from Pd^{II}Cl(Ar)(I^{t}Bu)$_2$ complexes produced by oxidative addition of Ar-Cl to Pd⁰(I^{t}Bu)$_2$ confirms the relevance of complexes of type **48** [65].

Under catalytic conditions, it was shown that two concurrent mechanisms based on monophosphane "Pd⁰(P^{t}Bu$_3$)" complexes were operating (Scheme 4.5); with either rate-limiting oxidative addition to [Pd⁰(P^{t}Bu$_3$)] or to anionic [Pd⁰(OR)(P^{t}Bu$_3$)]⁻ complexes **49** [OR⁻ = anion of an alkoxide base such as (Na/K)O^{t}Bu], yielding PdCl(L)(Ar) **48** or Pd(OR)L(Ar) **50**, respectively. Subsequent reaction of **50** with amine, and of **48** with amine and base yields a Pd^{II}(aryl)(amido)L complex **51**, which subsequently undergoes reductive elimination of the coupled aryl amine and regeneration of the [Pd⁰L] catalyst **47** [66]. The reductive elimination of coupled aryl amines from three-coordinate Pd^{II}(Ar)(NR$_2$)(P^{t}Bu$_3$) complexes of type **51** has been confirmed and shown to be facilitated by sterically demanding ligands [67]. NHC ligands probably behave similarly to highly basic phosphanes in this process. The nature of the base can discriminate between these two mechanisms with, for example, NaO^{t}Bu favoring the anionic, and NaO-CEt$_3$ the "conventional" route [66]. The operation of two concurrent mechanisms explains much of the observed variability with regards to the activity of the amination reaction when different catalysts, substrates, and bases are employed.

The major side-reaction encountered in aryl amination reactions is (hydro)dehalogenation of the aryl halide to the corresponding benzene derivative [60]. This process occurs through β-hydride elimination from the amido ligand of intermediate **51**, resulting in the formation of an arylhydridopalladium(II) complex (**52**) that subsequently reductively eliminates dehalogenated arene by-product and regenerates **47**.

A mechanism invoking monoligated Pd complexes is consistent with the observation of excess ligand slowing the amination reaction [68] and the poor results obtained using potentially chelating bis-imidazolium salt systems [69]. Where Pd(II) precatalyst complexes are employed, reduction of the Pd(II) center is necessary to allow formation of catalytically active "Pd⁰(NHC)" species. This process

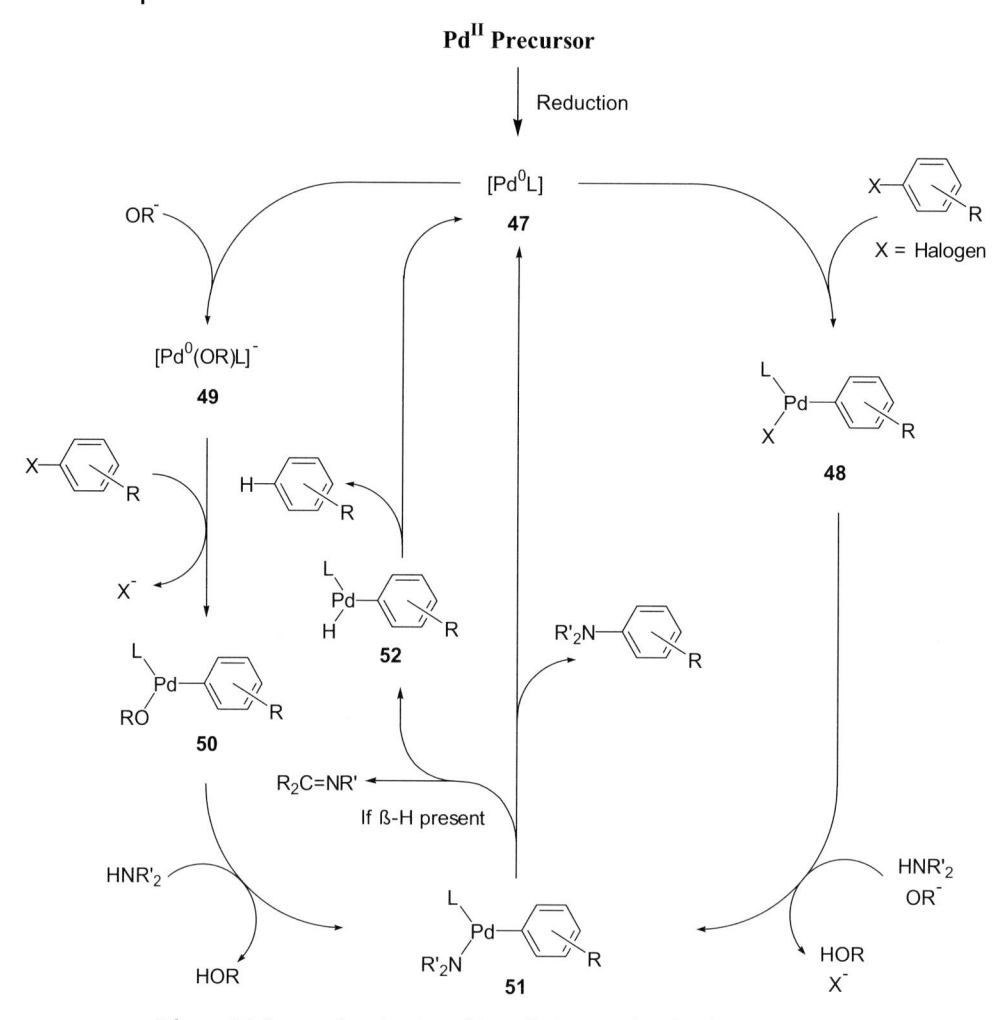

Scheme 4.5 Proposed mechanism of the palladium-catalyzed aryl amination reaction. (Adapted from Hartwig et al. [66].)

has been shown to be facilitated by alkoxide bases for several typical Pd(II) precursors, *vide infra*. For Pd(0) precatalyst complexes bearing NHC ligands, simple ligand dissociation yields the desired "Pd⁰(NHC)" catalysts. *In situ* catalyst systems employing imidazol(in)ium salts and a Pd(0/II) source must rely on the generation of NHC ligands through the action of base, or through redox processes involving the oxidative addition of the imidazolium C2–H bond to Pd(0) (Section 4.2.5).

4.3.3
Palladium-NHC Systems as Catalysts for Aryl Amination

It is useful to divide these into preformed Pd(0/II)(NHC) complexes and *in situ* systems generated from a Pd(0/II) precursor in combination with an imidazolium salt or free NHC. Whilst the former have the advantage of reproducibly generating defined catalyst complexes, the *in situ* protocols (particularly those employing the air- and moisture-stable imidazolium salts instead of the sensitive free NHCs) have significant advantages in terms of ease of synthesis and handling, and catalyst costs.

4.3.3.1 Application of Preformed Pd(0/II)(NHC) Complexes

Aryl amination reactions catalyzed by Pd(0/II) complexes typically require precatalyst loadings of the order of 1 mol% to give satisfactory conversions [70]. Noteworthy across the range of reported complexes is the generally superior performance of the IPr and SIPr ligated complexes over those bearing NHC ligands of lesser bulk (even IMes). The precatalyst complexes used fall into several main categories and will be addressed in turn. Most contain Pd:NHC associations in an ideal 1:1 ratio, with reduction of the Pd(II) precursors often facilitated by alkoxide bases.

I^{t}Bu, **53**: R = *tert*-butyl
ICy, **54**: R = cyclohexyl
IAd, **55**: R = adamant-1-yl
56: R = CH(Me)(Ph)

42: NHC = IMes (**17**)
43: NHC = IPr (**18**)
57: NHC = SIMes (**22**)
58: NHC = SIPr (**21**)
59: NHC = I^{t}Bu (**53**)
60: NHC = ICy (**54**)
61: NHC = IAd (**55**)
62: NHC = **56**

The air- and moisture-stable Pd^{II}Cl(η^3-allyl)(NHC) complexes **27** of Nolan et al. [44] have been extensively studied in the Pd-catalyzed amination of aryl chlorides, typically at Pd loadings of 1 mol% [71]. Reduction of precatalysts **27** to produce "Pd^0(NHC)" species in the presence of alkoxide bases has been confirmed by trapping experiments with phosphanes [44, 45]. The presence of sterically demanding NHC ligands in complexes **42**, **43**, **57–62** was found to correlate with high catalytic activity in the amination of aryl chlorides [44]. Specifically, complexes bearing the sterically demanding IPr, SIPr, IMes, SIMes, I^{t}Bu and IAd ligands showed significantly higher activities than the less bulky ICy and **56** [44]. Of the complexes bearing sterically demanding ligands, **43** and especially **58** were clearly superior, performing well even in the room temperature amination of aryl chlorides (Table 4.2,

Table 4.2 Representative comparative aryl chloride amination reactions catalyzed by Pd-NHC catalyst systems.

Entry	Catalyst	Ligand	Pd (mol%)	Pd:L	Reaction[a]	Conditions[b]	Time (h)	Yield (%)[c]	Ref.
1	42	IMes	1	1:1	A	DME, NaO^{t}Bu, 80 °C	2	100	[44]
2	43	IPr	1	1:1	A	DME, NaO^{t}Bu, 80 °C	0.25	100	[44]
3	58	SIPr	1	1:1	A	DME, NaO^{t}Bu, ambient	1.25	100	[44]
4	65	IPr	1	1:1	A	DME, KO^{t}Bu, 50 °C	0.5	97*	[75]
5	Pd(dba)$_2$-ImHBF$_4$	SIPr	1	1:1	A	DME, NaO^{t}Bu, ambient	3	100*	[84]
6	58	SIPr	1	1:1	B	DME, NaO^{t}Bu, ambient	1.3	93*	[45]
7	64	IPr	1	1:1	B	Dioxane, NaO^{t}Bu, 70 °C	0.5	100	[74]
8	67	IPr	1	1:1	B	DME, KO^{t}Bu, 50 °C	1	99	[70]
9	26	IPr	1	1:1	B	Dioxane, NaO^{t}Bu, 100 °C	16	96	[80]
10	31	IPr	1	1:1	B	Dioxane, NaO^{t}Bu, 100 °C	16	95	[80]
11	Pd(dba)$_2$-ImHCl	SIPr	1	1:1	B	DME, KO^{t}Bu, 165 °C	0.1	98*	[79]
12	Pd(SIPr)$_2$	SIPr	1	1:2	B	DME, KO^{t}Bu, 160 °C, mw	0.1	45*	[79]
13	Pd(dba)$_2$-ImHBF$_4$	SIPr	1	1:1	B	DME, KO^{t}Bu, 160 °C, mw	0.1	72*	[79]
14	Pd(dba)$_2$-ImHBF$_4$	SIPr	1	1:1	B	DME, NaO^{t}Bu, ambient	5	96*	[84]
15	Pd$_2$(dba)$_3$-ImHCl	IPr	2	1:2	B	Dioxane, KO^{t}Bu, 100 °C	30	80*	[82]
16	64	IPr	1	1:1	C	Dioxane, NaO^{t}Bu, 70 °C	2	85	[74]
17	PdCl$_2$(75)$_2$	75	1	1:2	C	IL, 50 °C	4	92	[77]
18	67	IPr	1	1:1	D	DME, KO^{t}Bu, 50 °C	1	98	[70]
19	Pd(SIPr)$_2$	SIPr	2	1:2	D	Dioxane, KO^{t}Bu, 100 °C	5	91*	[78]
20	Pd(SI^{t}Bu)$_2$	SI^{t}Bu	2	1:2	D	Dioxane, KO^{t}Bu, 100 °C	5	30*	[78]
21	Pd$_2$(dba)$_3$-ImHCl	SIPr	2	1:1	D	Dioxane, KO^{t}Bu, 100 °C	5	99*	[78]

a) A: *p*-chlorotoluene + morpholine; B: *p*-chloroanisole + morpholine; C: *p*-chloroanisole + aniline; D: *p*-chlorotoluene + aniline.
b) DME = dimethoxyethane; mw = microwave heating.
c) GC yields; * denotes isolated yield.

entries 1–3, 6) [44, 45]. The superiority of **43** over **42** and **59** was even more apparent in the amination of aryl triflates, which are convenient substrates readily prepared from widely available phenols [72]. In contrast, the SIMes complex **57** showed greatest activity (compared to **43** and **58**) in intramolecular aminations to produce *Cryptocarya* alkaloids from Ar-Br/Cl precursors in quantitative yield [73]. This switch in relative activities from the intermolecular reactions was ascribed to greater steric congestion at the reaction center in the intramolecular examples [73].

It has been proposed that the NHC-containing palladacycles **63** and **64** will also generate an active "Pd^0(NHC)" species through either the action of alkoxide or the aryl halide [74]. The IPr complex **64** (Table 4.2, entries 7 and 16) showed better activity than **63** in the coupling of aryl chlorides with various amines [74]. The Pd^{II}(O,O'-acac)X(IPr) (X = Cl, C-acac; acac = acetylacetonate) complexes **65** and **66** were tested in the amination of aryl chlorides, with the former found to perform significantly better (Table 4.2, entry 4) [75]. Reduction of Pd^{II}(O,O'-acac)Cl(IPr) to "Pd^0(NHC)" species via alkoxide attack on acac, with a key step being metathesis of the chloride ligand by alkoxide, was proposed [75]. The air- and moisture-stable dinuclear [PdCl(μ-Cl)(IPr)]$_2$ (**67**) complex was also active in the amination of aryl chlorides and bromides (Table 4.2, entries 8 and 18); in most cases the catalyst retained its activity when reactions were performed under air, in reagent grade solvents [70]. However, indoles were inert to coupling, and aryl triflates were not suitable substrates [70].

63: Ar = Mes
64: Ar = dipp
65: X = Cl
66: X = C-acac
67: Ar = 2,6-di-(isopropyl)phenyl

Various NHC ligand architectures (including acyclic N,N/O/S examples) in complexes **68–71**, obtained from oxidative addition of the respective 2-chloro-imidazolidinium or -amidinium salts to Pd(0), have been tested as precatalysts for the amination of aryl chlorides and bromides [76]. High conversions were observed using 1 mol% of precatalyst for the coupling of morpholine with bromobenzene at 70 °C and 2-chloropyridine at room temperature [76]. However, the observed activities may be attributable, at least in part, to the PPh$_3$ co-ligands. Several PdCl$_2$(NHC)$_2$ precatalysts (**72–76**) showed good performances for the amination of aryl chlorides when run in [Sbmim]BF$_4$ (Sbmim = 1-butyl-3-methylimidazolinium) ionic liquid (Table 4.2, entry 17) [77]. The solution of catalyst in ionic liquid could be reused several times with similar yields [77].

Complexes of type PdClXL(NHC)

68: X = L = PPh$_3$ (N,N'-dimethyl imidazolylidene)

69: X = Cl, L = PPh$_3$ (N-alkyl, N'-SPh imidazolylidene)

70: X = L = PPh$_3$

71: X = Cl, L = PPh$_3$ (N-alkyl, N'-OPh imidazolylidene)

Complexes of type PdCl$_2$(NHC)$_2$

72: Ar = C$_6$H$_2$(OMe)$_3$
73: Ar = Mesityl

74: R = R' = OMe
75: R = R' = Ph
76: R = OMe, R' = Ph

Palladium(0)-(NHC) amination precatalysts have taken the form Pd^0(NHC)$_2$, as employed by Cloke [78, 79], the commercially available Pd^0(NHC)(dvds) complexes **28** and **31**, and [Pd^0(NHC)(naphthoquinone)]$_2$ **26** (NHC = IMes, IPr) [80]. A potential drawback of Pd^0(NHC)$_2$ complexes is the Pd:NHC ratio of 1:2; mechanistic work indicates that dissociation of one NHC ligand prior to oxidative addition of the aryl halide is necessary [65, 66]. The activities of the commercially available IPr complexes (**26** and **31**) in the amination of aryl chlorides were significantly greater than the IMes analogues (Table 4.2, entries 9 and 10); furthermore, it was possible to substitute the expensive NaO^{t}Bu base with KOH with no loss of activity [80]. Beller et al. showed that SIMes complex **25** was superior to IMes complex **28**, but that *in situ* Pd$_2$(dba)$_3$-IPr.HCl systems generally showed higher activities than preformed complexes **25**, **26**, **28**, and **31** for the amination of a range of aryl bromide and chloride substrates [81]. At high temperature, Pd^0(SIPr)$_2$ was an efficient precatalyst for the amination of aryl chlorides, with results comparable to an *in situ* Pd$_2$(dba)$_3$-SIPr.HBF$_4$ catalyst system (Table 4.2, entry 11) [79]. However, when the same reactions were carried out using a microwave heating protocol, the *in situ* system performed significantly better (Table 4.2, entries 12 and 13) [79]. At lower temperatures, Pd^0(SIPr)$_2$ remains active, and more so than the Pd^0(SI^{t}Bu)$_2$ analogue, but less than the *in situ* system (Table 4.2, entries 19–21) [78].

4.3.3.2 *In situ* Pd Imidazolium Catalyst Systems

Combinations of Pd(0/II) complexes with imidazol(in)ium salts have been reported to produce effective catalyst systems for various C–C and C–N coupling reactions, including the aryl amination reaction [61]. Pd(0)-based methodologies generally make use of Pd(dba)$_2$ or Pd$_2$(dba)$_3$ as readily accessible metal sources, in combination with alkoxide bases. In addition to being consumed in the reaction, many bases are known to deprotonate imidazolium salts and generate NHC ligands, forming catalytically active "Pd^0(NHC)" species *in situ*. The presence of base is not strictly necessary for the formation of Pd^{II}(NHC) precursor complexes, given that imidazolium salts can oxidatively add to Pd(0) [36]. However, systems employing Pd(II) precursors, principally Pd(OAc)$_2$, rely on the alkoxide base to facilitate reduction to Pd(0).

The best performing Pd(0)-imidazol(in)ium systems can aminate aryl chloride substrates under mild conditions (often at room temperature) with catalyst loadings in the order of 1 mol%. The sterically demanding salts SIPr.HCl and IPr.HCl are usually pre-eminent and perform significantly better than IMes [82, 83] or I^{t}Bu/SI^{t}Bu salts [78]. Extensive testing of Pd$_2$(dba)$_3$-4(IPr.HCl) as a catalyst for the coupling of aryl bromides and chlorides with amines, imines and indoles has been conducted [82, 83]. The saturated analogue SIPr.HCl generally shows even higher activities, especially at lower temperatures, with Pd:L ratios of 1:1 (Table 4.2, entries 5, 11, 13, 14, 21) [78, 79, 82, 84]. At high temperature under convective heating a Pd(dba)$_2$-SIPr.HBF$_4$ system showed almost complete amination of aryl chlorides in only 5 min; under microwave heating the yield was somewhat reduced (Table 4.2, entries 11 and 13) [79]. Aryl bromides in conjunction with 10 mol% Pd$_2$(dba)$_3$-3(SIPr.HBF$_4$/IPr.HCl) have been used to functionalize amine resins [85]. Finally, *in situ* catalyst systems generated from Pd(dba)$_2$ in combination with various potentially chelating pyridine-, thiophene-, and furan-functionalized bis-imidazolium salts (**77**) were found to compare poorly with the analogous Pd(dba)$_2$-IPr.HCl system for the amination of aryl bromides, as could be expected from the mechanism shown in Scheme 4.5 [69].

77: Donor group = pyridyl, thiophene, furan
R = Me, Mesityl, ^{t}Bu

High-throughput screening of the aryl amination reaction using Pd(II)-imidazolium salt catalyst systems was demonstrated by incorporating fluorescence emitting and quenching moieties on the amination coupling partners, with generation of the coupled product measured as a decrease in fluorescence intensity [86]. Using this technique a range of 119 phosphane and NHC ligands (specifically IPr.HCl, SIPr.HCl/HBF$_4$, and SIMes.HBF$_4$) were evaluated in Pd-catalyzed aryl-bromide and -chloride amination reactions. The IPr and SIPr ligands were the most active across the entire study, with Pd(OAc)$_2$ a better Pd(II) source than PdCp(η^3-allyl) [86]. A Pd(OAc)$_2$-Me$_2$IPr system was used for the amination of aryl bromides, but no comparison with other NHC ligands was reported [87]. Intramolecular aryl chloride amination reactions were found to be efficiently catalyzed by Pd(OAc)$_2$ in the presence of SIPr.HCl. The optimum Pd:SIPr ratio was 1:2, with higher amounts of ligand destroying the activity [68]. An intramolecular aryl chloride amination has also been combined with a Sonogashira coupling in a one-pot indole synthesis; here a Pd(OAc)$_2$-IPr.HCl system outperformed the SIPr analogue, with IMes and SIMes showing no activity [88]. Again, the somewhat divergent ligand activities compared to intermolecular couplings highlights the inconsistent steric demands possible in intramolecular amination reactions.

4.4 Conclusions

Clearly, from studies to date and as reported in this chapter, the use of NHCs and related species in C–E coupling reactions, such as telomerization and amination, has enormous potential. Further detailed studies on catalyst design and on mechanistic aspects may well lead to important industrial developments.

References

1 W. Keim, A. Behr, M. Röper, Alkene and alkyne oligomerisation, cooligomerisation and telomerization reactions, in *Comprehensive Organometallic Chemistry*, Pergamon Press, Oxford, **1982**, vol. 8, p. 371–462.
2 A. Behr, in *Aspects of Homogeneous Catalysis*, ed. R. Ugo, Reidel, Dordrecht, **1984**, vol. 5, pp. 3–73; J. M. Takacs, Transition metal allyl complexes: Telomerization of dienes, in *Comprehensive Organometallic Chemistry II*, ed. L. S. Hegedus, Pergamon Press, Oxford, **1995**, vol. 12, pp. 785–796.
3 N. Yoshimura, Telomerization (hydrodimerization) of olefins, in *Applied Homogeneous Catalysis with Organometallic Compounds: A Comprehensive Handbook*, Special Workbench Edition, ed. B. Cornils, W. A. Herrmann, Wiley-VCH, Weinheim, **2000**, pp. 351–358.
4 F. Vollmüller, W. Mägerlein, S. Klein, J. Krause, M. Beller *Adv. Synth. Catal.* **2001**, *343*, 29–33.
5 S. M. Maddock, M. G. Finn *Organometallics* **2000**, *19*, 2684–2689.
6 W. Keim, M. Röper *J. Org. Chem.* **1981**, *46*, 3702–3707; W. Keim, K. R. Kurtz, M. Röper *J. Mol. Catal. A: Chem.* **1983**, *20*, 129–138.
7 P. W. Jolly *Angew. Chem. Int. Ed. Engl.* **1985**, *24*, 283–295.
8 F. Vollmüller, J. Krause, S. Klein, W. Mägerlein, M. Beller *Eur. J. Inorg. Chem.* **2000**, 1825–1832.
9 F. Hénin, A. Bessmertnykh, A. Serra-Muns, J. Muzart, H. Baillia *Eur. J. Org. Chem.* **2004**, 511–520.
10 W. Keim, A. Koehnes, T. Roethel *J. Organomet. Chem.* **1990**, *382*, 295–301.
11 V. Desvergnes-Breuil, C. Pinel, P. Gazellot *Green Chem.* **2001**, *3*, 175–177.
12 S. Takahashi, T. Shibano, N. Hagihara *Tetrahedron Lett.* **1967**, 2451–2453.
13 E. J. Smutny *J. Am. Chem. Soc.* **1967**, *89*, 6793–6794.
14 R. McCrindle, G. Ferguson, G. J. Arsenault, A. J. McAlees, D. K. Stephenson *J. Chem. Res. (S)* **1984**, 360–361.
15 W. A. Herrmann *Angew. Chem. Int. Ed.* **2002**, *41*, 1290–1309.
16 A. F. Littke, G. C. Fu *Angew. Chem. Int. Ed.* **2002**, *41*, 4176–4211; A. Zapf, M. Beller *Chem. Commun.* **2005**, 431–440.
17 D. Rottger, M. Beller, R. Jackstell, H. Klein, K.-D. Wiese US2005038273, WO2100803, **2005**.
18 R. Jackstell, S. Harkal, H. Jiao, A. Spannenberg, C. Borgmann, D. Röttger, F. Nierlich, M. Elliot, S. Niven, K. J. Cavell, O. Navarro, M. S. Viciu, S. P. Nolan, M. Beller *Chem. Eur. J.* **2004**, 3891–3900.
19 J. Krause, G. Cestaric, K.-J. Haack, K. Seevogel, W. Storm, K.-R. Pörschke *J. Am. Chem. Soc.* **1999**, *121*, 9807–9823.

20 R. Jackstell, M. G. Andreu, A. Frisch, K. Selvarkumar, A. Zapf, H. Klein, A. Spannenberg, D. Röttger, O. Briel, R. Karch, M. Beller *Angew. Chem. Int. Ed.* **2002**, *41*, 986–989.

21 R. Jackstell, A. Frisch, M. Beller, D. Röttger, M. Malaun, B. Bildstein *J. Mol. Catal. A: Chem.* **2002**, *185*, 105–112.

22 M. S. Viciu, F. K. Zinn, E. D. Stevens, S. P. Nolan *Organometallics* **2003**, *22*, 3175–3177.

23 T. Prinz, B. Driessen-Hölscher *Chem. Eur. J.* **1999**, *5*, 2069–2076.

24 T. Prinz, W. Keim, B. Driessen-Hölscher *Angew. Chem. Int. Ed. Engl.* **1996**, *35*, 1708–1710.

25 N. M. Scott, S. P. Nolan *Eur. J. Inorg. Chem.* **2005**, 1815 1828.

26 A. C. Hillier, W. J. Sommer, B. S. Yong, J. L. Petersen, L. Cavallo, S. P. Nolan *Organometallics* **2003**, *22*, 4322–4326.

27 A. Behr, G. v. Ilsemann, W. Keim, C. Krüger, Y.-H. Tsay *Organometallics* **1986**, *5*, 514–518.

28 H. M. Büch, P. Binger, R. Benn, C. Krüger, A. Rufinska *Angew. Chem. Int. Ed. Engl.* **1983**, *22*, 774–775.

29 A. Döhring, P. W. Jolly, R. Mynott, K.-P. Schick, G. Wilke *Z. Natursforsch., Teil B.* **1981**, *36*, 1198–1199.

30 U. Storzer, O. Walter, T. Zevaco, E. Dinjus *Organometallics* **2005**, *24*, 514–520.

31 A. Döhring, R. Goddard, G. Hopp, P. W. Jolly, N. Kokel, C. Krüger *Inorg. Chim. Acta* **1994**, *222*, 179–192.

32 R. Benn, B. Büssemeier, S. Holle, P. W. Jolly, R. Mynott, I. Tkatchenko, G. Wilke *J. Organomet. Chem.* **1985**, *279*, 63–86.

33 K. J. Szabó *Chem. Eur. J.* **2000**, *6*, 4413–4421.

34 S. Harkal, R. Jackstell, F. Nierlich, D. Ortmann, M. Beller *Org. Lett.* **2005**, *7*, 541–544.

35 B. Åkermark, B. Hansson, B. Krakenberger, A. Vitagliano, K. Zetterberg *Organometallics* **1984**, *3*, 679–682; B. Åkermark, K. Zetterberg, S. Hansson, B. Krakenberger, A. Vitagliano *J. Organomet. Chem.* **1987**, *335*, 133–142.

36 K. J. Cavell, D. S. McGuinness *Coord. Chem. Rev.* **2004**, *248*, 671–681.

37 M. Beller, R. Jackstell, H. Klein, D. Röttger, K.-D. Weise, D. Maschmeyer, A. Tuchlenski, A. Kaizik, S. S. Fernandez US2004242947, WO03031379, **2004**.

38 M. Beller, R. Jackstell, H. Klein, D. Röttger, D. Maschmeyer, S. S. Fernandez US2005065387, WO03031378, **2005**.

39 M. Beller, Personal communication, **2005**.

40 D. S. McGuinness, B.Sc. Honours Thesis, University of Tasmania, Hobart, **1997**.

41 N. D. S. Clément, PhD Thesis, Cardiff University, **2005**.

42 A. M. Magill, K. J. Cavell, B. F. Yates *J. Am. Chem. Soc.* **2004**, *126*, 8717–8724.

43 L. R. Titcomb, S. Caddick, F. G. N. Cloke, D. J. Wilson, D. McKerrecher *Chem. Commun.* **2001**, 1388–1389.

44 M. S. Viciu, O. Navarro, R. F. Germaneau, R. A. I. Kelly, W. J. Sommer, N. Marion, E. D. Stevens, L. Cavallo, S. P. Nolan *Organometallics* **2004**, *23*, 1629–1635.

45 M. S. Viciu, R. F. Germaneau, O. Navarro-Fernandez, E. D. Stevens, S. P. Nolan *Organometallics* **2002**, *21*, 5470–5472.

46 P. Wasserscheid, W. Keim *Angew. Chem. Int. Ed.* **2000**, *39*, 3772–3789; J. Dupont, R. F. de Souza, P. A. Z. Saurez *Chem. Rev.* **2002**, *102*, 3667–3692; T. Welton *Coord. Chem. Rev.* **2004**, *248*, 2459–2477.

47 H. Lebel, M. K. Janes, A. B. Charette, S. P. Nolan *J. Am. Chem. Soc.* **2004**, *126*, 5046–5047.

48 L. Xu, W. Chen, J. Xiao *Organometallics* **2000**, *19*, 1123–1127.

49 C. J. Mathews, P. J. Smith, T. Welton, A. J. P. White, D. J. Williams *Organometallics* **2001**, *20*, 3848–3850.

50 N. A. Hamill, C. Hardacre, S. E. J. McMath *Green Chem.* **2002**, *4*, 139–142.

51 N. D. S. Clément, K. J. Cavell, C. Jones, C. J. Elsevier *Angew. Chem. Int. Ed.* **2004**, *43*, 1277–1279.

52 D. S. McGuinness, K. J. Cavell, B. F. Yates, B. W. Skelton, A. H. White *J. Am. Chem. Soc.* **2001**, *123*, 8317–8328.

53 D. Bacciu, K. J. Cavell, I. A. Fallis, L. Ooi *Angew. Chem. Int. Ed.* **2005**, *44*, 5282–5284.

54 D. Bacciu, K. J. Cavell, unpublished results.
55 J. Dupont, J. Spencer *Angew. Chem. Int. Ed.* **2004**, *43*, 5296–5297.
56 L. Magna, Y. Chauvin, G. P. Niccolai, J.-M. Basset *Organometallics* **2003**, *22*, 4418–4425.
57 Y. Chauvin, L. Magna, G. P. Niccolai, J.-M. Basset EP 1201634 A1, **2002**.
58 G. S. Fonseca, R. F. de Souza, J. Dupont *Catal. Commun.* **2002**, *3*, 377–380.
59 J. E. L. Dullius, P. A. Z. Saurez, S. Einloft, R. F. de Souza, J. Dupont, J. Fischer, A. De Cian *Organometallics* **1998**, *17*, 815–819.
60 B. H. Yang, S. L. Buchwald *J. Organomet. Chem.* **1999**, *576*, 125–146.
61 A. C. Hillier, G. A. Grasa, M. S. Viciu, H. M. Lee, C. Yang, S. P. Nolan *J. Organomet. Chem.* **2002**, *653*, 69–82.
62 A. S. Guram, R. A. Rennels, S. L. Buchwald *Angew. Chem. Int. Ed. Engl.* **1995**, *34*, 1348–1350; J. Louie, J. F. Hartwig *Tetrahedron Lett.* **1995**, *36*, 3609–3612.
63 M. Kosugi, M. Kameyama, T. Migita *Chem. Lett.* **1983**, 927–928.
64 V. Farina *Adv. Synth. Catal.* **2004**, *346*, 1553–1582; U. Christmann, R. Vilar *Angew. Chem. Int. Ed.* **2005**, *44*, 366–374.
65 A. K. d. K. Lewis, S. Caddick, F. G. N. Cloke, N. C. Billingham, J. Leonard *J. Am. Chem. Soc.* **2003**, *125*, 10066–10073.
66 L. M. Alcazar-Roman, J. F. Hartwig *J. Am. Chem. Soc.* **2001**, *123*, 12905–12906.
67 M. Yamashita, J. F. Hartwig *J. Am. Chem. Soc.* **2004**, *126*, 5344–5345.
68 R. Omar-Amrani, R. Schneider, Y. Fort *Synthesis* **2004**, 2527–2534.
69 D. J. Nielsen, K. J. Cavell, M. S. Viciu, S. P. Nolan, B. W. Skelton, A. H. White *J. Organomet. Chem.* **2005**, *690*, 6133–6142.
70 M. S. Viciu, R. M. Kissling, E. D. Stevens, S. P. Nolan *Org. Lett.* **2002**, *4*, 2229–2231.
71 N. Marion, O. Navarro, R. A. I. Kelly, S. P. Nolan *Synthesis* **2003**, 2590–2592.
72 O. Navarro, H. Kaur, P. Mahjoor, S. P. Nolan *J. Org. Chem.* **2004**, *69*, 3173–3180.
73 S. S. Cämmerer, M. S. Viciu, E. D. Stevens, S. P. Nolan *Synlett* **2003**, 1871–1873.
74 M. S. Viciu, R. A. I. Kelly, E. D. Stevens, F. Naud, M. Studer, S. P. Nolan *Org. Lett.* **2003**, *5*, 1479–1482.
75 O. Navarro, N. Marion, N. M. Scott, J. González, D. Amoroso, A. Bell, S. P. Nolan *Tetrahedron* **2005**, *61*, 9716–9722.
76 D. Kremzow, G. Seidel, C. W. Lehmann, A. Fürstner *Chem. Eur. J.* **2005**, *11*, 1833–1853.
77 I. Özdemir, S. Demir, Y. Gök, E. Çetinkaya, B. Çetinkaya *J. Mol. Catal. A: Chem.* **2004**, *222*, 97–102.
78 K. Arentsen, S. Caddick, F. G. N. Cloke *Tetrahedron* **2005**, *61*, 9710–9715.
79 A. J. McCarroll, D. A. Sandham, L. R. Titcomb, A. K. d. K. Lewis, F. G. N. Cloke, B. P. Davies, A. P. de Santana, W. Hiller, S. Caddick *Mol. Diversity* **2003**, *7*, 115–123.
80 L. J. Gooßen, J. Paetzold, O. Briel, A. Rivas-Nass, R. Karch, B. Kayser *Synlett* **2005**, 275–278.
81 A. C. Frisch, A. Zapf, O. Briel, B. Kayser, N. Shaikh, M. Beller *J. Mol. Catal. A: Chem.* **2004**, *214*, 231–239.
82 J. Huang, G. Grasa, S. P. Nolan *Org. Lett.* **1999**, *1*, 1307–1309.
83 G. A. Grasa, M. S. Viciu, J. Huang, S. P. Nolan *J. Org. Chem.* **2001**, *66*, 7729–7737.
84 S. R. Stauffer, S. Lee, J. P. Stambuli, S. I. Hauck, J. F. Hartwig *Org. Lett.* **2000**, *2*, 1423–1426.
85 K. Weigand, S. Pelka *Org. Lett.* **2002**, *4*, 4689–4692.
86 S. R. Stauffer, J. F. Hartwig *J. Am. Chem. Soc.* **2003**, *125*, 6977–6985.
87 T. Wenderski, K. M. Light, D. Ogrin, S. G. Bott, C. J. Harlan *Tetrahedron Lett.* **2004**, *45*, 6851–6853.
88 L. T. Kaspar, L. Ackermann *Tetrahedron* **2005**, *61*, 11311–11316; L. Ackermann *Org. Lett.* **2005**, *7*, 439–442.

5
Metal-mediated and -catalyzed Oxidations Using N-Heterocyclic Carbene Ligands

Mitchell J. Schultz and Matthew S. Sigman

5.1
Introduction

Recent years have seen the emergence of N-heterocyclic carbenes (NHCs) as a valuable new ligand class for metal-based catalysis. However, their use as ligands in catalytic oxidations has received scarce attention. Since phosphine ligands cannot be generally used in most metal-catalyzed oxidations due to their susceptibility to oxidation, ligands in oxidation catalysis have typically been restricted to nitrogen- or oxygen-based ligands. With this limitation in mind, NHCs may provide a valuable alternative ligand class to both improve current methods and explore novel oxidative strategies.

The stability of free NHCs to molecular oxygen has been established, although the carbenes are highly sensitive to water [1, 2]. In contrast, the tolerance of metal–NHC complexes to oxidative conditions was not reported until recently. Fortunately, metal–NHC complexes have been shown to indeed survive under oxidative conditions. Specifically, Pd-catalyzed oxidation reactions using NHC ligands have provided some of the most active homogenous catalysts for the aerobic oxidation of alcohols reported to date. This chapter overviews recent progress in two areas: (1) the use of NHC metal complexes to study fundamental steps in aerobic oxidations and (2) the development of NHC–metal complexes as oxidation catalysts.

5.2
Metal–NHC-mediated Activation of Molecular Oxygen

5.2.1
Co

Metals containing tripodal ligands have been shown to be useful for the activation and functionalization of small molecules [3]. Meyer and coworkers have reported the development of a chelating tripodal NHC ligand [4] and, in a recent publication, prepared a Co-complex to test the reactivity of such complexes with molecular oxygen [5]. Addition of 1 equivalent of NaBPh$_4$ to **1** followed by exposure to molecular oxygen afforded the side-on peroxo species **2** (Fig. 5.1). DFT calculations performed on the Co–peroxo complex reveal a high-lying, filled π^* orbital on dioxygen, which suggests that **2** is a nucleophilic peroxo complex.

Fig. 5.1 O$_2$ activation by a tripodal-Co complex.

To test this idea, **2** was allowed to react with the electron-poor olefin tetracyanoethylene (TCNE) (Fig. 5.2). Upon reaction, complex **3** was isolated (the organic products were not identified), demonstrating that the Co–peroxo species is indeed nucleophilic. Complex **2** also reacts with benzyolidenemalonitrile to yield **3** and benzaldehyde. Additionally, reacting **2** with benzoyl chloride also results in the formation of **3**, along with phenyl benzoate.

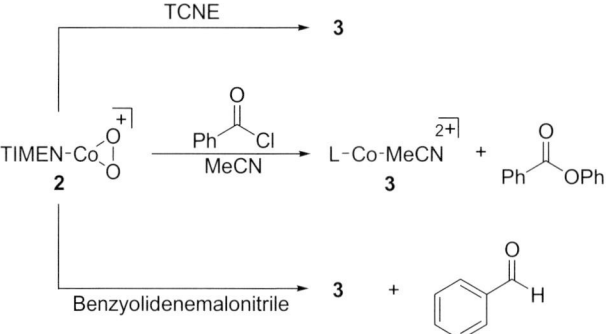

Fig. 5.2 Reactivity of **2**.

^{18}O-labeling studies revealed complete isotopic incorporation into the organic product, thus confirming the oxygen atoms originate from **2**. The authors propose a mechanism consistent with this labeling where the Co–peroxo species **2** attacks benzoyl chloride to form a Co(III)–peroxyphenylacetate (**5**) (Scheme 5.1). Complex **5** undergoes homolytic cleavage of the Co–O bond to form Co(II) and a peroxybenzoate radical (**7**). The radical decomposes to produce CO_2 and a phenoxy radical (**8**). The phenoxy radical then reacts with another equivalent of benzoyl chloride to generate the observed phenyl benzoate.

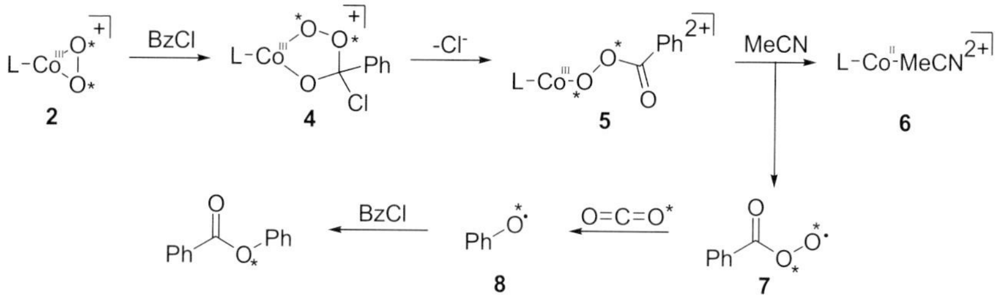

Scheme 5.1 Proposed mechanism for phenyl benzoate formation from benzoyl chloride.

5.2.2
Ni

While significant effort has been put forth developing catalytic aerobic oxidations involving Cu [6] and Pd [7], very little work on similar Ni-catalyzed aerobic oxidations is represented in the literature. To investigate the viability of Ni complexes for use in catalytic oxidations, 1:1 Ni:NHC complexes have been synthesized by Dible and Sigman [8]. Starting with $Ni(cod)_2$ (cod = cycloocta-1,5-diene), (π-allyl)-Ni(NHC)Cl complexes of type **9** were prepared (Fig. 5.3). Upon exposing **9** to molecular oxygen, a bis-μ-hydroxonickel(II) dimer (**10**) was formed. The isolated organic products were cinnamaldehyde (**11**) and phenyl vinyl ketone (**12**).

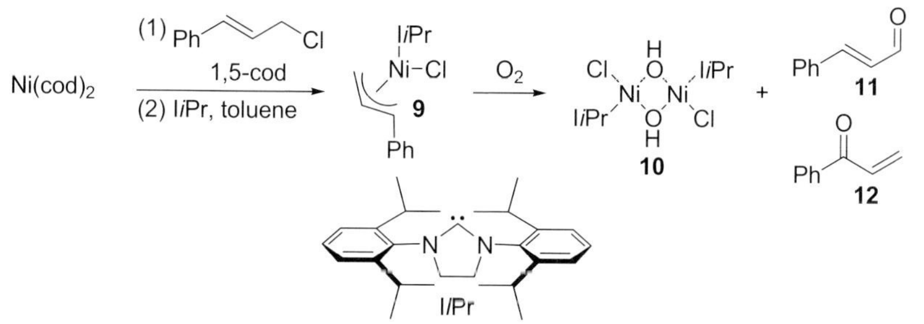

Fig. 5.3 Synthesis of bis-μ-hydroxonickel(II) dimer **10**.

To understand the fate of the allyl group and to gain insight into the oxidation mechanism, several experiments were performed. Oxidation with ^{18}O-labeled molecular oxygen lead to complete isotopic incorporation into the carbonyl products **11** and **12** and also a shift in the OH stretching frequency of the resulting bis-μ-hydroxonickel(II) dimer **10**, thus confirming the presence of ^{18}O. Additionally, kinetic studies revealed saturation in $[O_2]$, which is consistent with reversible O_2 binding followed by rate-limiting decomposition. A mechanism that accounts for these observations was proposed wherein O_2 binds to **13** followed by rate-limiting hydrogen abstraction from **14** and decomposition of **15** to form the carbonyl product and the bis-μ-hydroxonickel(II) dimer (Scheme 5.2).

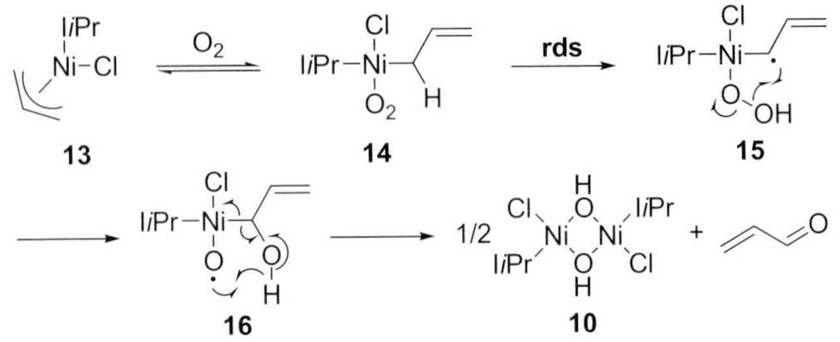

Scheme 5.2 Proposed mechanism for the formation of **10**.

While attempting to synthesize a (π-allyl)Ni(NHC)Cl species using 4-chloro-2,4-dimethylpent-2-ene, Dible and Sigman observed a complex mixture of products with the major product identified as a bis-μ-choloronickel(I)NHC dimer **17** [9]. A higher yielding method to synthesize these complexes was developed by mixing Ni(cod)$_2$, Ni(dme)Cl$_2$, and IiPr or SIiPr in toluene to form complexes of type **17** (Fig. 5.4). Exposing these complexes to oxygen led to the formation of bis-μ-hydroxonickel(II) dimers **18** containing one NHC ligand that had been dehydrogenated. This was confirmed by decomposing the complexes and analyzing the resulting NHC salts, which revealed a 1:1 mixture of IiPr/SIiPr (**19**) and dehydrogenated IiPr/SIiPr (**20**).

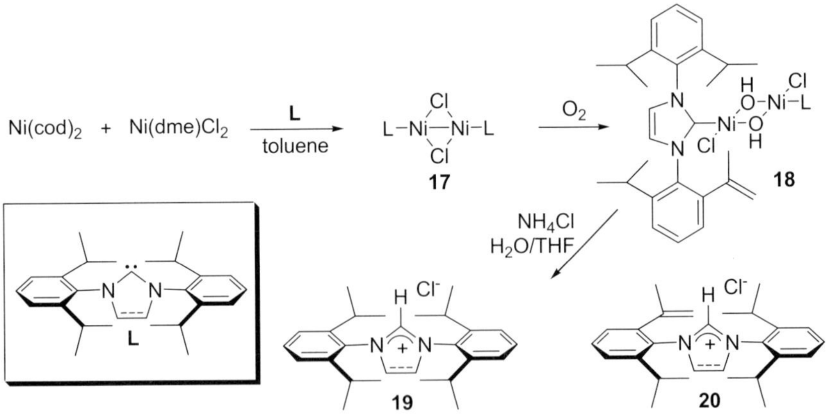

Fig. 5.4 O_2 activation/ligand dehydrogenation by bis-μ-choloronickel(I)NHC dimers.

5.2.3
Pd

Aerobic oxidations utilizing Pd complexes is a rapidly developing field in catalysis [7]. However, how O_2 promotes catalyst turnover remains a key fundamental question. There are two primary mechanistic proposals for catalyst turnover with O_2 to reform the active Pd(II)-catalyst: (1) direct insertion of O_2 into a Pd–hydride bond to form a Pd–hydroperoxo species that can be protonated to form hydrogen peroxide and the active Pd(II)–catalyst and (2) the formation of Pd(0) followed by direct oxidation by O_2 to form a Pd–peroxo species that can react with 2 equivalents of acid to form hydrogen peroxide and regenerate the catalyst.

To evaluate these possibilities, Stahl and coworkers have synthesized Pd(IMes)$_2$ (21) and evaluated its reactivity with molecular oxygen (Fig. 5.5) [10]. Exposing 21 to molecular oxygen yields the η^2-peroxo complex 22. Addition of 1 equivalent of acid to 22 gave the isolable Pd–hydroperoxide complex 23. Addition of a second equivalent of acid results in the extremely slow formation of hydrogen peroxide and Pd(IMes)$_2$(OAc)$_2$ (24). This slow second step is proposed to arise from the steric bulk associated with having two NHCs on Pd. While this reaction sequence demonstrates that a Pd(0) complex can be oxidized to Pd(II) by molecular oxygen, it still does not rule out the direct insertion of O_2 into a Pd–H bond under catalytic conditions. Additionally, computational work on this problem has been reported using metal complexes that do not contain NHC ligands [11, 12].

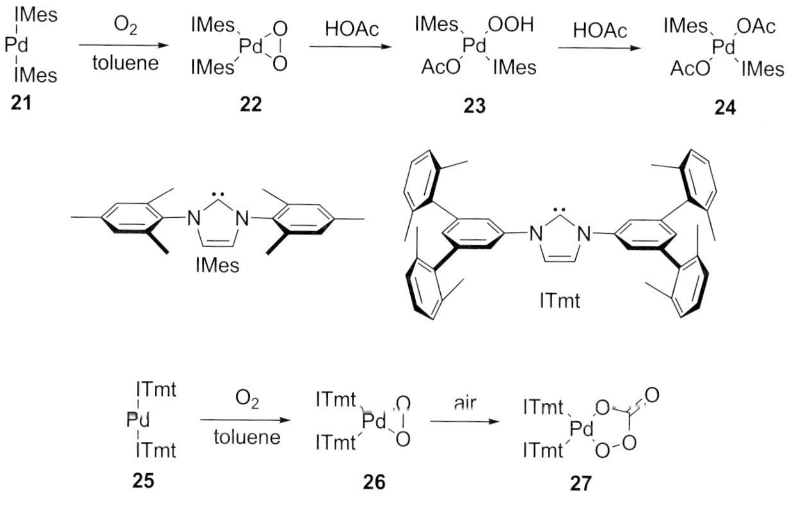

Fig. 5.5 Reactions of Pd(NHC)$_2$ with O_2/air.

In a similar study, Kawashima and coworkers found that by substituting ITmt for IMes and exposing the resultant complex to molecular oxygen in the solid state that an equivalent of Pd–peroxo species (26) is formed [13]. However, upon exposing 26 to air, the η^2-peroxo complex reacts with CO_2 to form a Pd–peroxycarbonate complex (27). In a control experiment, the authors found that Pd(ITmt)$_2$ does not react directly with CO$_2$ but must first form **26** to access **27**. Additionally, the authors tested Pd(IMes)$_2$O$_2$ and found that it does not react with CO$_2$ to form the corresponding Pd–peroxycarbonate. This is presumably due to the subtly increased steric hindrance about the complex with Pd(IMes)$_2$ as opposed to Pd(ITmt)$_2$.

5.3
Metal-catalyzed Oxidations, Pd

5.3.1
Methane Oxidation

The efficient oxidation of methane to methanol is a crucial challenge in catalysis that has received significant attention [14]. In 2002, Herrmann and coworkers published a report on a Pd–NHC catalyst that succeeds in this transformation (Fig. 5.6) [15]. This was the first report of a Pd-catalyzed oxidation using an NHC as a ligand. In this system, a chelating bis-carbene ligand is used with PdBr$_2$ and K$_2$S$_2$O$_8$ in a TFA/TFAA solvent mixture to produce methyl trifluoroacetate with TONs up to 30. Interestingly, several different halide counter-ions were tested and all, except iodine, produced catalysts with similar activity. Of note, initial DFT calculations do not reveal the origin of this difference in reactivity [15b].

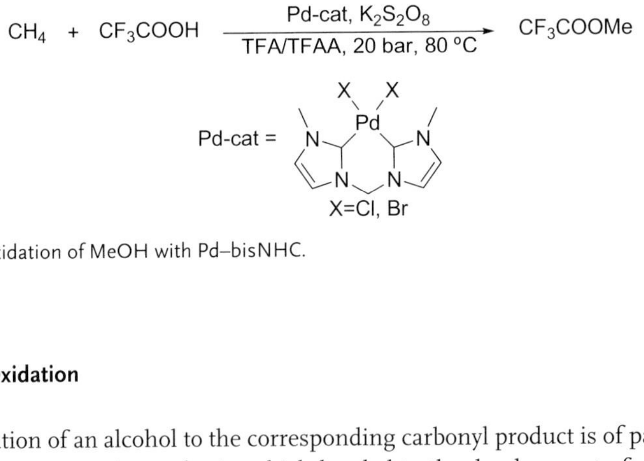

$$CH_4 + CF_3COOH \xrightarrow[\text{TFA/TFAA, 20 bar, 80 °C}]{\text{Pd-cat, K}_2\text{S}_2\text{O}_8} CF_3COOMe$$

Pd-cat = [Pd complex with X=Cl, Br]

Fig. 5.6 Oxidation of MeOH with Pd–bisNHC.

5.3.2
Alcohol Oxidation

The oxidation of an alcohol to the corresponding carbonyl product is of paramount importance in organic synthesis, which has led to the development of numerous methods to accomplish this transformation [16]. Recently, Sigman and coworkers have reported several studies on Pd–NHC-catalyzed aerobic oxidations of alcohols.

In their first report, Jensen and Sigman replaced (−)-sparteine with an NHC as a ligand for Pd in the Pd-catalyzed aerobic oxidative kinetic resolution of secondary alcohols (Fig. 5.7) [17]. While using (−)-sparteine as a chiral Brønsted base, several achiral NHCs were tested for the oxidative kinetic resolution. I*i*Pr was the best NHC, leading to k_{rel} values of up to 14.3 for the oxidation of secondary

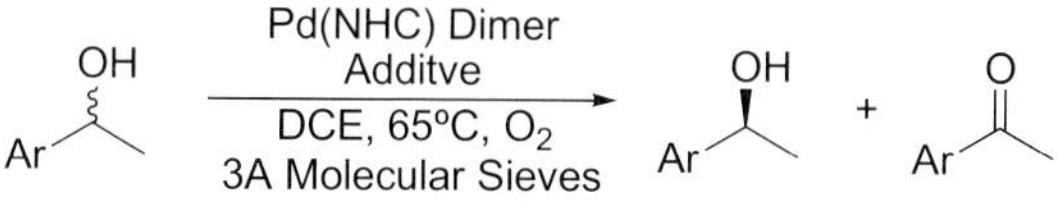

5.3 Metal-catalyzed Oxidations, Pd

Entry	NHC	R	Additive	k_{rel}
1	I/Pr	Ph	(−)-sparteine	11.6
2	I/Pr	p-MeO-Ph	(−)-sparteine	14.3
3	IMes	Ph	(−)-sparteine	6.1
4	(R,R)-28	Ph	(−)-sparteine	4.5
5	(S,S)-28	Ph	(−)-sparteine	11.8
6	29	Ph	AgOAc	3.3

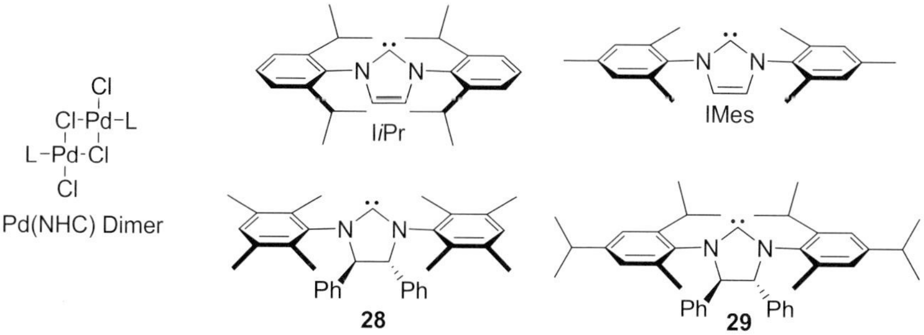

Fig. 5.7 Pd-NHC-catalyzed oxidative kinetic resolution of secondary alcohols.

benzylic alcohols. When using a chiral carbene (28) as a ligand along with (−)-sparteine as the base, a "matched" and "mismatched" relationship between the two chiral elements was demonstrated. Finally, by eliminating (−)-sparteine and testing several chiral NHC–Pd complexes with an achiral base, the highest k_{rel} value reported is 3.3, using 29 as the ligand [18].

Sigman and coworkers have also reported using Pd–NHCs as catalysts for the general aerobic oxidation of alcohols [19]. Upon testing different Pd–NHC catalysts, it was found that Pd(I/Pr)(OAc)$_2$(H$_2$O) (30) afforded an extremely active catalyst (Fig. 5.8). Using small amounts of added AcOH or ^{n}Bu$_4$NOAc, this oxidation was successful for the transformation of both primary and secondary benzylic, aliphatic, and allylic alcohols to the corresponding carbonyl products. With further optimization, these oxidations could also be performed using as little as 0.1 mol% of 30 as well as under an ambient air atmosphere. Additionally, continued evaluation of this catalyst system illustrated that the oxidation can be applied to both 1,2- and 1,3-mono-protected diols [20].

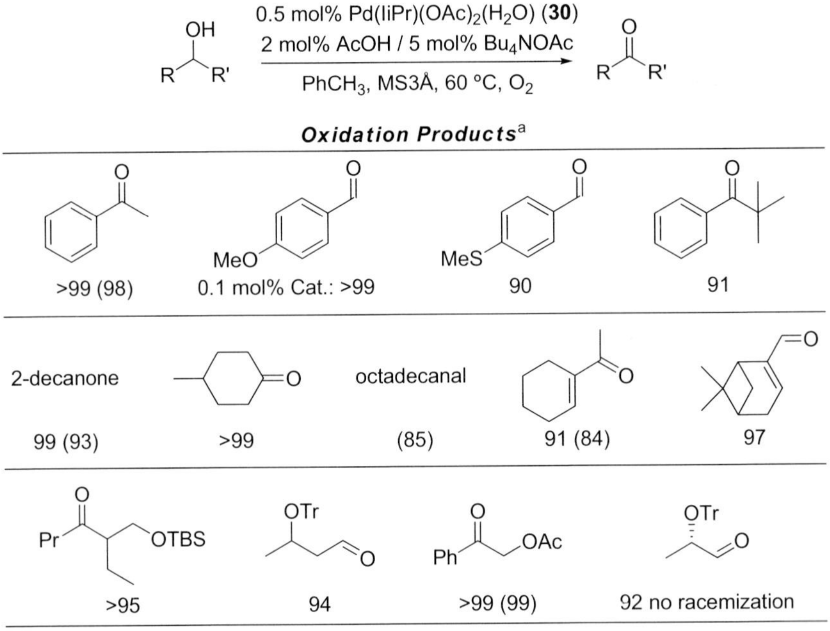

5 Metal-mediated and -catalyzed Oxidations Using N-Heterocyclic Carbene Ligands

$$\text{R-CH(OH)-R'} \xrightarrow[\text{PhCH}_3, \text{MS3Å}, 60\,°C, O_2]{\substack{0.5\ \text{mol\%}\ \text{Pd(I}i\text{Pr)(OAc)}_2(\text{H}_2\text{O})\ (\mathbf{30})\\ 2\ \text{mol\%}\ \text{AcOH} / 5\ \text{mol\%}\ \text{Bu}_4\text{NOAc}}} \text{R-CO-R'}$$

Oxidation Products[a]

acetophenone	4-MeO-benzaldehyde	4-MeS-benzaldehyde	PhC(O)tBu
>99 (98)	0.1 mol% Cat.: >99	90	91

2-decanone	cyclohexanone	octadecanal	1-acetylcyclohexene	pinene-aldehyde
99 (93)	>99	(85)	91 (84)	97

Pr-CH(Et)C(O)-OTBS	CH₃CH(OTr)CH₂CHO	PhC(O)CH₂OAc	CH₃CH(OTr)CHO
>95	94	>99 (99)	92 no racemization

[a] Conversions measured by GC, Isolated yield in parenthesis.

Fig. 5.8 Aerobic oxidation of alcohols catalyzed by **30**.

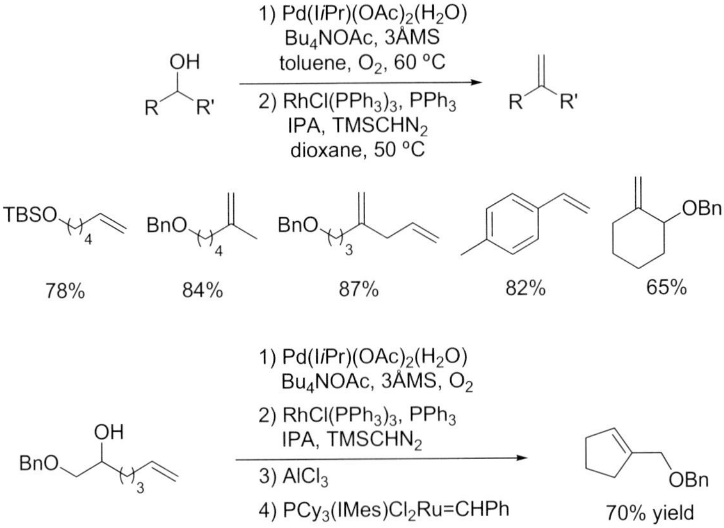

$$\text{R-CH(OH)-R'} \xrightarrow[\substack{2)\ \text{RhCl(PPh}_3)_3, \text{PPh}_3\\ \text{IPA, TMSCHN}_2\\ \text{dioxane, 50\,°C}}]{\substack{1)\ \text{Pd(I}i\text{Pr)(OAc)}_2(\text{H}_2\text{O})\\ \text{Bu}_4\text{NOAc, 3ÅMS}\\ \text{toluene, } O_2, 60\,°C}} \text{R=CH-R'}$$

TBSO(CH₂)₄CH=CH₂	BnO(CH₂)₄C(CH₃)=CH₂	BnO(CH₂)₃CH=CH₂	4-vinyltoluene	methylenecyclohexyl-OBn
78%	84%	87%	82%	65%

BnO-CH₂CH(OH)CH₂CH₂CH=CH₂
$$\xrightarrow[\substack{3)\ \text{AlCl}_3\\ 4)\ \text{PCy}_3(\text{IMes})\text{Cl}_2\text{Ru=CHPh}}]{\substack{1)\ \text{Pd(I}i\text{Pr)(OAc)}_2(\text{H}_2\text{O})\\ \text{Bu}_4\text{NOAc, 3ÅMS, } O_2\\ 2)\ \text{RhCl(PPh}_3)_3, \text{PPh}_3\\ \text{IPA, TMSCHN}_2}} \text{cyclopentenyl-CH}_2\text{OBn}$$

70% yield

Fig. 5.9 Multi-catalytic olefination of alcohols.

Lebel and Paquet have recently used the Pd(I*i*Pr)(OAc)$_2$(H$_2$O) catalyst system in a one-pot multi-catalytic transformation of alcohols into olefins (Fig. 5.9) [21]. In this reaction, aliphatic or benzylic alcohols were first submitted to an aerobic oxidation using **30**, followed by a Rh-catalyzed olefination. This one-pot procedure works with both primary and secondary alcohols, resulting in good yields with no apparent negative interactions between the Pd and Rh catalysts. Additionally, this method was applied to a four-step sequence transforming an alcohol into a cyclopentene ring in a 70% overall yield.

Sigman's group have also performed a mechanistic study of the Pd(I*i*Pr)(OAc)$_2$-(H$_2$O)-catalyzed aerobic alcohol oxidation [22]. Several features from this study are notable. First, upon growing a single crystal of **30**, X-ray crystallographic analysis revealed a water molecule bound to the Pd center. Interestingly, this molecule of water was hydrogen-bonded to the carbonyl oxygen of each acetate ligand. This water binding can be thought of as an alcohol surrogate and provides evidence supporting intramolecular deprotonation of the alcohol, as originally proposed in the design of the catalyst (Scheme 5.3). In the mechanistic study, an unusually large kinetic isotope effect (KIE) for β-hydride elimination of 5.5 ± 0.1 was measured. This large KIE led to the interpretation of a late transition state for β-hydride elimination with significant Pd-hydride character.

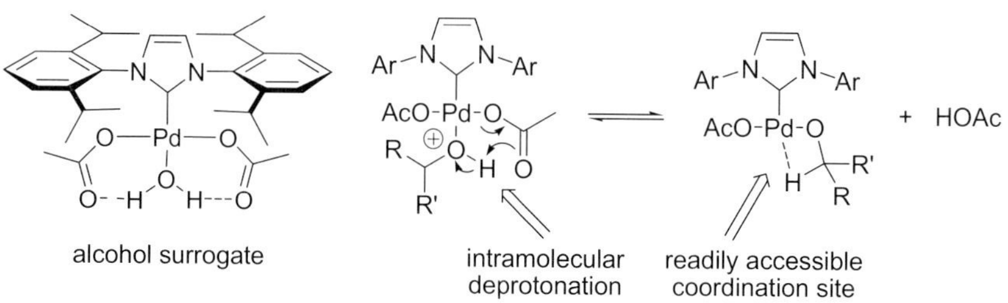

Scheme 5.3 Pd(I*i*Pr)(OAc)$_2$(H$_2$O) as an alcohol surrogate.

Additionally, increasing the basicity of the carboxylate ligand or the pK_a of the carboxylic acid added to the reaction resulted in increased catalytic activity. Based on this observation, a catalyst system was developed using pivalate as the anionic ligand for Pd. This system was successful for the aerobic oxidation of alcohols using relatively low catalyst loadings (1 mol%) at ambient temperature and an air atmosphere (Fig. 5.10) [15]. While this method represents the mildest aerobic oxidation to date, it has a more limited substrate scope – primary aliphatic alcohols and sterically encumbered alcohols are not successfully oxidized.

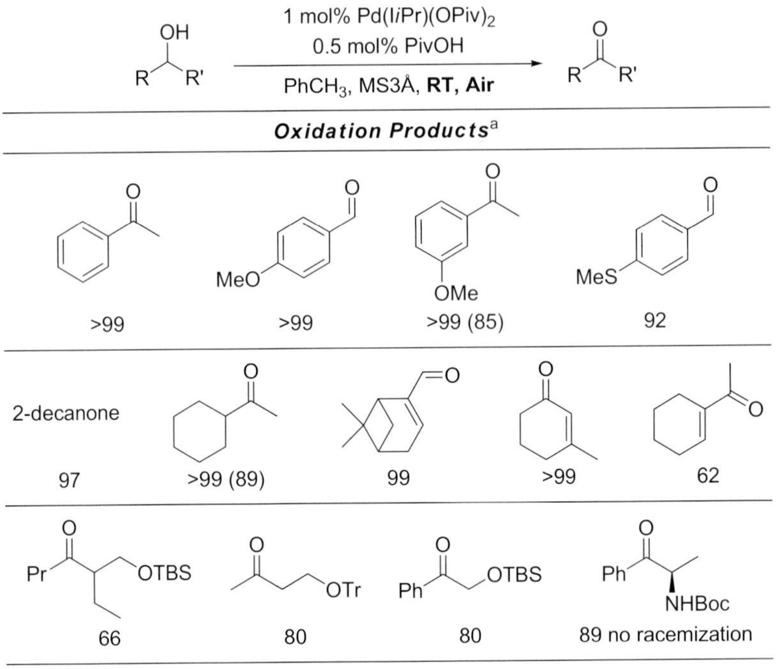

^a Conversion measured by GC, isolated yield in parenthesis

Fig. 5.10 Pd(I*i*Pr)(Pivalate)$_2$-catalyzed aerobic oxidation of alcohols.

5.3.3
Wacker-type Oxidations

The use of Pd–NHC catalysts has also been recently applied to Wacker-type oxidations [23]. One example is the Pd-catalyzed aerobic Wacker cyclization of allyl-phenol derivatives (Fig. 5.11) [24, 25]. Several NHC ligands were evaluated for this process, with IMes in combination with $Pd(TFA)_2$ producing superior results. The optimized conditions use 5 mol% $Pd(IMes)(TFA)_2$ generated *in situ* at 80 °C in toluene under an oxygen atmosphere.

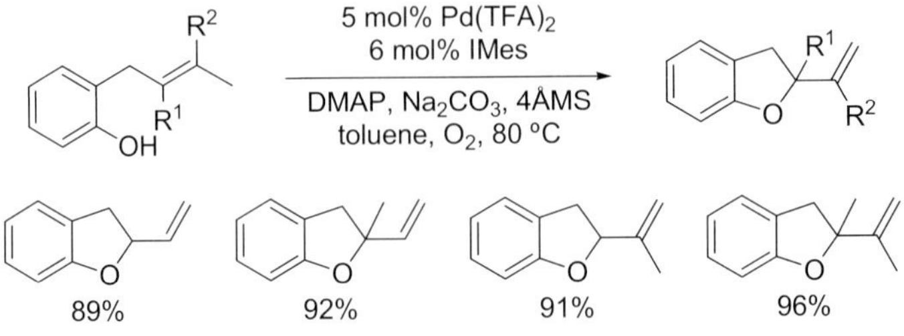

Fig. 5.11 Pd–NHC-catalyzed Wacker cyclization of allyl phenols.

It has also been demonstrated that Pd–NHC complexes can catalyze the Wacker oxidation of styrene derivates using aqueous TBHP (tert-butyl hydroperoxide) as the oxidant [26]. Using Pd(IiPr)(OTf)$_2$ prepared *in situ* from **31** and AgOTf, various styrenes were oxidized to the corresponding ketones with good selectivity and very little aldehyde formation (Fig. 5.12). This oxidation represents an extremely mild method for oxidizing styrene derivates that in most other cases do not undergo high-yielding Wacker oxidations. Kinetic studies of this oxidation revealed a zero-order dependence in both [TBHP] and [styrene]. Additionally, using anhydrous TBHP the reaction has an inverse first-order dependence on [H$_2$O]. This implies a rate-limiting step of water dissociation from the precatalyst. Using ^{18}O labeled H$_2$O, it was found that there was very little ^{18}O incorporation into the product at low conversions. Additionally, deuterium labeling at the a-position on styrene resulted in a high incorporation of deuterium in the methyl group on the product. Together, this data implicates that TBHP acts as a nucleophile attacking the Pd-bound olefin, followed by an a-hydride shift [27].

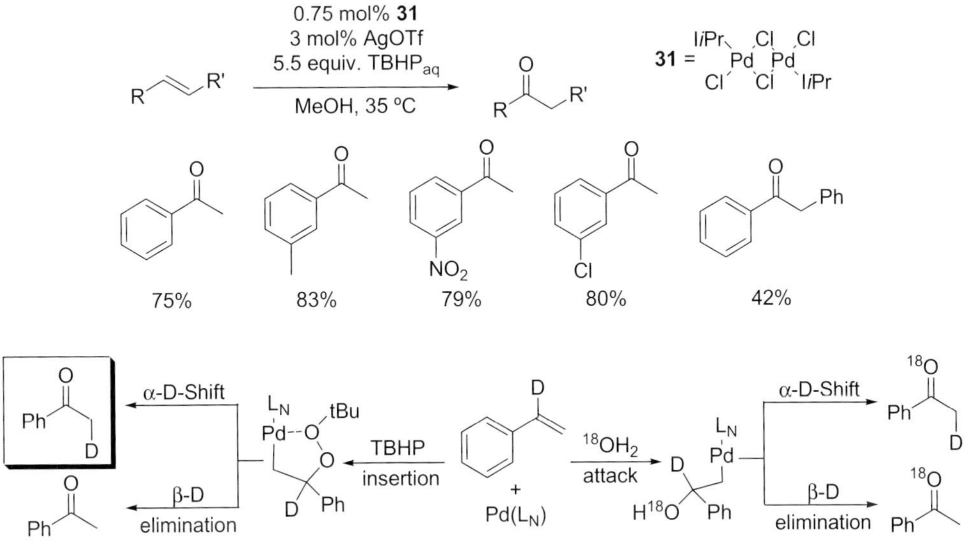

Fig. 5.12 Pd–NHC-catalyzed Wacker oxidation of styrenes.

5.3.4
Oxidative Carbonylation

Diphenyl carbonates (DPCs) are a common starting material for the production of aromatic polycarbonates (PCs), which formulate an important class of plastics [28]. The use of Pd, for the oxidative carbonylation of phenols and carbon monoxide, is a new option that has been recently explored to improve upon the synthesis of DPCs and aromatic PCs [29]. Recently, Sugiyama and coworkers have published several reports on the use of Pd–NHC catalyzed oxidative carbonylation using molecular oxygen as the stoichiometric oxidant to produce diphenyl carbonate and aromatic polycarbonates.

In their first report, Sugiyama and coworker's found that Pd–NHC complexes with bulky ligands containing electron-donating groups attached to the nitrogen produced the best results [30]. Under optimized conditions, a mixture of a Pd(NHC) catalyst, Ce(TMHD)$_4$ [31], ^{n}Bu$_4$NBr, hydroquinone, and 3Å molecular sieves in CH$_2$Cl$_2$ under a gaseous mixture of CO and O$_2$ was used to produce a 45% yield of diphenyl carbonate with a TOF of 51 h^{-1}. Only small amounts of by-products, including phenyl salicylate, phenoxy phenols, and biphenyl, were formed under these reaction conditions. Mechanistically this reaction is believed to proceed by substitution of the Br ligand with phenoxide, followed by CO insertion to form **33**, and reductive elimination to form the DPC and a Pd(0) species (**34**) that is then reoxidized to form the active catalyst (**32**) (Scheme 5.4).

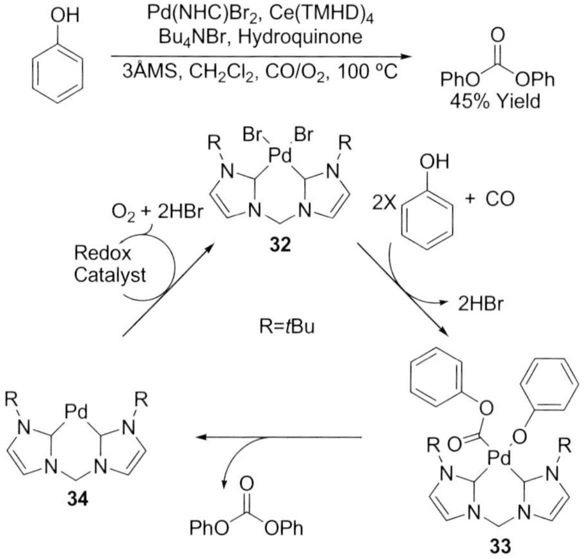

Scheme 5.4 Synthesis and mechanism of diphenyl carbonate (DPC) catalyzed by Pd–NHC.

Shortly following this report, Sugiyama also described two related Pd–NHC catalyst systems for the synthesis of polycarbonate from bisphenol A (**35**) (Fig. 5.13). In the first report, identical conditions to those used for the synthesis of diphenyl carbonate were found to be optimal for the synthesis of aromatic PCs [32]. This system produced aromatic PC with a M_n of 9600 and an M_w of 24 000. In the second report, the authors use a solid supported Pd–NHC catalyst (**37**) to synthesize both DPCs and aromatic PCs [33]. Using a NHC covalently attached to a support allowed for a synthesis of DPCs in high selectivity and TON (5100) but lower yields than those reported for the unsupported catalyst system. This same supported catalyst was used for the synthesis of an aromatic PC using bisphenol A. This system produced the polycarbonate in 95% yield with an M_n of 9200 and an M_w of 22 400.

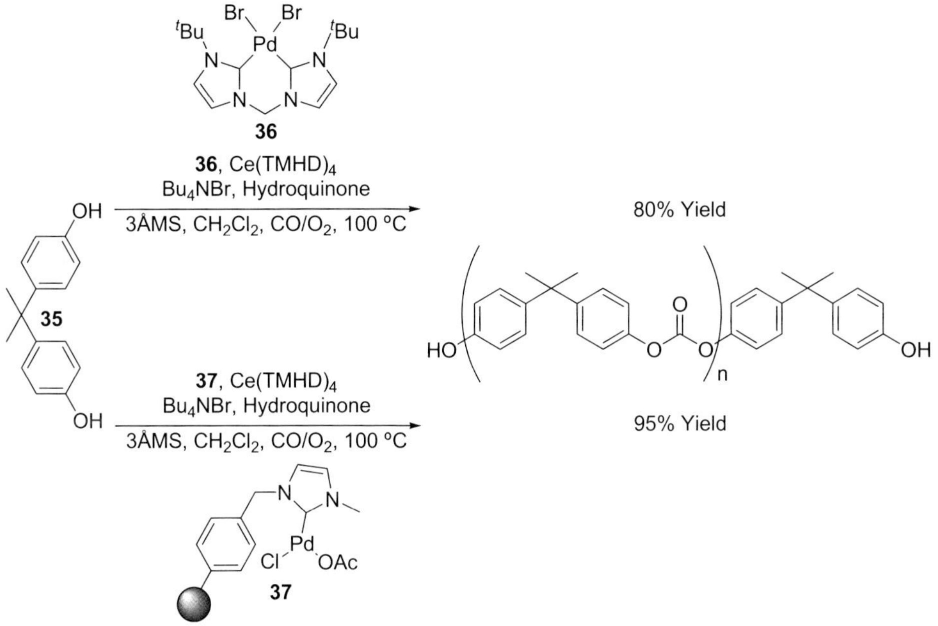

Fig. 5.13 Pd–NHC-catalyzed polymerization of bisphenol A.

5.4
Ir-catalyzed Oppenauer Oxidation of Alcohols

Yamaguchi and coworkers have described recently the first report of an Ir–NHC-catalyzed Oppenauer oxidation [34] of both primary and secondary alcohols [35]. Of several catalysts tested, [Cp*Ir(NHC)(MeCN)$_2$](OTf)$_2$ (**38**) proved to be the most active, achieving a TON of 6400 for the oxidation of cylcopentanol (Fig. 5.14). This catalyst was successful for the oxidation of aliphatic and benzylic alcohols, producing the corresponding carbonyl compounds in good to excellent yields.

Mechanistically this oxidation is thought to proceed via the accepted Oppenauer oxidation pathway (Scheme 5.5) [36]. The authors found no significant KIE for the β-hydride on sec-phenethyl alcohol, which implies that β-hydride elimination to form the Ir–hydride species (**40**) is not rate limiting. In addition, the observed rate dependence on [alcohol] pointed towards rate-limiting alkoxy exchange to form **39**. Studying the reaction by ^{1}H-NMR, an Ir–hydride species was observed, providing evidence that the reaction proceeds via the Ir–monohydride and not a Ir–dihydride. Finally, **38** reacted in stoichiometric fashion with sec-phenethyl alcohol to afford a Cp*Ir(NHC)(OTF)$_2$ (**42**) and a dinuclear Ir–NHC complex (**43**). The authors found that **43** was not active for the oxidation and that dimerization of **40** represented a deactivation pathway in the catalytic cycle.

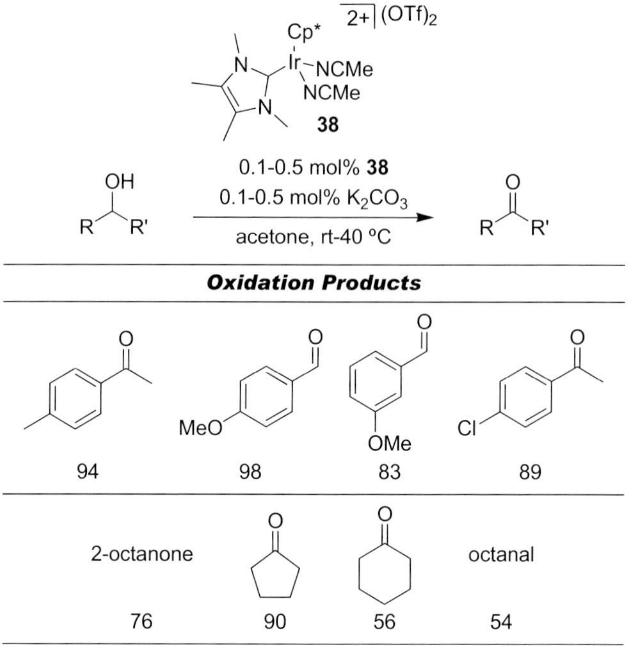

116 | *5 Metal-mediated and -catalyzed Oxidations Using N-Heterocyclic Carbene Ligands*

Oxidation Products

(4-methylphenyl methyl ketone)	MeO-(4-formylphenyl)	3-OMe-benzaldehyde	Cl-(4-acetylphenyl)
94	98	83	89
2-octanone	cyclopentanone	cyclohexanone	octanal
76	90	56	54

Fig. 5.14 Ir–NHC-catalyzed Oppenauer oxidation.

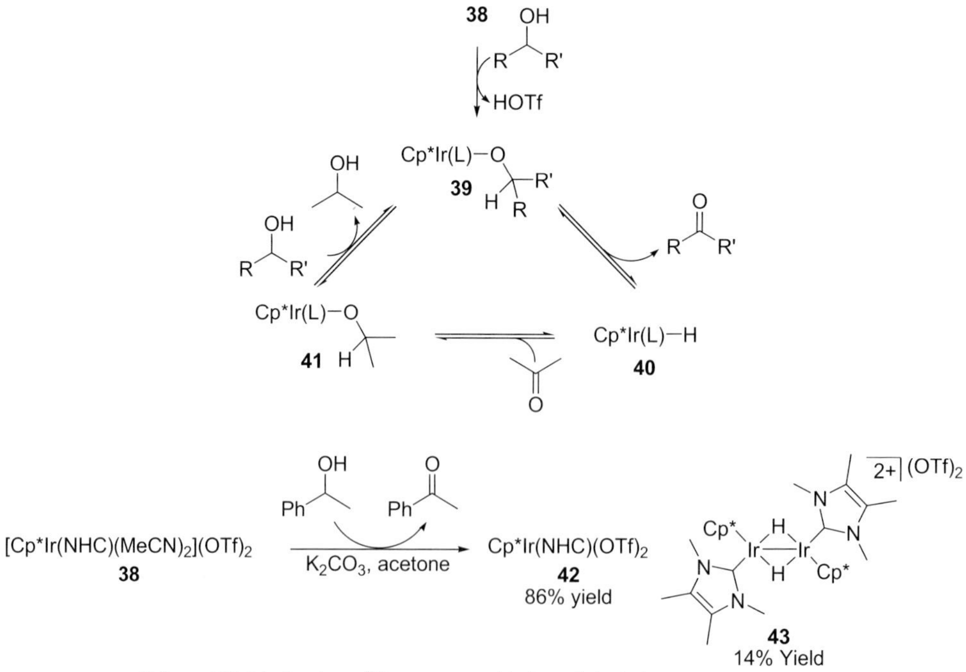

Scheme 5.5 Mechanism of Oppenauer oxidation of alcohols.

5.5
Conclusion

The use of N-heterocyclic carbenes as ligands for metals, especially Pd, in oxidative reactions has already proven successful. As demonstrated with Pd-based systems, NHCs produce highly robust complexes, which can withstand forcing, oxidative environments. Considering that the use of NHCs in oxidative reactions is only in its early stage of development, this chapter highlights the extremely bright potential for their continued use in oxidative catalysis.

References

1 For studies on the stability of carbenes, see: D. Bourissou, O. Guerret, P. G. François, G. Bertrand *Chem. Rev.* **2000**, *100*, 39–91 and references therein.
2 For studies on the stability of metal-NHC complexes, see: C. M. Crudden, D. P. Allen *Coord. Chem. Rev.* **2004**, *248*, 2247–2273.
3 For examples, see: (a) D. V. Yandulov, R. R. Schrock *Science* **2003**, *301*, 76–78. (b) S. D. Brown, T. A. Betley, J. C. Peters *J. Am. Chem. Soc.* **2003**, *125*, 322–323. (c) C. E. MacBeth, A. P. Golombek, V. G. Young, C. Yang, K. Kuczera, M. P. Hendrich, A. S. Borovick *Science* **2000**, *289*, 938–941.
4 X. Hu, I. Castro-Rodrigues, K. Meyer *J. Am. Chem. Soc.* **2003**, *125*, 12 237–12 245.
5 X. Hu, I. Castro-Rodrigues, K. Meyer *J. Am. Chem. Soc.* **2004**, *126*, 13 464–13 473.
6 For recent examples of Cu-catalyzed oxidations, see: (a) P. Gamez, I. W. C. E. Arends, R. A. Sheldon, J. Reedijk *Adv. Synth. Catal.* **2001**, *316*, 805–811. (b) I. E. Markó, A. Gautier, R. Dumeunier, K. Doda, F. Phillippart, S. M. Brown, C. J. Urch *Angew. Chem. Int. Ed.* **2004**, *43*, 1588–1591 and references therein.
7 For a review on Pd-catalyzed aerobic oxidations, see: S. S. Stahl *Angew. Chem. Int. Ed.* **2004**, *43*, 3400–3420.
8 B. R. Dible, M. S. Sigman *J. Am. Chem. Soc.* **2003**, *125*, 872–873.
9 B. R. Dible, M. S. Sigman, A. M. Arif *Inorg. Chem.* **2005**, *44*, 3774–3776.
10 M. M. Konnick, I. A. Guzei, S. S. Stahl *J. Am. Chem. Soc.* **2004**, *126*, 10 212–10 213.
11 T. Privalov, C. Linde, K. Zetterberg, C. Moberg *Organometallics* **2005**, *24*, 885–893.
12 C. R. Landis, C. M. Morales, S. S. Stahl *J. Am. Chem. Soc.* **2004**, *126*, 16 302–16 303.
13 M. Yamashita, K. Goto, T. Kawashima *J. Am. Chem. Soc.* **2005**, *127*, 7294–7295.
14 R. H. Crabtree *Chem. Rev.* **1995**, *95*, 987–1007.
15 (a) M. Muehlhofer, T. Strassner, W. A. Herrmann *Angew. Chem. Int. Ed.* **2002**, *41*, 1745–1747. (b) T. Strassner, M. Muehlhofer, A. Zeller, E. Herdtweck, W. A. Herrmann *J. Organomet. Chem.* **2004**, *689*, 1418–1424.
16 R. C. Larock *Comprehensive Organic Transformations*, 2nd edn., VCH, New York, **1999**, pp. 1234–1250.
17 D. R. Jensen, M. S. Sigman *Org. Lett.* **2003**, *5*, 63–65.
18 G. J. Mercer, M. Sturdy, D. R. Jensen, M. S. Sigman *Tetrahedron* **2005**, *61*, 6418–6424.
19 D. R. Jensen, M. J. Schultz, J. A. Mueller, M. S. Sigman *Angew. Chem. Int. Ed.* **2003**, *42*, 3810–3813.
20 M. J. Schultz, S. S. Hamilton, D. R. Jensen, M. S. Sigman *J. Org. Chem.* **2005**, *70*, 3343–3352.

21 H. Lebel, V. Paquet *J. Am. Chem. Soc.* **2004**, *126*, 11 152–11 153.
22 J. A. Mueller, C. P. Goller, M. S. Sigman *J. Am. Chem. Soc.* **2004**, *126*, 9724–9734.
23 J. M. Takacs, X.-T. Jiang *Curr. Org. Chem.* **2003**, *7*, 369–396.
24 K. Muñiz *Adv. Synth. Catal.* **2004**, *346*, 1425–1428.
25 Stahl has also stated that an NHC can be substituted for triethylamine for the Pd-catalyzed oxidative amination of sytrenes, see: V. I. Timokhin, N. R. Anastasi, S. S. Stahl *J. Am. Chem. Soc.* **2003**, *125*, 12 996–12 997.
26 C. N. Cornell, M. S. Sigman *J. Am. Chem. Soc.* **2005**, *127*, 2796–2797.
27 This is similar to the mechanism proposed by Mimoun, see: (a) H. Mimoun, R. Charpentier, A. Mitschler, J. Fischer, R. Weiss *J. Am. Chem. Soc.* **1980**, *45*, 1047–1054. (b) M. Roussel, H. Mimoun *J. Org. Chem.* **1980**, *45*, 5387–5390.
28 D. Freitag, U. Grigo, P. R. Muller, W. Nouvertne *Encyclopedia of Polymer Science and Engineering*, Wiley, New York, **1988**, vol. 11.
29 Y. Ono *Pure Appl. Chem.* **1996**, *68*, 367–375.
30 K. Okuyama, J. Sugiyama, R. Nagahata, M. Asai, M. Ueda, K. Takeuchi *J. Mol. Cat. A* **2003**, *203*, 21–27.
31 TMHD = 2,2,6,6-tetramethylheptane-3,5-dionate.
32 K. Okuyama, J. Sugiyama, R. Nagahata, M. Asai, M. Ueda, K. Takeuchi *Macromolecules* **2003**, *36*, 6953–6955.
33 K. Okuyama, J. Sugiyama, R. Nagahata, M. Asai, M. Ueda, K. Takeuchi *Green Chem.* **2003**, *5*, 563–566.
34 C. F. de Graauw, J. A. Peters, H. van Bekkum, J. Huskens *Synthesis* **1994**, 1007–1017.
35 (a) F. Hanasaka, K. Fujita, R. Yamaguchi *Organometallics* **2004**, *23*, 1490–1492. (b) F. Hanasaka, K. Fujita, R. Yamaguchi *Organometallics* **2005**, *24*, 3422–3433.
36 J.-E. Bäckvall *J. Organomet. Chem.* **2002**, *652*, 105–111.

6
Efficient and Selective Hydrosilylation of Alkenes and Alkynes Catalyzed by Novel N-Heterocyclic Carbene Pt⁰ Complexes

Guillaume Berthon-Gelloz and István E. Markó

6.1
Introduction

The hydrosilylation reaction, the addition of a Si-H unit **2** onto a carbon–carbon double or triple bond **1** to generate an alkyl- or a vinyl-silane **3**, is of paramount importance both for academia and for the silicon industry [1]. This transformation, which is highly atom-economical [2, 3], is the second most important C–Si bond-forming process, after the Rochow process [4]. It is at the heart of the production of silicon polymers, the basic ingredients for the manufacture of commodities such as lubricating oils, resins, pressure-sensitive adhesives, liquid injection molding products and paper release coatings [5]. Moreover, the alkyl- and vinyl-silanes produced are important synthetic intermediates [6, 7]. For example, they can be transformed into the corresponding alcohols, aldehydes and ketones under Tamao–Fleming conditions [8–11] or into alkyl- and vinyl-halides and chalcogenides by reaction with the appropriate electrophiles [12, 13]. Furthermore, the vinylsilanes can be engaged in Hiyama-type coupling reactions, leading regio- and stereo-selectively to aryl-, vinyl- and alkynyl-substituted alkenes [14–20].

Whilst the hydrosilylation of alkenes and alkynes can be initiated in numerous ways [21], the use of metal complexes has been most successful in promoting this transformation. Amongst the plethora of metals employed, platinum derivatives occupy a cardinal position [22]. Karstedt complex **8** [23, 24] and the Speier system (H_2PtCl_6/^{i}PrOH) [25] are the most commonly operating industrial catalysts. However, over the years, Speier's catalyst has been gradually superseded by the more active Karstedt derivative. Unfortunately, the hydrosilylation of alkenes catalyzed by complex **8** leads, besides the desired adduct **3**, to a mixture of by-products, among which the isomerized alkene **4**, the saturated derivative **5**, the secondary alkylsilane **6** and the vinylsilane **7** are the main components. Hydrosilylation of alkynes typically affords mixtures of regioisomers and, sometimes, even geometric isomers (Scheme 6.1).

N-Heterocyclic Carbenes in Synthesis. Edited by Steven P. Nolan
Copyright © 2006 WILEY-VCH Verlag GmbH & Co. KGaA, Weinheim
ISBN: 3-527-31400-8

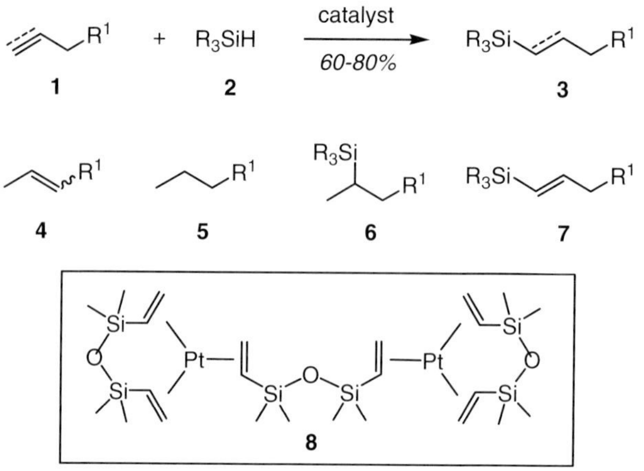

Scheme 6.1 Hydrosilylation of alkenes and alkynes.

Furthermore, colloidal platinum species are always generated under these conditions, leading to coloration of the reaction medium [26, 27]. These colloids are also believed to participate in the formation of some of the side products [28]. Moreover, the extreme reactivity of **8** precludes its application with substrates bearing sensitive functionalities, such as epoxides. To overcome some of these limitations, we launched a few years ago a research program aimed at the synthesis of novel platinum complexes that would display high reactivity, enhanced selectivity and increased functional group tolerance whilst, at the same time, being readily available and insensitive towards air and moisture. In addition, these new catalysts should not lead to the formation of platinum colloids. Elsevier et al. have, independently, investigated a similar approach [29–31].

6.2
Initial Results

At the onset we investigated the hydrosilylation of the model alkene **10** with the commercially available monomeric silane **9** (MD^{H}M, Me$_3$SiOSi(H)(Me)OSiMe$_3$, a suitable polysiloxane backbone mimic) (Scheme 6.2). Whereas Karstedt catalyst proved to be highly active (30 ppm Pt, 72 °C, 30 min, 80% yield), the products **4–7** were also generated in significant quantities (up to 20% overall). The appearance of these impurities follows the formation of the desired adduct **11** and coincides with the rapid generation of colloidal Pt species [27, 32].

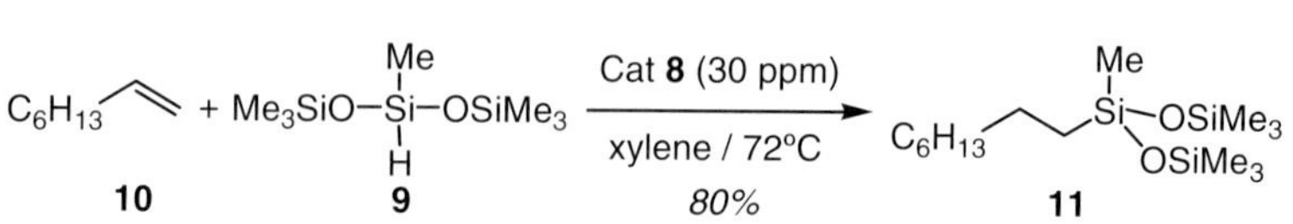

Scheme 6.2 Model hydrosilylation with various platinum(0) catalysts.

6.2 Initial Results

These experiments suggest that the divinyltetramethylsiloxane (dvtms) ligand is particularly labile, dissociating readily under the reaction conditions and leading rapidly to Pt colloids that catalyze the formation of the undesired side products. To stabilize the catalyst and avoid colloid procreation, it was decided to append more strongly binding, electron-rich ligands onto the platinum. Accordingly, preparation of the phosphine-containing complexes **12** became our prime objective (Scheme 6.3) [33–35].

Product	R	Yield (%)
12a	Phenyl	54
12b	Cyclohexyl	49
12c	Furyl	90
12d	*tert*-Butyl	60
12e	*ortho*-Tolyl	55

Scheme 6.3 Synthesis of (phosphine)Pt(dvtms) complexes.

Organometallic derivatives **12a–12e** were readily obtained by treating a toluene solution of Karstedt catalyst with an equimolar amount of the corresponding phosphine. The desired, highly soluble monophosphine-dvtms complexes could be purified by crystallization in moderate to good yields. Figure 6.1 displays the ORTEP X-ray structure of the tris(2-furyl)phosphine adduct **12c**.

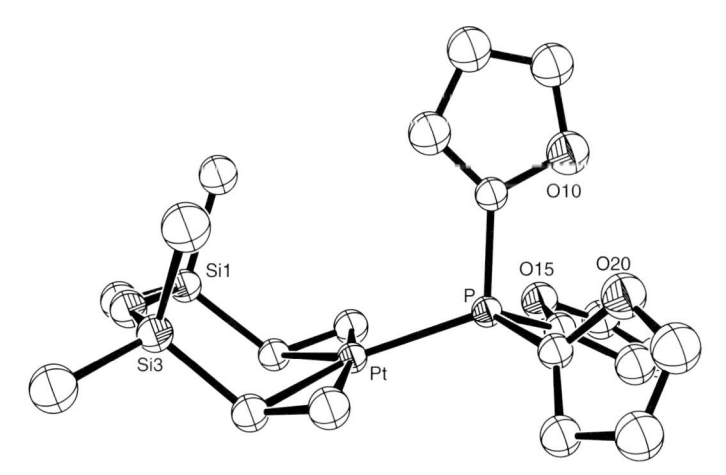

Fig. 6.1 ORTEP view of the tris(2-furyl)phosphine-Pt(dvtms) complex (**12c**).

These complexes were then tested in the model hydrosilylation reaction and compared with the Karstedt catalyst. Some selected results are collected in Fig. 6.2.

As can be seen, the Karstedt catalyst is superior to all complexes **12** in terms of rate of reaction. However, the conversion never exceeds 80% and the amount of by-products is higher with **8** (up to 18%) than with any of the derivatives **12a** to **12e**. Although the best conversion is reached using **12a**, containing a triphenylphosphine ligand, the fastest rate of reaction is achieved by complex **12d**, bearing a tris(*tert*-butyl)phosphine substituent. The least reactive catalyst appears to be the tris(2-furyl)phosphine-containing derivative **12c**. Examination of the amount of by-products (mostly internal alkenes) reveals an interesting trend: as the reactivity of the catalyst increases, so does the amount of undesired side-products. Moreover, with complexes **8**, **12a** and **12d**, it can be seen that, after reaching a maximum value, the amount of isomerized olefins decreases over time. The disappearance of the internal alkenes coincides with the formation of platinum colloids and with the appearance of the corresponding alkane **6**, indicating that these colloidal species are responsible, at least in part, for the hydrogenation of the substrate and its isomerized derivatives. In all cases, and depending upon the nature of the phosphine ligand, colloidal platinum species are still produced, implying that the phosphorous substituent is displaced to some extent during the reaction.

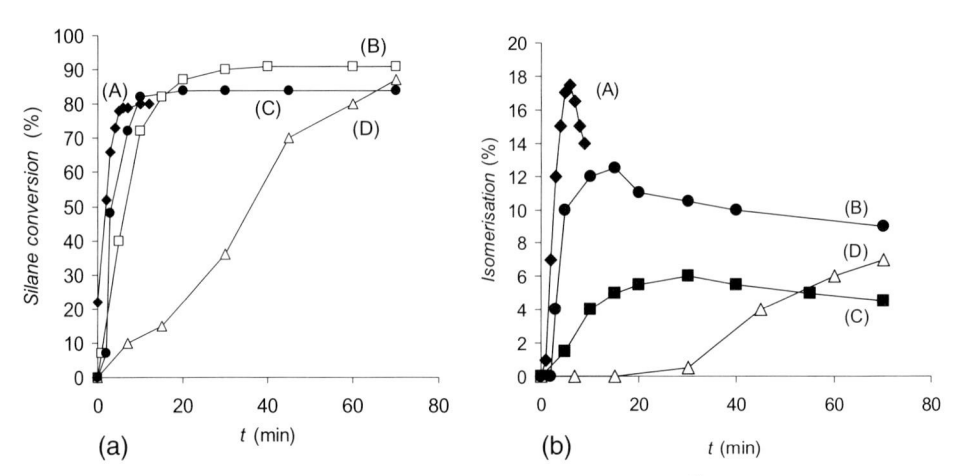

Fig. 6.2 (a) Reaction profile for the hydrosilylation of oct-1-ene by MD^{H}M catalyzed by phosphine-Pt(dvtms) complexes. (b) Isomerization reaction. Reaction conditions: MD^{H}M (0.5 mol L^{-1}), oct-1-ene (0.5 mol L^{-1}), [Pt] (2.5 × 10^{-5} mol L^{-1}), *o*-xylene (25 ml), 60 °C. Karstedt catalyst **8** (A), (Ph$_3$P)Pt(drtms) **12a** (B), (*t*-Bu$_3$P)Pt(drtms) **12d** (C) and **12c** (D).

6.3
Synthesis, Structure and Reactivity of (NHC)Pt(dvtms) Complexes

6.3.1
(Alkyl-NHC)Pt(dvtms) Complexes

To circumvent the above problem, we decided to use more robust σ-donor ligands and selected the corresponding N-heterocyclic carbenes [36–38]. The desired Pt-carbene complexes [31, 39–43] were readily obtained by treatment of the Karstedt catalyst with the N-heterocyclic carbenes, generated from the corresponding imidazolium salts **13** by treatment with NaH or KOBu^t [44–46]. Exclusive displacement of the internal, more labile, bridging dvtms ligand occurred, affording organometallics **14** in high yields and purity (Scheme 6.4).

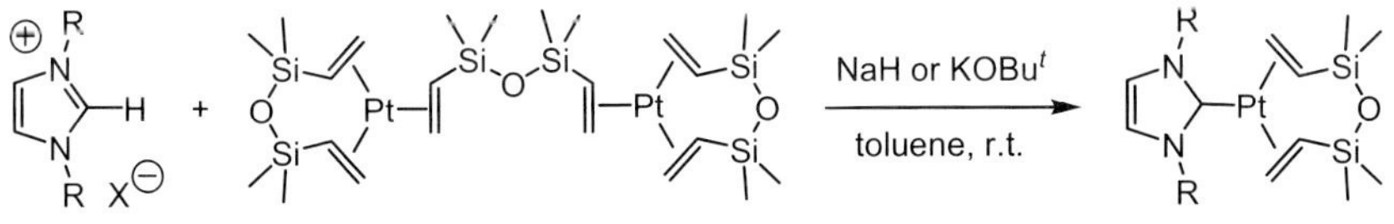

Complex	Yield(%)	Complex	Yield(%)
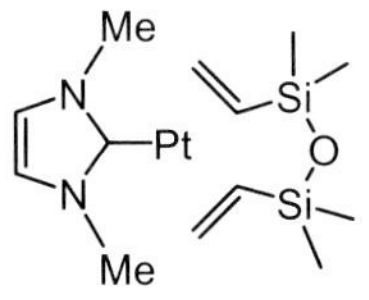 14a (Me, Me) 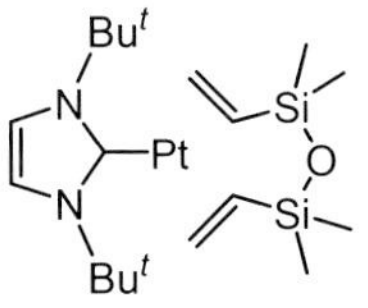	77	14c (Bu^t, Bu^t)	65
14b (Cy, Cy)	90	14d (Ad, Ad)	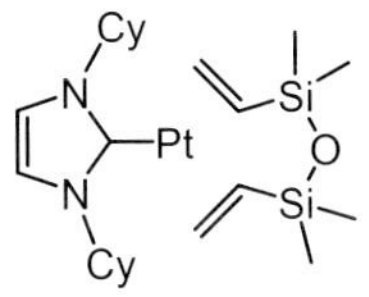 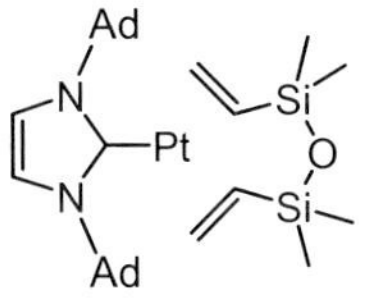 66

Scheme 6.4 Synthesis of alkyl-substituted N-heterocyclic carbene platinum(0) complexes.

These complexes are beautifully crystalline solids, insensitive to air and moisture. When protected from light, they can be stored for prolonged periods without noticeable loss in activity. The structure of complexes **14a–d** was determined by NMR spectroscopy and single-crystal X-ray diffraction analysis. The ^{1}H NMR region of the vinylsilane fragment is a very sensitive indicator of the coordination of the dvtms ligand to platinum. The shift of the vinylsilane protons, from the 5–6 ppm region to the 1–2.5 ppm, is indicative of the binding of the alkenes to the metal center. Another salient feature of the ^{1}H NMR spectrum is the splitting of the SiMe$_2$ singlet into two distinct signals for the pseudo-equatorial and pseudoaxial positions, at 0.3 and −0.5 ppm, respectively. This behavior, which points to the formation of a chelate, has been reported previously for complexes bearing phosphine ligands [47]. Another important indication of the coordination of the alkene residues to platinum is the coupling of the ^{195}Pt nuclei with several protons. Thus, a $^4J_{Pt-H}$ coupling of ca. 10 Hz is observed with the imidazol-2-ylidene backbone protons and an average $^2J_{Pt-H}$ coupling of 54 Hz for the =CH$_2$ protons. At room temperature, all the complexes, apart from **14a**, display diastereotopic N-CH$_2$ protons in their ^{1}H NMR spectrum, revealing the absence of rotation around the carbene–metal bond on the NMR time scale [48].

^{13}C NMR is also indicative of the coordination of the NHC and dvtms fragments to Pt, as evidenced through the ^{13}C/^{195}Pt couplings (Table 6.1). The ^{13}C NMR signal of the carbenic carbon appears on average at 182 ppm. The carbene carbons are strongly coupled with ^{195}Pt with a $^1J_{Pt-C}$ = 1365 Hz. The coupling constants between the ^{195}Pt and the carbons of the vinylic fragment of the dvtms ligand are $^1J_{Pt-C}$ = 166 Hz for the internal carbon and 116 Hz for the terminal carbon.

The ^{195}Pt NMR spectra of these complexes also provide a valuable and sensitive probe of the electronic environment of the Pt. An average chemical shift of −5350 ppm for platinum was observed (Table 6.2). This chemical shift is indicative of a Pt^0 complex (for reference, the chemical shift of Karstedt's catalyst is −6156 ppm) [24, 33]. The ^{195}Pt values for the related (R$_3$P)Pt(dvtms) for R = Ph, Cy and ^{t}Bu are −5598, −5633 and −5735 ppm respectively [33, 49]. The ^{195}Pt shifts of the (NHC)Pt(dvtms) complexes result from the combination of two factors. The

Table 6.1 ^{13}C carbene shift, J_{Pt-C} coupling, and J_{Pt-H}.

Entry	Complex	δc Pt–C$_{carb}$ (ppm)	^1J Pt–C$_{carb}$ (Hz)	^3J Pt–C$_{Im}$ (Hz)	^1J Pt–C$_{Vi}$ (Hz)	^1J Pt–C$_{Vi}$ (Hz)	^4J Pt–H$_{Im}$ (Hz)	^2J Pt–H$_{vi}$ (Hz)
1	(IMe)Pt(dvtms) (**14a**)	179.8					11.8	52.8
2	(ICy)Pt(dvtms) (**14b**)	180.0	1350.0	24.3	157.9	119.0	12.6	53.1
3	(IAd)Pt(dvtms) (**14d**)	180.3			185.4	123.6	12.0	53.2
4	(ItBu)Pt(dvtms) (**14c**)	181.2	1360.8	23.3	184.2	123.2	11.8	54.0
5	(IMes)Pt(dvtms) (**17a**)	184.2		41.8	165.7	117.7	9.2	55.0
6	(IPr)Pt(dvtms) (**17b**)	186.4		42.0	166.5	120.9	7.8	54.1
7	(BInPr)Pt(dvtms) (**19b**)	198.3	1373.3	41.6	158.2	113.6		54.1
8	(BImAllyl)Pt(dvtms) (**19c**)	198.3	1398.3	41.6	155.4	113.6		54.3
9	(BIMe)Pt(dvtms) (**19a**)	199.5	1377.9	39.0	155.7	113.7		54.1
10	(BIneoP)Pt(dvtms) (**20**)	201.9	1337.0	40.7	173.3	114.0		55.0
11	(SIMes)Pt(dvtms) (**17c**)	211.0		48.0	166.3	109.7		55.5
12	(SIPr)Pt(dvtms) (**17d**)	213.3		48.4	169.2	112.8		50.6

Table 6.2 ^{195}Pt shift of (NHC)Pt(dvtms) complexes.

Entry	Complex	δ ^{195}Pt (ppm)	Entry	Complex	δ ^{195}Pt (ppm)
1	(IAd)Pt(dvtms) (14d)	−5306	6	(ICy)Pt(dvtms) (14b)	−5343
2	(ItBu)Pt(dvtms) (14c)	−5333	7	(SIPr)Pt(dvtms) (17d)	−5361
3	(IMes)Pt(dvtms) (17a)	−5339	8	(SIMes)Pt(dvtms) (17c)	−5365
4	(IPr)Pt(dvtms) (17b)	−5340	9	(BIMe)Pt(dvtms) (19a)	−5379
5	(IMe)Pt(dvtms) (14a)	−5343	10	(BInPr)Pt(dvtms) (19b)	−5383

first is the σ-donating ability of the ancillary ligand: the more electron rich the ligand, the more downfield the ^{195}Pt shift. The second factor is back-donation of the platinum(0) center to the π* orbital of the chelating alkene ligand. With extended back donation, the electron density on the platinum decreases and results in an upfield shift. In (NHC)Pt(dvtms) derivatives, the NHC ligand is almost solely σ-donating, resulting in high electron density on the platinum center, which redistributes this density onto the olefins. The net result is a higher ^{195}Pt shift than the phosphine analogues and an increased stability. Hence, the ^{195}Pt shift reflects the overall reactivity of such platinum(0) complexes.

The structures of (IMe)Pt(dvtms) (14a), (ICy)Pt(dvtms) (14b) and (I^{t}Bu)Pt(dvtms) (14c) have been unambiguously elucidated by single-crystal X-ray diffraction analysis (Fig. 6.3) [50–52]. In all these organometallic derivatives, the platinum occupies the center of a trigonal planar arrangement, a coordination mode characteristic of (L)Pt^0(alkene)$_2$ complexes. This trigonal planar conformation around the metal provides a better overlap between the Pt d-orbitals and the olefinic π* systems, thus improving the back-bonding [53]. The dvtms ligand wraps around the platinum in a pseudo-chair conformation that nicely accommodates the required trigonal planar geometry.

Examination of the Pt–C$_{Carbenic}$ distance in various members of this (NHC)Pt(dvtms) family reveals only a small variation in bond length, which remains close to the mean value of 2.05 Å (Table 6.3) [54]. Such a bond length between platinum and the carbenic center correlates best with a single-bond character, in good accordance with the almost exclusive σ-donor properties of these ligands [55, 56].

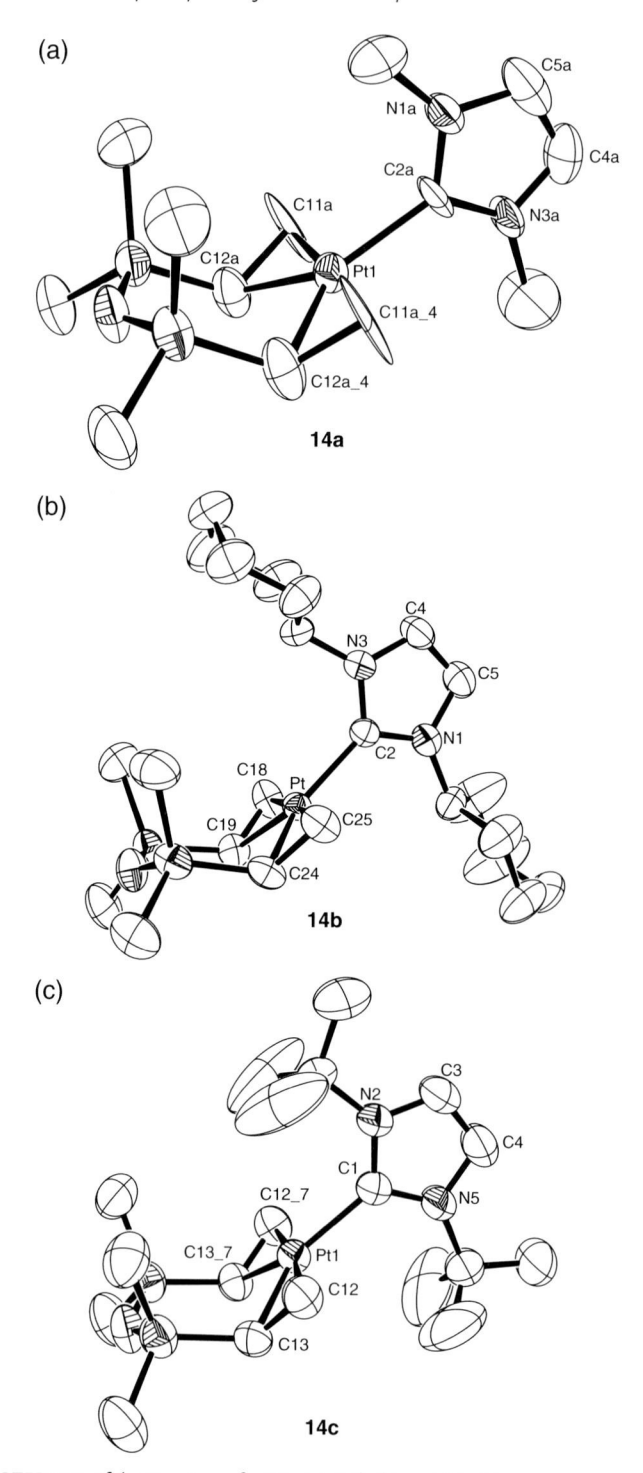

Fig. 6.3 ORTEP view of the structures for **14a** (a), **14b** (b) and **14c** (c).

Table 6.3 Selected structural data (NHC)Pt(dvtms) complexes characterized by X-ray crystallography.

Bond length (Å)	IMe (14a)	ICy (14b)	ItBu (14c)	IMes (17a)	SIPr (17d)	IBMe (19a)	BImAllyl (19c)	BIneoPent (20)
Pt–C$_{carbene}$	2.050(11)	2.026(5)	2.085(5)	2.046(4)	2.053(4)	2.035(4)	2.042(4)	2.058(4)
Pt–C$_{C=C}$[a]	2.103(15)	2.120(5)	2.130(6)	2.138(5)	2.138(6)	2.116(4)	2.127(5)	2.144(4)
Pt–C$_{C=C}$[b]	2.178(11)	2.136(5)	2.150(4)	2.148(5)	2.151(7)	2.143(5)	2.146(4)	2.149(4)
C$_{carbene}$–N[c]	1.34(2)	1.358(7)	1.361(7)	1.359(5)	1.347(6)	1.360(5)	1.351(6)	1.365(5)
C$_{Im}$–N[d]	1.35(2)	1.385(8)	1.371(9)	1.377(6)	1.468(6)	1.390(6)	1.398(6)	1.394(5)
C$_{Im}$–C$_{Im}$[e]	1.34(3)	1.335(10)	1.329(10)	1.332(7)	1.499(7)	1.384(6)	1.398(7)	1.387(6)
C$_{C=C}$–C$_{C=C}$[f]	1.491(13)	1.436(8)	1.433(6)	1.436(7)	1.433(7)	1.438(7)	1.421(7)	1.432(6)
C$_{C=C}$–C$_{C=C}$[f]	1.490(13)	1.420(9)	1.433(6)	1.419(7)	1.419(7)	1.428(6)	1.411(8)	1.415(7)
Angle (°)								
N–C$_{carbene}$–N	104.8(10)	103.8(5)	105.3(5)	103.1(3)	107.0(4)	105.8(3)	106.0(3)	106.0(3)
Torsion (°)								
N–C$_{carbene}$–Pt–C$_{C=C}$[g]	87.9(6)	82.4(9)	88.9(14)	63.8(6)	52.1(6)	88.2(4)	83.8(4)	70.3(5)
N–C$_{Im}$C$_{Im}$N[h]	0.0(7)	1.9(6)	0.9(12)	0.49(6)	18.2(5)	0.49(5)	0.9(4)	3.0(6)
C$_{C=C}$–C$_{C=C}$–C$_{C=C}$–C$_{C=C}$[i]	0.0(6)	2.4(5)	0.0(13)	4.3(6)	5.9(6)	1.1(4)	3.1(3)	1.5(5)

a) Distance of platinum to the terminal carbon.
b) Distance of platinum to the internal carbon.
c) Distance between the carbenic carbon and nitrogen.
d) Distance between the carbon of the imidazol-2-ylidene backbone and nitrogen.
e) Distance between the carbons of the imidazol-2-ylidene backbone.
f) Distance between the two olefinic carbons.
g) Dihedral angle between the plane of the imidazol-2-ylidene backbone and the trigonal plane formed by the dvtms ligand around the platinum center.
h) Refers to the distortion of the imidazol-2-ylidene backbone.
i) Distortion from planarity of the trigonal planar arrangement of the dvtms ligand.

The coordination features of the dvtms ligand are also excellent indicators of the strong electron-donating nature of the NHC substituents. For example, the length of the C=C bond is a good measure of the extent of back-bonding interactions [57]. Indeed, the mean C=C bond distance for dvtms ligand in (NHC)Pt(dvtms) complexes has an average value of 1.43 Å. This value, which is halfway between a double and a single bond, indicates that the olefinic linkage is particularly elongated as a result of intensive back-bonding. The longest bond [1.491(13) Å] occurs in (IMe)Pt(dvtms) (**14a**), the least sterically hindered complex. This correlation between the extent of back-bonding and the average bond length

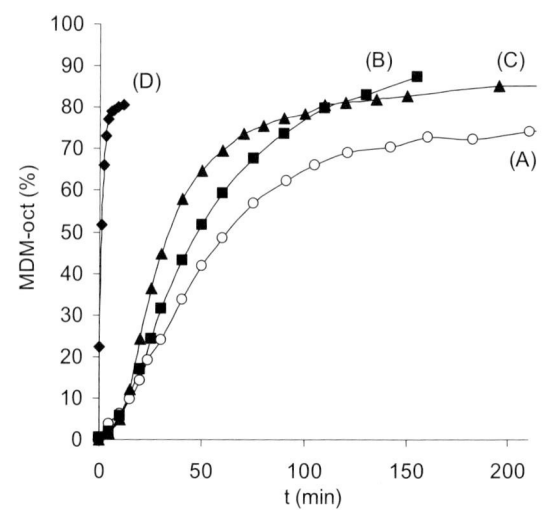

Fig. 6.4 Hydrosilylation of oct-1-ene by MD^{H}M catalyzed by (IMe)Pt(dvtms) (**14a**) curve (A), (ICy)Pt(dvtms) (**14b**) (B), (I^{t}Bu)Pt(dvtms) (**14c**) (C) and the Karstedt catalyst (**8**) (D).

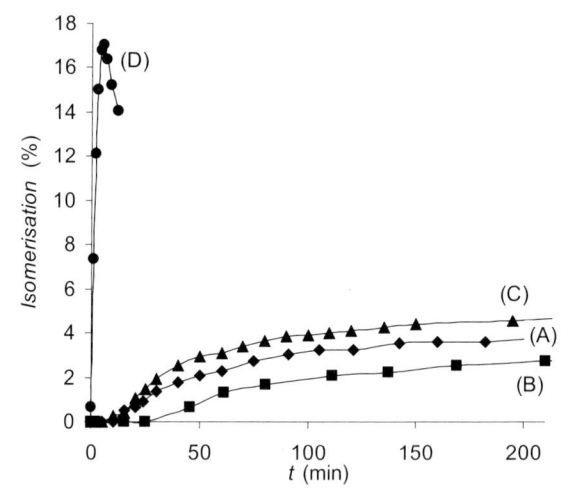

Fig. 6.5 Isomerization (sum of all isomerized alkenes) of oct-1-ene catalyzed by (IMe)Pt(dvtms) (**14a**) curve (A), (ICy)Pt(dvtms) (**14b**) (B), (I^{t}Bu)Pt(dvtms) (**14c**) (C) and the Karstedt catalyst (**8**) (D).

of the alkene ligand is further substantiated by literature data obtained for related (L)Pt^0(dvtms) complexes [24, 34, 35, 58–61].

The platinum(0) catalysts **14a–d** were initially tested in our model system, using oct-1-ene as the substrate, 30 ppm of **14**, and MD^{H}M (**9**) as the silylating agent, in xylene, at 72 °C. The results of these experiments are collected in Fig. 6.4.

As can be seen from these kinetic curves, the (NHC)Pt(dvtms) complexes **14** are all less reactive than Karstedt catalyst. Within this series, the activity of the platinum species increases as the steric bulk of the substituent on nitrogen becomes larger. Thus, the (I^{t}Bu)Pt(dvtms) catalyst **14c** hydrosilylates oct-1-ene faster than (ICy)Pt(dvtms) (**14b**), which is more reactive than (IMe)Pt(dvtms) (**14a**). However, interestingly, the conversions are higher with complexes **14b** and **14c** than with Karstedt catalyst. This behavior can be easily understood by examining the amount of isomerized alkenes produced in this reaction (Fig. 6.5).

Whilst Karstedt catalyst is the most reactive, it produces also a large amount of isomerized alkenes (up to 17%). In contrast, the slower-reacting Pt-carbene derivatives only generate small quantities of the internal alkenes (up to 4%). Thus, the selectivity appears to be inversely proportional to reactivity, and the formation of a minimum of by-products is best achieved by using a moderately active catalyst. Moreover, whereas platinum colloids are engendered in the hydrosilylation catalyzed by **8**, as evidenced by the appearance of a yellow color and by the formation of octane, no such species are produced in the reactions using the (NHC)Pt(dvtms) complexes [59].

To further optimize this process, an inverse addition protocol was performed [62]. Accordingly, silane **9** was added slowly to a solution of oct-1-ene (**10**) containing catalyst **14b**. Gratifyingly, the desired adduct **11** could be isolated in 96% yield; less than 1% of the isomerized alkenes could be detected in the crude reaction mixture (Scheme 6.5) [63].

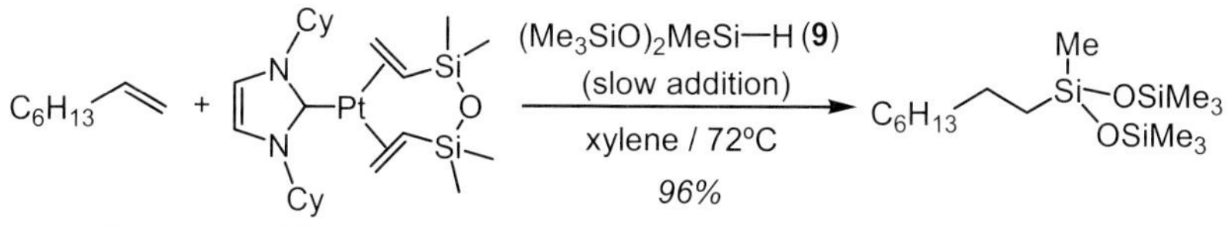

Scheme 6.5 Inverse addition protocol.

The platinum carbene complexes also proved to be far more tolerant towards a wide range of functionalities and protecting groups than the Karstedt catalyst. Table 6.4 gives some pertinent results.

As illustrated in Table 6.4, catalytic hydrosilylation of all the terminal olefins proceeds smoothly and leads in excellent yields to the desired addition products (entries 1–7). In all cases, complete regiocontrol is exercised and only the anti-Markovnikov adduct is obtained. Internal alkenes are completely inert under these conditions and the starting material is recovered quantitatively (entry 8). The catalyst tolerates various functionalities and protecting groups. Thus, hydrosilylation of olefins containing a tetrahydropyranyl ether (entry 1) or a *tert*-butyl dimethylsilyl ether (entry 2) affords the desired adducts in high yields. Notably, free alcohol functions are unaffected under these conditions (entry 3). No trace of the corresponding silylether could be detected in the crude reaction mixture [64]. In stark contrast to Karstedt catalyst **8**, which leads to significant decomposition of epox-

Table 6.4 Hydrosilylation of functionalized alkenes catalyzed by **14b**.

R⁀⁀ + Me₃SiO–Si(Me)(H)–OSiMe₃ →[14b (30 ppm), xylene / 72°C] R–CH₂–CH(Me)–Si(OSiMe₃)₂(OSiMe₃)
15 **9** **16**

Entry	R	Product	Yield (%)[a, b]
1	THPO⁀⁀	THPO–(CH₂)₄–Si(Me)(OTMS)(OTMS)	92
2	TBSO⁀⁀	TBSO–(CH₂)₄–Si(Me)(OTMS)(OTMS)	81
3	HO⁀⁀	HO–(CH₂)₄–Si(Me)(OTMS)(OTMS)	96
4	epoxide-(CH₂)₂-CH=CH₂	epoxide-(CH₂)₅-Si(Me)(OTMS)(OTMS)	90
5	cyclohexene epoxide with vinyl	corresponding product	92
6	pivaloyl-CH₂-CH=CH₂	pivaloyl-(CH₂)₃-Si(Me)(OTMS)(OTMS)	78[c]
7	EtO₂C-CH₂-CH=CH₂	EtO₂C-(CH₂)₃-Si(Me)(OTMS)(OTMS)	80
8	internal alkene (hex-2-ene)	–	–[d]

a) All yields are for isolated, pure compounds. Unless otherwise mentioned, all the conversions are quantitative.
b) In all cases, the use of Karstedt catalyst **8** leads to a mixture of products and to the formation of colloidal platinum species.
c) The reaction was stopped after 80% conversion.
d) No reaction was observed and the starting materials were recovered unchanged.

ide-containing substrates, the carbene complex **14b** tolerates the presence of this sensitive functionality (entries 4 and 5) and the hydrosilylated adducts are obtained in excellent yields [65]. Finally, complete chemoselectivity in favor of the hydrosilylation of the terminal alkene is observed in the presence of a ketone or an ester function (entries 6 and 7, respectively) [66, 67].

6.3.2
(Aryl-NHC)Pt(dvtms) Complexes

With the desire to improve the reactivity of these unique catalysts, whilst retaining their selectivity, it was decided to prepare (NHC)Pt(dvtms) complexes bearing aromatic substituents on the imidazolyl nitrogen atoms. Compounds **17a** and **17b** were readily synthesized using the protocol depicted in Scheme 6.6. The saturated analogues **17c** and **17d** were obtained in a similar manner, though in lower yields [68]. Figure 6.6 displays the ORTEP structures of **17a** and **17d**.

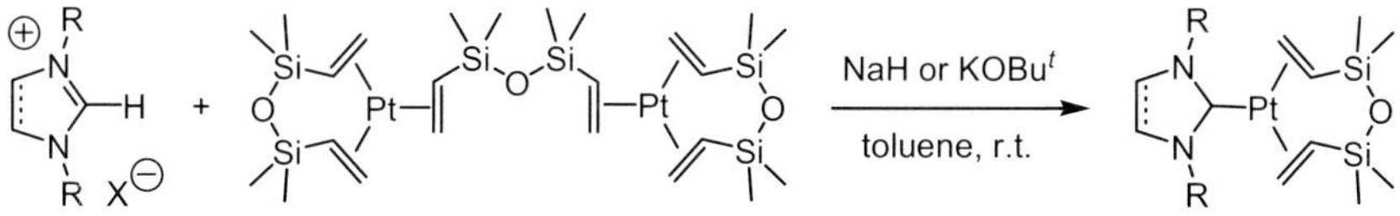

Complex	Yield(%)	Complex	Yield(%)
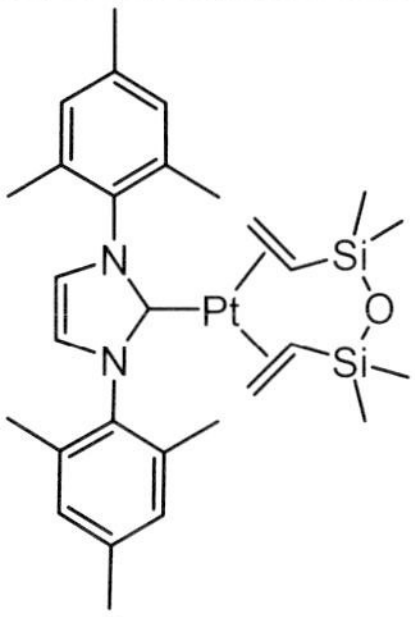 17a	85	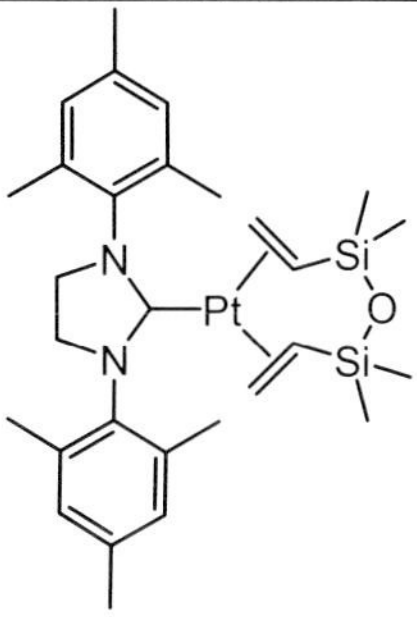 17c	50
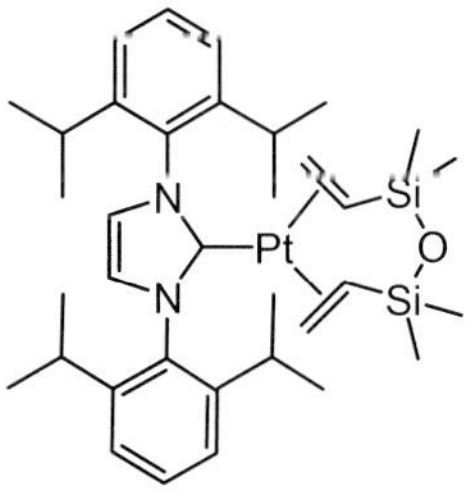 17b	70	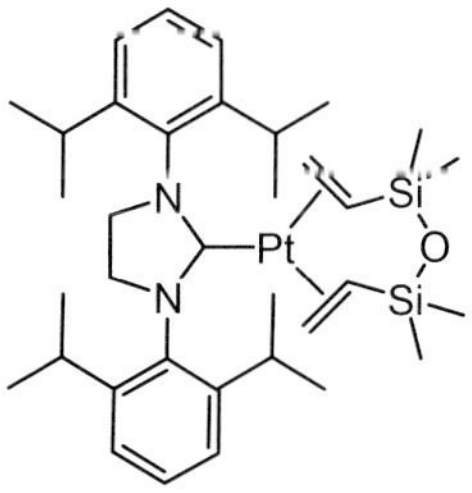 17d	50

Scheme 6.6 Synthesis of aryl-substituted (NHC)Pt(dvtms) complexes.

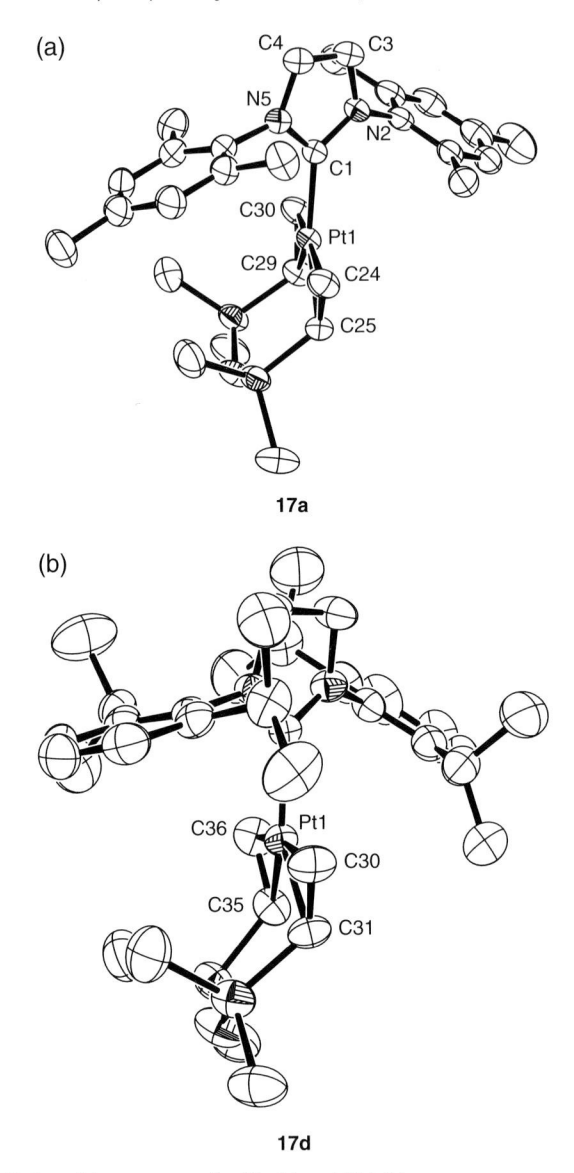

Fig. 6.6 ORTEP view of the structures for **17a** (a) and **17d** (b).

The aryl substituted (NHC)Pt(dvtms) catalysts possess a structure similar to that of their aliphatic analogues, with the platinum occupying the center of a trigonal planar arrangement. The aryl substituents adopt a "propeller" conformation around the imidazolyl backbone. Thus, in complex **17a**, the mesityl groups are tilted by 77.6(6)° and −79.2(7)°. This particular surrounding around the platinum center creates an unusual steric crowding that results in enhanced reactivity and selectivity.

These organometallic species were then assayed in the standard hydrosilylation of oct-1-ene. Figures 6.7 and 6.8 display the results.

As seen in Fig. 6.7, all of these catalysts display greater reactivity than their alkyl substituted (NHC)Pt(dvtms) counterparts. However, in contrast to derivatives **14**, which reacted immediately with the alkene (Fig. 6.4), complexes **17** require variable induction periods. After this initial latency time, hydrosilylation proceeds

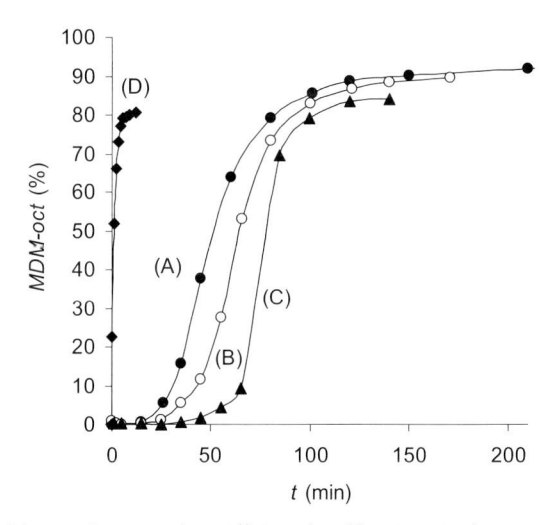

Fig. 6.7 Hydrosilylation of oct-1-ene by MD^{H}M catalyzed by (IMes)Pt(dvtms) (**17a**) curve (A), (IPr)Pt(dvtms) (**17b**) (B), (SIPr)Pt(dvtms) (**17d**) (C) and the Karstedt catalyst (**8**) (D).

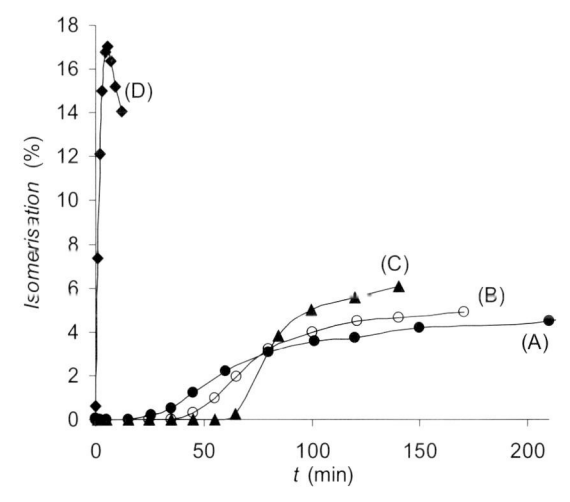

Fig. 6.8 Isomerization (sum of all isomerized alkenes) of oct-1-ene catalyzed by (IMes)Pt(dvtms) (**17a**) curve (A), (IPr)Pt(dvtms) (**17b**) curve (B), (SIPr)Pt(dvtms) (**17d**) curve (C) and the Karstedt (**8**) catalyst curve (D).

with a remarkably high rate, almost similar to that of the powerful Karstedt catalyst. Interestingly, even the most active platinum-carbene complex generates only minute amounts of isomerized olefins (ca. 6%, Fig. 6.8). Consequently, organometallic species **17** appear to provide a good comprise between reactivity and selectivity. Finally, colloidal platinum is never generated in these reactions.

6.3.3
(Benzimidazolyl-NHC)Pt(dvtms) Complexes

Although a plethora of imidazolyl ligands have been prepared and used in homogeneous catalysis [69], the synthesis and applications of their benzimidazolyl counterparts have remained relatively unexplored, though these derivatives display several interesting features [70–72]. Electronically, they possess a σ-donating ability that is intermediate between that of imidazolylidene and imidazolinylidene carbenes. In addition, benzimidazolylidene carbenes exhibit the topology of unsaturated N-heterocyclic carbenes but show structural properties and reactivity akin to those of carbenes containing a saturated N-heterocyclic ring system [73].

Complexes **19a–d** were synthesized in good to high yields, by *in situ* deprotonation of the corresponding benzimidazolium salts, in the presence of ^{t}BuOK, followed by reaction of the carbene with Karstedt's catalyst (Scheme 6.7).

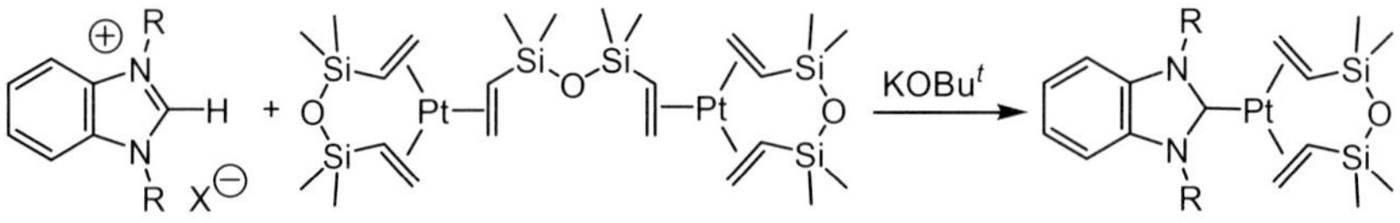

Complex	Yield	Complex	Yield
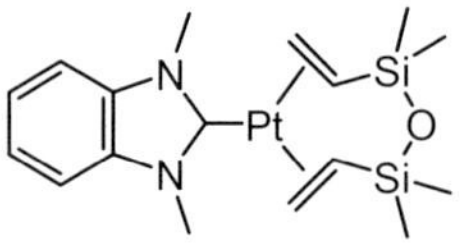 19a	80%	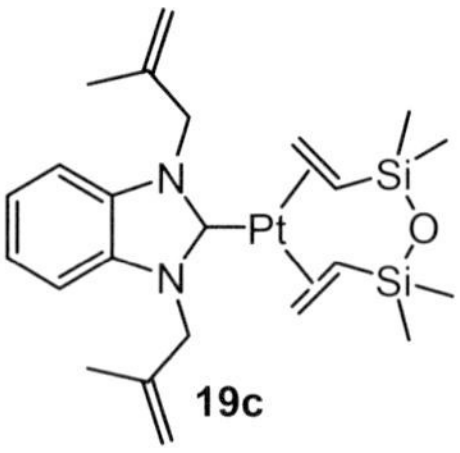 19c	78%
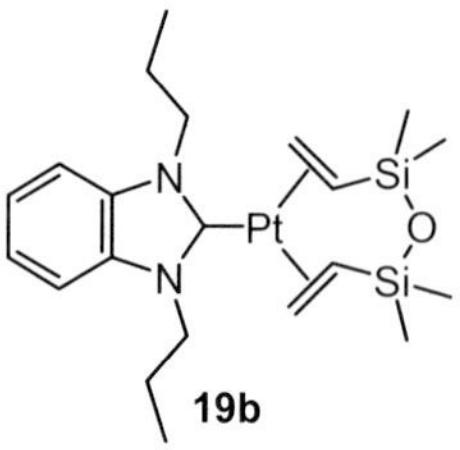 19b	60%	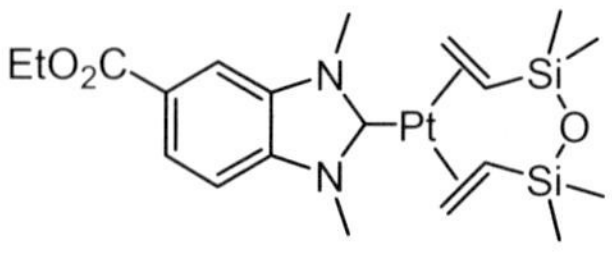 19d	77%

Scheme 6.7 Synthesis of benzimidazolyl platinum(0) carbenes.

Complex **19e** was prepared in a different manner. The benzimidazolylidene carbene was generated from the corresponding thiourea by reduction over Na/K amalgam, in toluene, for 20 days and then reacted with Karstedt's catalyst (Fig. 6.9) [73–75].

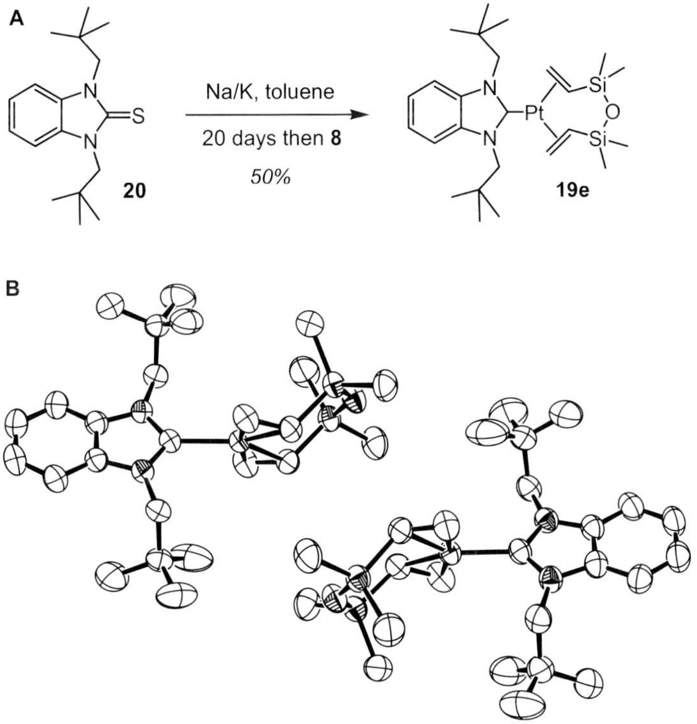

Fig. 6.9. (A) Synthesis of benzimidazolyl platinum(0) carbene **19e**.
(B) ORTEP view of the complex of the two enantiomers of **19e**.

Crystals of complex **19e**, suitable for X-ray diffraction analysis, were grown by slow evaporation of a dichloromethane solution. The molecular structure is shown in Fig. 6.9(B). Compound **19e** crystallises as a pair of enantiomers. The folding of the *neo* pentyl arms around the benzimidazole core defines a C_2 symmetric environment. This complex owes its chirality to the fact that the C_2 symmetry of the carbene ligand is broken by the pseudo-chair conformation adopted by the dvtms ligand. The benzimidazole carbene is tilted by 68° (C20–Pt1–C1–N9 torsion angle) from the plane defined by the Pt center and the alkene ligands.

Having obtained the desired carbene-platinum complexes, we next investigated their catalytic activity by subjecting them to our standard hydrosilylation conditions. The reaction was performed with a catalyst loading of 5×10^{-3} mol%. Figure 6.10 displays representative curves of the catalytic activity of **19a–c** and **19e**.

Figure 6.10 reveals that, apart from **19e**, the benzimidazolylidene carbene complexes **(19a–d)** are more active than the parent (IMe)Pt(dvtms) derivative. For example, after 100 min reaction, only 65% conversion is observed for the latter catalyst (Fig. 6.4) whilst 83% conversion is obtained with **19b** and **19c**. The initial reaction rate is also faster for the benzimidazole-containing complexes **19a–d** than for their imidazole counterparts. Interestingly, reagent **19b**, which displayed the highest activity, contains the *n*-propyl substituted benzimidazolylidene ligand. This high activity is, seemingly, due to a good compromise between the steric and the electronic properties of the NHC substituent.

Of the complexes described above, both **19c** and **19e** display a slightly different catalytic behavior, involving a prolonged induction period. For **19c**, this induction period can be attributed to the presence of the pendant alkene side-chains, which coordinate to the platinum center after the loss of the dvtms ligand and efficiently inhibit the approach of the reactants. Hydrosilylation of these olefinic substituents is thus required before the catalyst could exert its full activity. For **19e**, the bulky *neo*-pentyl groups wrap around the platinum, effectively blocking both sides and significantly increasing the induction period. Even after departure of the dvtms ligand, the *neo*-pentyl substituents are sufficiently bulky to reduce considerably the rate of the hydrosilylation of oct-1-ene. These experiments suggest that the loss of the dvtms ligand might be the rate-determining step of all these reactions.

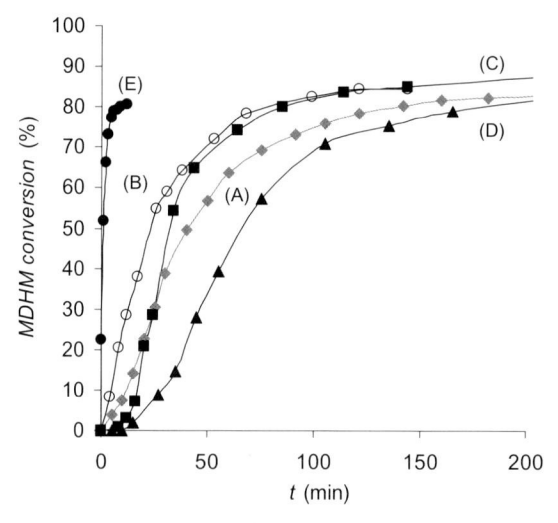

Fig. 6.10 Hydrosilylation of oct-1-ene by MD^{H}M catalyzed by (BIMe)Pt(dvtms) **(19a)** curve (A), (BI*n*Pr)Pt(dvtms) **(19b)** (B), (BI*m*Allyl)Pt(dvtms) **(19c)** (C), (BI*neo*Pent)Pt(dvtms) **(19e)** (D), and Karstedt's catalyst **(8)** (E).

6.4 Kinetic and Mechanistic Studies

At this stage, a detailed mechanistic understanding of the mode of action of these NHC-bearing catalysts became mandatory and an investigation of the overall reaction kinetics was performed, including the measurement of primary kinetic isotope effects and the isolation of key intermediates.

The hydrosilylation of oct-1-ene by bis(trimethylsilyloxy)methylsilane (MD^HM) catalyzed by the (ICy)Pt(dvtms) complex (**14b**) was selected as the model reaction (Scheme 6.2). The color of the solution remained perfectly clear throughout the reaction, indicating the lack of colloid formation. The consumption of the substrates as a function of time is sigmoidal (Fig. 6.11), with a direct relationship between substrate expenditure and product formation. Only minute amounts (<3%) of isomerized alkenes are formed [50]. Below 50 °C, the hydrosilylation becomes extremely sluggish, with reaction times over 12 h.

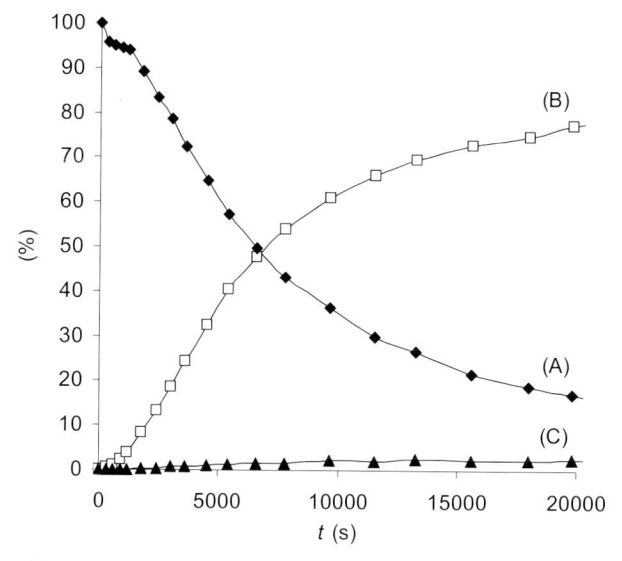

Fig. 6.11 Typical reaction profile for the hydrosilylation of oct-1-ene by MD^HM catalyzed by (ICy)Pt(dvtms) (**14b**). Reaction conditions: MD^HM (0.5 mol L^{-1}), oct-1-ene (0.5 mol L^{-1}), (ICy)Pt(dvtms) (**14b**) (2.5 × 10^{-5} mol L^{-1}), o-xylene, 60 °C. Curve (A) oct-1-ene, (B) MDM-oct, (C) sum of isomerization byproducts.

Recent studies on Karstedt and closely related Pt^0 catalysts have investigated in detail the homogeneous vs. heterogeneous nature of the active species formed during the hydrosilylation reaction. This research produced strong evidence in favor of a purely homogeneous nature of the catalytic active species. The formation of platinum colloids concured with a low concentration of alkene and proceeded after the hydrosilylation reaction [27, 59]. In none of the hydrosilylation reactions catalyzed by (NHC)Pt(dvtms) catalysts could visible platinum colloids be observed in the reaction media. Therefore, it was assumed that these transformation were also homogeneous.

At the onset of our study, we examined the mechanism of formation of the active catalyst from the (ICy)Pt(dvtms) precatalyst (**14b**). The sigmoidal reaction profile hinted at the presence of a prior equilibrium between the precatalyst and the active species. To assess qualitatively if the formation of the catalyst was a rate-determining step, we added fresh reactants (1:1 mixture of oct-1-ene and MD^{H}M) (Fig. 6.12). Surprisingly, we observed a marked increase in reaction rate ($\times$ 8.9) relative to the initial velocity. The activity increased even further ($\times$ 16) upon addition of a third batch of fresh reactants. This key experiment revealed the importance of the activation of the precatalyst. It also indicated that complete generation of the active species was significantly longer than the actual hydrosilylation process.

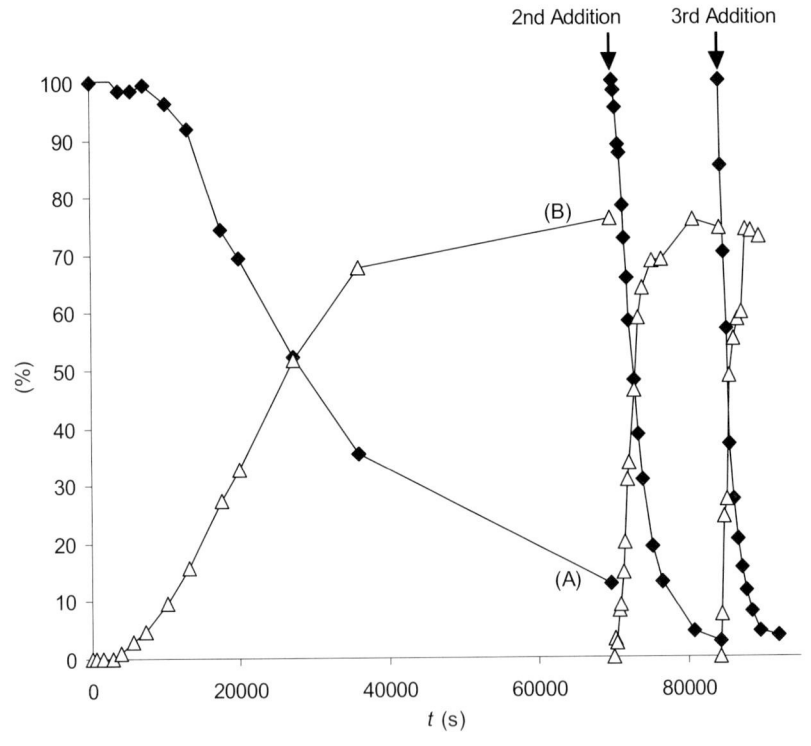

Fig. 6.12 Effect of the addition of fresh reactants on the reaction profile of the hydrosilylation of oct-1-ene by MD^{H}M. Reaction conditions: MD^{H}M (0.05 mol L^{-1}), oct-1-ene (0.05 mol L^{-1}), (ICy)Pt(dvtms) (**14b**) (2.5 $\times$ 10^{-6} mol L^{-1}), o-xylene, 70 °C. (A) Conversion of oct-1-ene. (B) Formation of MDM-oct. On the 2nd addition and 3rd addition, MD^{H}M (0.05 mol L^{-1}), oct-1-ene (0.05 mol L^{-1}) were added to the reaction mixture. For clarity, the reaction profile for the formation of MDM-oct (**11**) is corrected by subtraction of the previously formed product.

A reasonable mechanism for this process involves the initial decoordination of one alkene ligand, followed by subsequent hydrosilylation of the bound olefin of **21** to yield the catalytically active intermediate [NHC-Pt] **22** (Scheme 6.8).

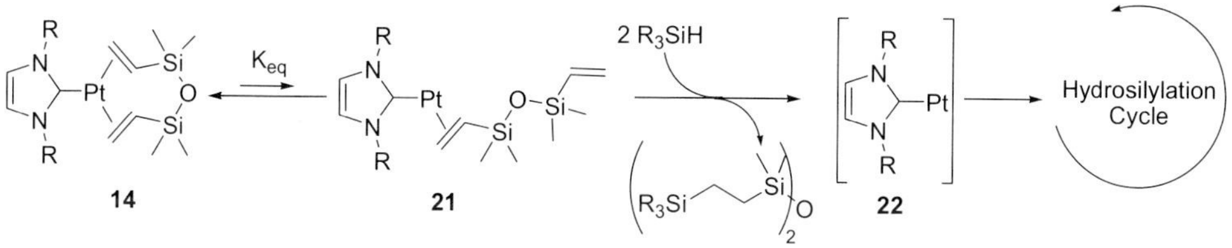

Scheme 6.8 Proposed mechanism for the activation of (NHC)Pt(dvtms) complexes.

To establish if such a process is indeed operative, the coordinating ability of the bis-alkene ligand was modulated through the synthesis of various (NHC)Pt(alkene)$_2$ derivatives in which the divinylsiloxane backbone was sequentially replaced by an allylic fragment, leading to complexes **23** and **24** (Fig. 6.13).

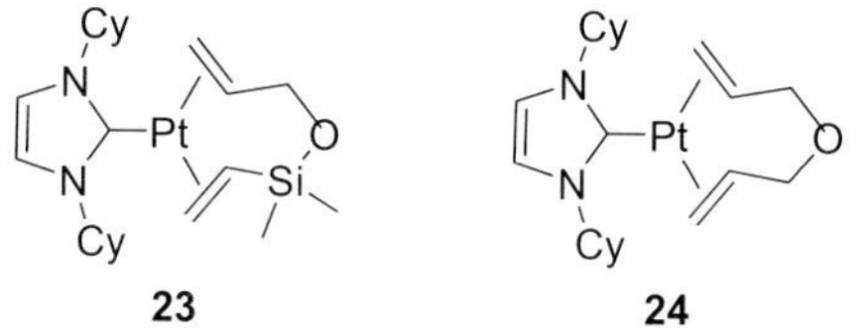

Fig. 6.13 Modified (ICy)Pt(dvtms) complexes.

The catalytic activity of the modified (ICy)Pt(dvtms) species **23** and **24**, in the model hydrosilylation reaction, was strikingly magnified (Fig. 6.14). The replacement of one vinylsilane unit by its whole carbon analogue provided a 4.5-fold rise in the reaction rate, whilst the second substitution led to an eight-fold increase. These experiments clearly demonstrated that the hydrosilylation activity is directly linked to the lability of the alkene ligands and that the rate-determining step in the initiation process is the initial decoordination of the olefins and not the subsequent hydrosilylation.

Examination of the amount of isomerized alkenes formed in the kinetic runs using **23** and **24** reveals that an increase in the hydrosilylation rate results in a concomitant rise in these by products (Fig. 6.15). Thus, it appears that the rate of hydrosilylation and isomerization are intimately coupled.

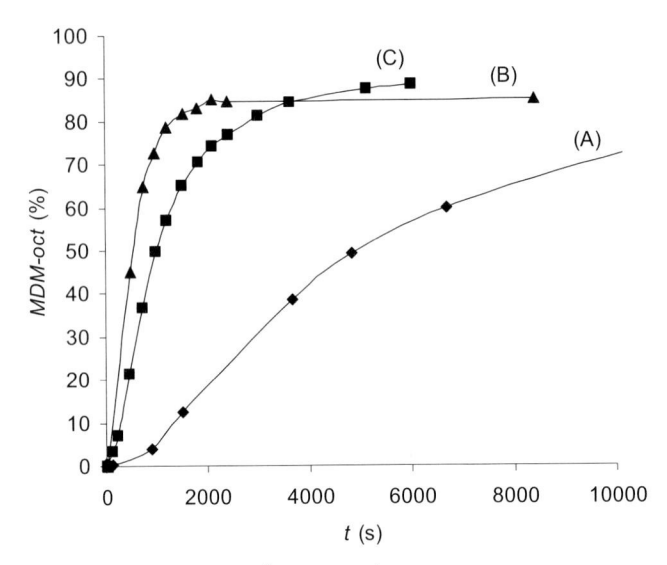

Fig. 6.14 Hydrosilylation with (ICy)Pt(alkene)$_2$ complexes. Reaction conditions: MD^{H}M (0.5 mol L^{-1}), oct-1-ene (0.5 mol L^{-1}), [Pt] (2.5 × 10^{-5} mol L^{-1}), o-xylene, 70 °C. **14b** (A), **23** (B), **24** (C).

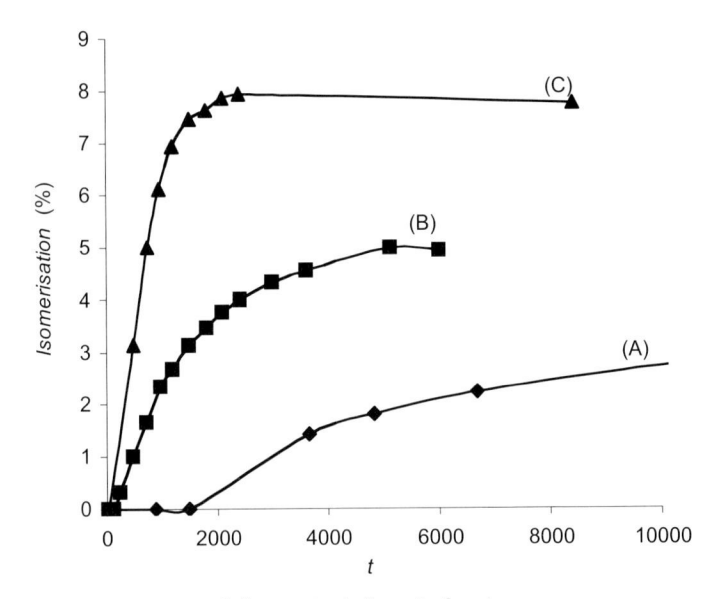

Fig. 6.15 Isomerization (sum of all isomerized alkenes) of oct-1-ene. Reaction conditions: MD^{H}M (0.5 mol L^{-1}), oct-1-ene (0.5 mol L^{-1}), [Pt] (2.5 × 10^{-5} mol L^{-1}), o-xylene, 70 °C. **14b** (A), **23** (B), **24** (C).

6.4 Kinetic and Mechanistic Studies

In an effort to isolate reaction intermediates, (ICy)Pt(dvtms) (**14b**) was treated with 5 equivalents of MD^{H}M, affording the dimeric complex **25** (Scheme 6.9). This platinum derivative was fully characterized and its structure unambiguously established by X-ray diffraction analysis [76].

Scheme 6.9 Synthesis of platinum dimer **25**.

The kinetics of the hydrosilylation of oct-1-ene catalyzed by **25**, followed by ^{1}H NMR spectroscopy, displayed first-order kinetics in (NHC)Pt(dvtms) complex (Fig. 6.16). At 70 °C, the observed rate constant was $10^{-4} \pm 5 \times 10^{-5}$ s^{-1}. The analogous phosphine derivatives have been characterized by Stone et al. and found to catalyze efficiently the hydrosilylation of alkenes and alkynes at room temperature [77–79]. Surprisingly, the dimeric hydride **25** displayed a slightly lower catalytic activity than the (ICy)Pt(dvtms) complex **14b**.

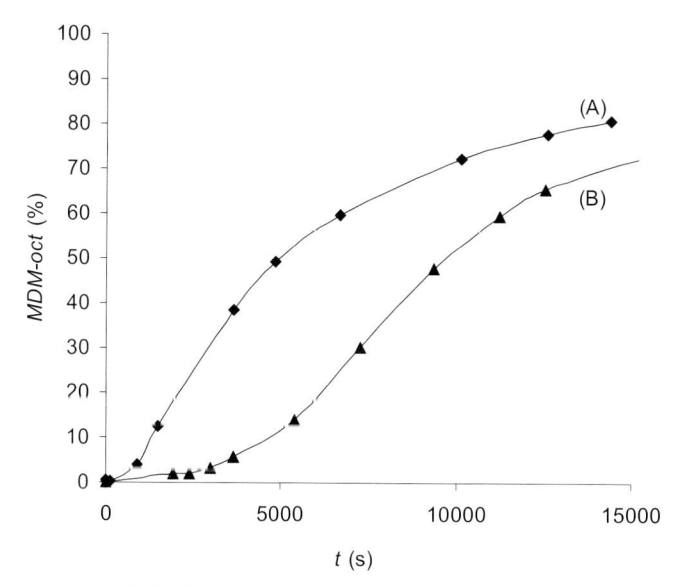

Fig. 6.16 Reaction profile for the dimeric complex (**25**) and (ICy)Pt(dvtms) (**14b**). Reaction conditions: MD^{H}M (0.5 mol L^{-1}), oct-1-ene (0.5 mol L^{-1}), [Pt] (2.5 × 10^{-5} mol L^{-1}), o-xylene, 70 °C. (ICy)Pt(dvtms) (**14b**) (A). Dimer (**25**) (B).

Examination of the reaction profile reveals a prolonged induction period, indicating very slow formation of the active species. This behavior can be tentatively explained by the strength of the Pt–H–Pt bond, which must undergo dissociation to release the (NHC)Pt(SiH) intermediate **26** to catalyze the hydrosilylation reaction (Scheme 6.10).

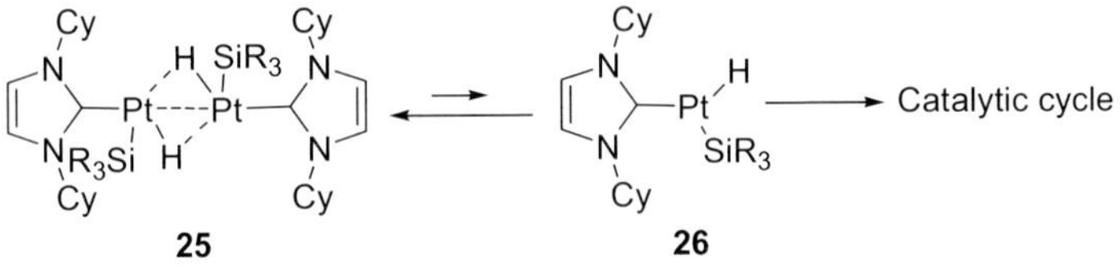

Scheme 6.10 Proposed equilibrium for the formation of the active species from dimer **25**.

The implication of dimer **25** in the catalytic cycle was further probed by reacting **25** with 10 eq. of oct-1-ene at 70 °C in [d_8]-toluene. After 4 h, complete conversion of **25** into (ICy)Pt(octene)$_2$ **27**, along with the formation of the hydrosilylated adduct **11**, were observed by ^{1}H NMR (Scheme 6.11). This experiment demonstrates that an equilibrium can be established between **25** and the monomeric form **26**, which can complete the hydrosilylation reaction without any added silane present. Unfortunately, the (ICy)Pt(octene)$_2$ complex **27** was too unstable to be isolated from the reaction mixture.

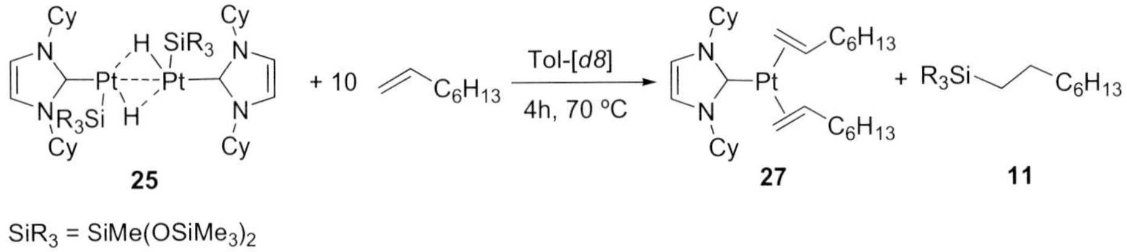

SiR$_3$ = SiMe(OSiMe$_3$)$_2$

Scheme 6.11 Reaction of dimer **25** with excess oct-1-ene.

The greater strength of the Pt–H–Pt interaction in **25** relative to the phosphine analogues is seemingly due to the higher propensity of the NHC-ligand to increase the electron density onto the metal center. Thus, on the reaction time scale, the formation of dimer **25** can be considered as a deactivation pathway.

Unless activation of the precatalyst can be made so fast as to become negligible on the time scale of the overall reaction, the initial rate measurement under conventional kinetic conditions is impossible. In the specific case of (NHC)Pt(dvtms) complexes, preactivation of the catalyst cannot be accomplished without formation of dimer **25**. Furthermore, we wanted to perform a precise kinetic analysis while keeping every parameter as close as possible to the "real" system, i.e., with low catalyst loading (S/C = 20 000) and no pseudo-first order conditions in order to avoid artifacts. At the onset of this study, we chose a rudimentary semi-quantitative analysis of the rate constant by measuring the tangent of the active phase of

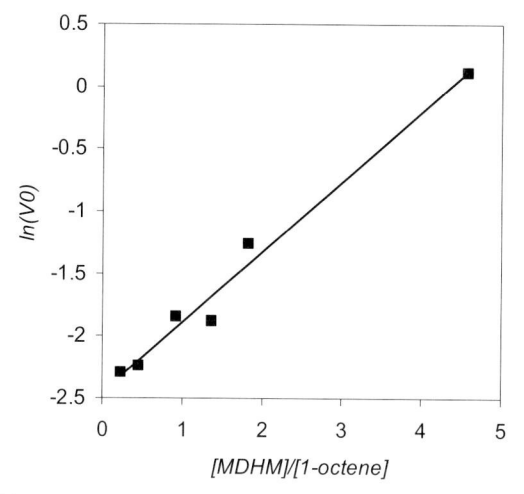

Fig. 6.17 Effect of the variation of silane. Reaction conditions: oct-1-ene (0.5 mol L^{-1}), **14b** (2.5 × 10^{-5} mol L^{-1}), o-xylene, 70 °C.

the kinetic curve: "V_0". Although V_0 has no clear physical meaning, it enables the quantitative comparison between the different kinetic curves. Hence, using this approach, a tentative first order in silane was measured, which is fully coherent with literature precedent (Fig. 6.17) [27, 80].

The reaction is also first order in oct-1-ene, for low oct-1-ene to silane ratios ([oct-1-ene]/[MD^{H}M] < 3). At higher alkene concentration, the rate decreases (Fig. 6.18). Such a behavior is often observed with other hydrosilylation catalysts in the presence of strongly coordinating alkenes [81, 82]. This substrate inhibition is most likely due to reversible catalyst deactivation by formation of a (ICy)Pt(oct-1-ene)$_2$ complex (**27**, Scheme 6.11).

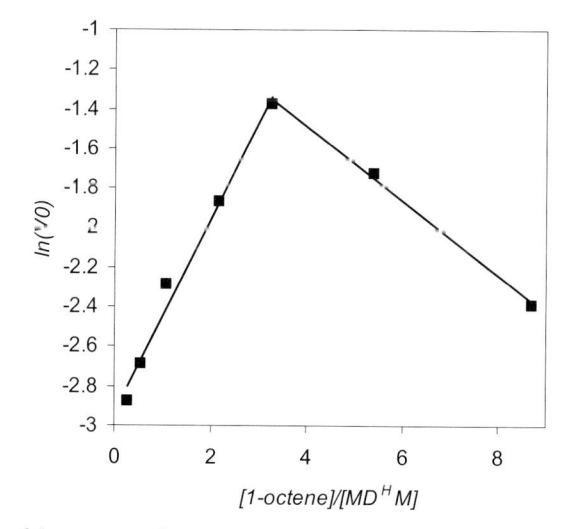

Fig. 6.18 Effect of the variation of concentration of alkene. Reaction conditions: MD^{H}M (0.5 mol L^{-1}), **14b** (2.5 × 10^{-5} mol L^{-1}), o-xylene, 70 °C.

The first order in silane and in octene (for low [octene]/[MD^{H}M] ratio) would indicate that the two reactants are present in the rate-limiting step of the catalytic cycle. The effect of the platinum concentration on the rate of hydrosilylation was measured similarly. Interestingly, a plateau at 5×10^{-3} mol% was observed, suggesting that the solution becomes "saturated" in catalyst. One plausible explanation for this phenomenon is that a high Pt concentration strongly favors the formation of dimer **25**, whilst the initiation step, which is unimolecular, remains unaffected. Therefore, the concentration of active catalyst does not rise significantly with the increase in the amount of precatalyst (Fig. 6.19).

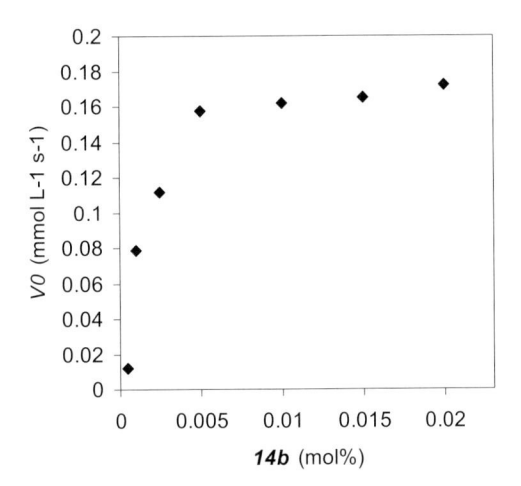

Fig. 6.19 Effect of catalyst loading on the reaction rate. Reaction conditions: MD^{H}M (0.5 mol L^{-1}), oct-1-ene (0.5 mol L^{-1}), o-xylene, 70 °C.

An in-depth analysis of the isomerization process vs. the hydrosilylation reaction led us to compare $V_{0\,Hydro}$ versus $V_{0\,Iso}$ for different platinum-based catalysts under otherwise identical conditions. This correlation led to an excellent linear relationship between $\ln(V_{0\,Hydro})$ and $\ln(V_{0\,Iso})$ (Fig. 6.20).

As can be seen from Fig. 6.20, the isomerization process appears to be clearly coupled with the hydrosilylation reaction by a common intermediate, most likely a Pt-H species. This relationship is seemingly independent of the nature of the Pt precatalyst. For example, (Ph$_3$P)Pt(dvtms), PtCl$_2$(COD) and Pt(norbonene)$_3$ also correlate well with the NHC-based platinum derivatives. The direct implication of this correlation, which can be expressed by the empirical power law of Eq. (1), is that the more active the catalyst, the greater the isomerization to the power 1.4!

$$\frac{V_{0Iso}}{V_{0REF}} = 0.03 \times \left(\frac{V_{0Iso}}{V_{0REF}}\right)^{1.4} \tag{1}$$

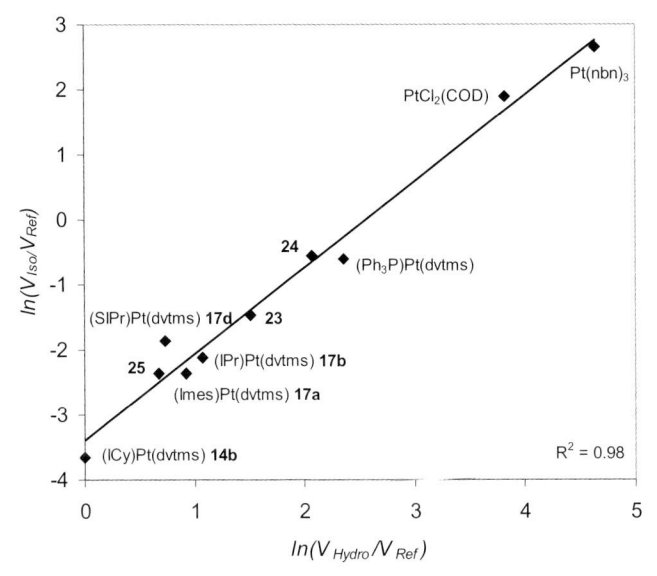

Fig. 6.20 Ln–ln relationship between the rate of hydrosilylation and the rate of isomerization (nbn = norbornene). Reaction conditions: MD^{H}M (0.5 mol L^{-1}), oct-1-ene (0.5 mol L^{-1}), [Pt] (2.5 × 10^{-5} mol L^{-1}), o-xylene, 70 °C.

Thus, it transpires that the minute amount of isomerization observed with (NHC)Pt(dvtms) complexes is due to their low hydrosilylation activity and high thermal stability.

To overcome the obvious limitations of this semi-quantitative approach, however informative, we turned our attention to numerical kinetic modeling with the use of the chemical kinetic software ReactOp. This software, designed for chemical engineering, enables the resolution of a set of differential equations, representing a proposed catalytic cycle, for the given kinetic data. The basic principle behind this approach is that numerical integration of the system of differential equations corresponding to the elementary transformations can be used to determine the elementary rate constants by least-squares fitting [83, 84]. In practice, each data point used with this method is equivalent to an individual experiment where only the initial rate was measured. Unlike traditional kinetic analysis, a high number of kinetic runs can be used to fit the model, thus providing rate constants with higher precision. Importantly, this mathematical treatment can be used to deconvolute the initiation rate constant of the active catalyst from the actual rate constant of the catalytic cycle, thereby obviating the need for parameters outside the actual regime of the catalytic reaction.

To determine a working kinetic model, every reasonable mechanism based upon our experimental studies was analyzed. Discrimination between each of them was performed by examining their fit with the experimental kinetic data. Thus, 15 different models were confronted with the kinetic data represented by 16 independent experiments, making up 900 data points. If two or more mechanisms fitted closely, further experiments were undertaken to eliminate one possibility. Based upon this method, we were able to identify a working kinetic model in good agreement with the kinetic data. Figure 6.21 gives a representative example of the fit obtained for the temperature variation. Scheme 6.12 depicts the catalytic cycle corresponding to the working kinetic model.

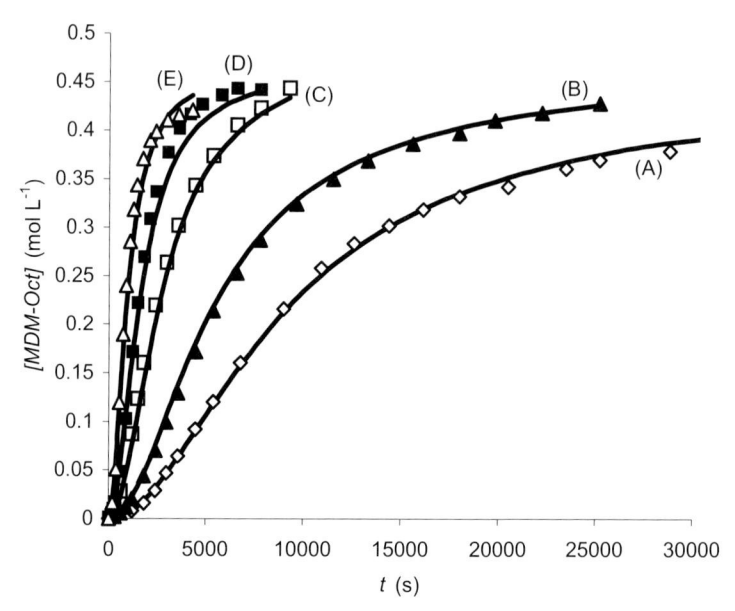

Fig. 6.21 Representative fit of the kinetic model with experimental curves (hydrosilylated product formation) at different reaction temperatures. Reaction conditions: MD^{H}M (0.5 mol L^{-1}), oct-1-ene (0.5 mol L^{-1}), (ICy)Pt(dvtms) (**14b**) (2.5 × 10^{-5} mol L^{-1}), o-xylene. Kinetic model (solid line), T (°C) = (A) 51, (B) 59, (C) 70, (D) 80 and (E) 90.

In this kinetic model, the release of the active [ICy-Pt] intermediate **28** from the precatalyst **14b** appears to be a first-order activation step. Moreover, this process is independent of both silane and alkene (models having a silane and/or an alkene in the activation step resulted in a poorer fit). The simulation is coherent with the observation that the slow step in the activation of the catalyst is the initial decoordination of the alkene. The intermediate **28** subsequently reacts with the alkene and the silane in a concerted oxidative addition – 1,2 migratory insertion (transition state **29**), leading directly to the (ICy)Pt(alkyl)(silyl) intermediate **30**, which, upon reductive elimination, yields the hydrosilylated product. Notably, this kinetic model fitted significantly better than the equivalent epitome based upon the classical Chalk and Harrod mechanism and embodying two kinetically distinct steps for the oxidative addition and 1,2 insertion [80]. Moreover, this concerted hydrosilylation mechanism agrees well with the first-order dependence on both alkene and silane. Inhibition of the catalyst at high alkene concentration was modeled

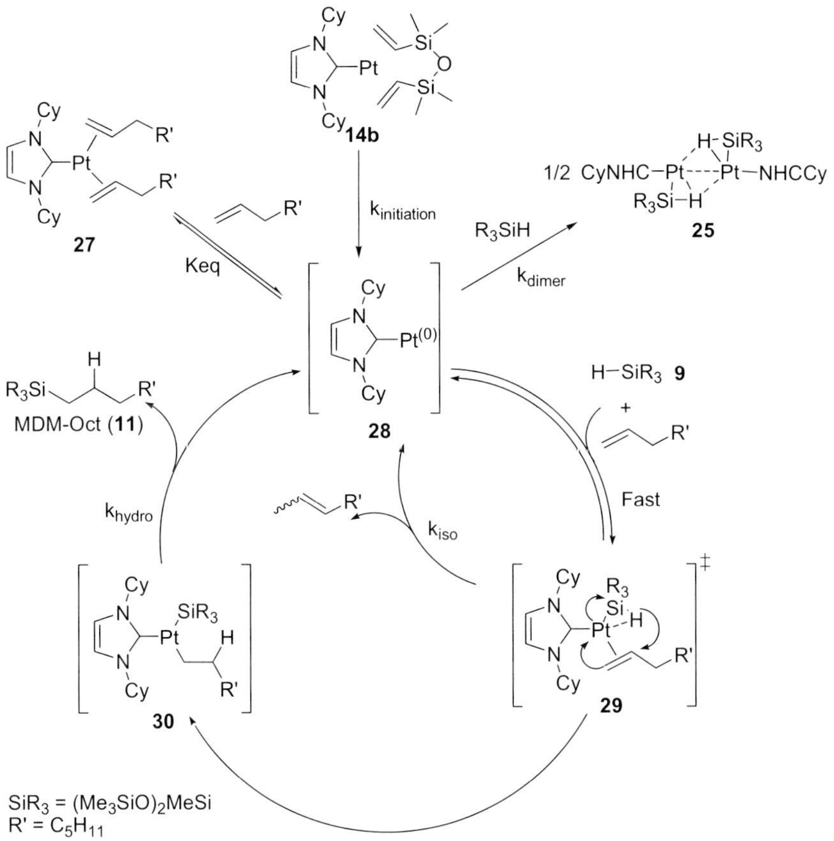

Scheme 6.12 Proposed catalytic cycle.

Table 6.5 Rate constants determined by fitting the kinetic model to the experimental data.

Constant	Fitting	Error
k_{ini}	$1.4 \times 10^{-4}\,s^{-1}$	$2 \times 10^{-6}\,s^{-1}$
k_{Hydro}	$127\,L^2\,mol^{-2}\,s^{-1}$	$1\,L^2\,mol^{-2}\,s^{-1}$
k_{Iso}	$8.3\,L^2\,mol^{-2}\,s^{-1}$	$0.3\,L^2\,mol^{-2}\,s^{-1}$
k_{dimer}	$15.6\,L\,mol^{-1}\,s^{-1}$	$0.3\,L\,mol^{-1}\,s^{-1}$
K_{eq}	$0.32\,L^2\,mol^{-2}$	$0.02\,L^2\,mol^{-2}$

by an equilibrium constant between [ICy-Pt] (**28**) and (ICy)Pt(octene)$_2$ (**27**). The formation of dimer **25** was considered to be an irreversible process, since its activity is lower than that of the precatalyst **14b**. Table 6.5 gives the numerical values for the rate constants obtained using this model.

The calculated rate constant for the first-order initiation process is in good agreement with the constant determined by ^{1}H NMR experiments. These calculated values reveal that the rate constant for the activation step is 900,000 times smaller than that for the hydrosilylation step.

With this working kinetic model in hand, we were able to simulate the catalyst distribution thoughout the reaction and under various conditions. Figure 6.22 gives a representative simulation for our standard transformation. The figure shows that the first-order decrease in precatalyst concentration results in the formation of the active [NHC-Pt] derivative **28**, the highest amount of this species coinciding with the highest catalytic activity. The increase in active species also results in its deactivation though the formation of the dimer **25**. Noticeably, under these conditions, the generation of the (ICy)Pt(octene)$_2$ complex **27** is negligible.

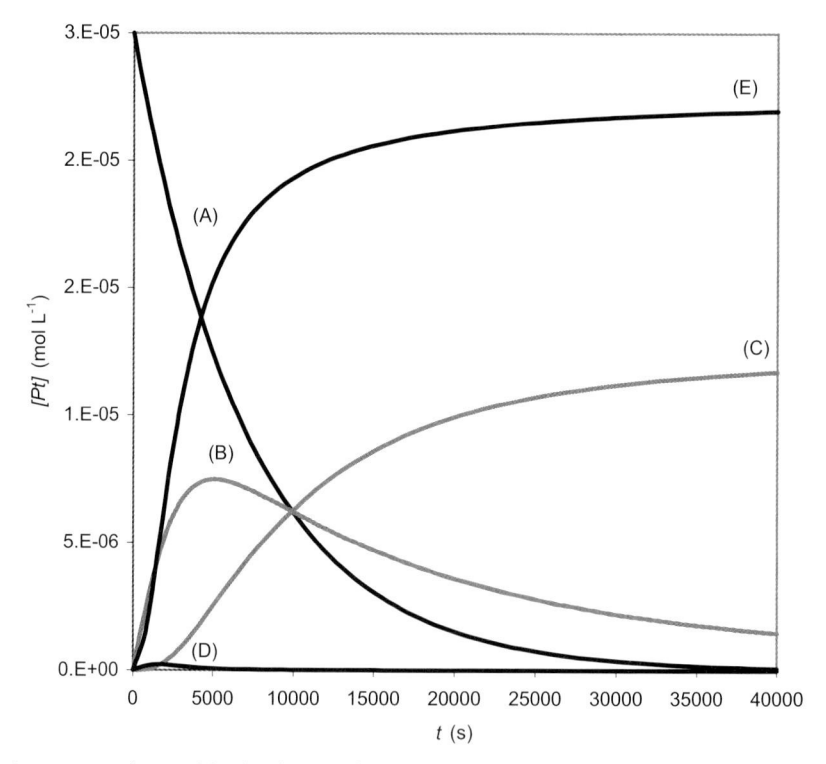

Fig. 6.22 Simulation of the distribution of platinum species during the hydrosilylation reaction under standard conditions. MD^{H}M (0.5 mol L^{-1}), oct-1-ene (0.5 mol L^{-1}), (ICy)Pt(dvtms) (**14b**) (2.5 × 10^{-5} mol L^{-1}), T = 70 °C. (ICy)Pt(dvtms) (**14b**) (A), active species (ICy)Pt^0 (**28**) (B), dimer (**25**) (C), (ICy)Pt(octene)$_2$ (**27**) (D), [MDM-oct] × 10^{-5} (E).

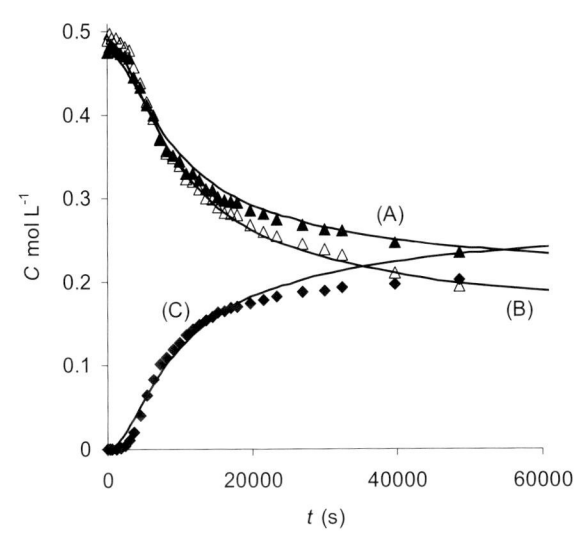

Fig. 6.23 Fit of the kinetic model with the experimental curves obtained for the primary kinetic isotope effect. Reaction conditions: MD^{D}M (0.5 mol L^{-1}), oct-1-ene (0.5 mol L^{-1}), (ICy)Pt(dvtms) (**14b**) (2.5 × 10^{-5} mol L^{-1}), o-xylene, 70 °C. Kinetic model (solid line), MD^{H}M curve (A), oct-1-ene curve (B), MDM-oct curve (C).

It is also instructive to know that during the peak of catalytic activity only 30% of the total platinum engaged actually catalyzes the reaction.

To further substantiate the unconventional concerted mechanism for the proposed key step in the hydrosilylation of oct-1-ene by (NHC)Pt catalysts, the primary kinetic isotope effect was measured. The model reaction was performed in the presence of the deuterated silane [27], at 70 and 80 °C. Figure 6.23 displays the kinetic curves.

Much to our surprise, the reaction with MD^{D}M at 70 °C does not go to completion and stops at a conversion of 43%. The rate constants were extracted from these runs, through the use of the kinetic model and a satisfactory fit was obtained (Fig. 6.23). Analysis of the rate constants revealed that the hydrosilylation step is greatly reduced in the presence of the deuterated silane whilst the initiation step remains unaffected (within experimental error) (Table 6.6).

The formation of the dimer is affected to a lesser extent than the hydrosilylation, leading to an increased production of **25** relative to adduct **11**. Therefore, in the presence of the deuterated silane, the active catalyst is siphoned into the inert dimeric form **25**, resulting in incomplete conversion. The large kinetic isotope effect suggests a "loose" transition state, with the hydride possessing a maximum degree of freedom. This observation is coherent with the proposed concerted mechanism in which the Si–H bond is only partially broken. (Fig. 6.24).

Table 6.6 (a) Rate constants for the primary kinetic isotope effect determined by fitting the kinetic model to experimental data obtained with the deuterated silane. (b) Primary kinetic isotope effect determined by fitting the kinetic model to experimental data.

(a)			(b)		
Constant	Fitting	Error	Constant	Fitting	Error
k_{ini}	$2.2 \times 10^{-4}\,s^{-1}$	$8 \times 10^{-5}\,s^{-1}$	$k_{H\text{-}ini}/k_{D\text{-}ini}$	0.6	0.2
k_{Hydro}	$7\,L^2\,mol^{-2}\,s^{-1}$	$0.6\,L^2\,mol^{-2}\,s^{-1}$	$k_{H\text{-}hydro}/k_{D\text{-}hydro}$	18	1.7
k_{Iso}	$1.7\,L^2\,mol^{-2}\,s^{-1}$	$0.1\,L^2\,mol^{-2}\,s^{-1}$	$k_{H\text{-}iso}/k_{D\text{-}iso}$	4.8	0.2
k_{dimer}	$4.4\,L\,mol^{-1}\,s^{-1}$	$0.1\,L\,mol^{-1}\,s^{-1}$	$k_{H\text{-}dimer}/k_{D\text{-}dimer}$	3.5	0.1
K_{eq}	$4.4\,L^2\,mol^{-2}$	$0.1\,L^2\,mol^{-2}$	$K_{H\text{-}eq}/K_{D\text{-}eq}$	0.073	0.003

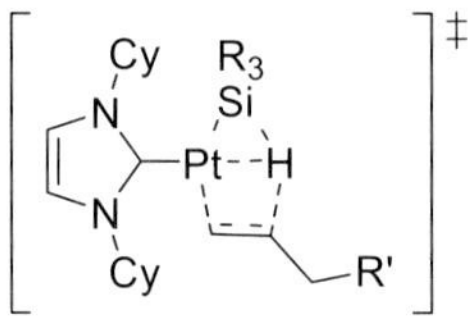

Fig. 6.24 Proposed transition state for the rate-determining step.

This incursion into the mechanistic study of the hydrosilylation of alkenes catalyzed by (NHC)Pt(dvtms) complexes afforded us a better understanding of this catalytic process. It transpires that the hydrosilylation and isomerization cycles are coupled and that high hydrosilylation activity inexorably leads to elevated alkene isomerization. The complex (ICy)Pt(dvtms) (**14b**) acts merely as a reservoir of the catalytically active Pt^0 species **28** and its slow release is a key prerequisite for minimizing the amount of isomerized olefins.

6.5
Hydrosilylation of Alkynes

Although the hydrosilylation of alkynes is the most efficient and atom-economical method for the synthesis of vinylsilanes, this reaction yields three regioisomers [the α, the β-(E) and the β-(Z) isomers], depending upon the metal, the ligand, the alkyne and the silane employed (Scheme 6.13). This selectivity issue has been studied extensively but few catalytic systems have enabled the selective synthesis of each regioisomer.

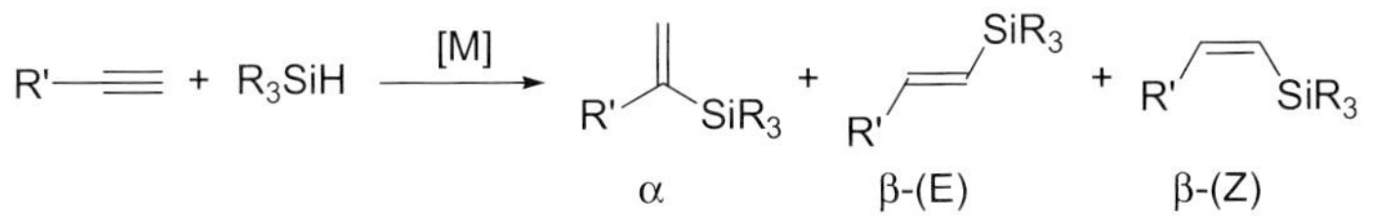

6.5 Hydrosilylation of Alkynes

$$R'\text{—}\equiv\text{—} + R_3SiH \xrightarrow{[M]} \underset{\alpha}{R'\text{—C(=CH}_2\text{)—SiR}_3} + \underset{\beta\text{-}(E)}{R'\text{—CH=CH—SiR}_3} + \underset{\beta\text{-}(Z)}{R'\text{—CH=CH—SiR}_3}$$

Scheme 6.13 Hydrosilylation of alkynes.

For example, whilst the α-isomer can be obtained in high yield using a [Cp*Ru]-based catalyst [85–87], neutral rhodium and ruthenium complexes can lead selectively to the (Z)-alkenylsilane [88–91]. When alkynes are hydrosilylated in the presence of platinum complexes, such as H_2PtCl_6 and **8**, only the α and β-(E) regioisomers are produced, with a low selectivity. Seminal studies by Stone and Tsipis have paved the way for the use of bulky phosphine ligands to obtain high β-(E) regioselectivities [79, 92]. The [$^tBu_3$P-Pt]-based complexes were first reported by Procter et al. for the hydrosilylation of propargylic alcohols. The desired (E)-vinylsilanes were typically produced with high regiocontrol [93]. An elegant application of the ($^tBu_3$P)Pt(dvtms) [33] complex is the sequential hydrosilylation/cross-coupling reaction described by Hiyama and thoroughly investigated by Denmark et al. [16, 94]. Yoshida achieved similar results using dimethyl(pyridyl)silane [95]. Although this procedure is highly efficient, with β-(E)/α ratios generally >99:1, the expensive, air-sensitive and pyrophoric $^tBu_3$P can be problematic when the reaction is performed on a large scale. Recently, Pt-complexes containing bulky phosphatrane ligands have been employed in the hydrosilylation of alkynes. Although exceptional regioselectivities and broad functional group tolerance have been observed, high catalyst loading (1 mol%) is required and only silanes of limited utility (Et_3SiH and Ph_3SiH) have been reported [96].

The remarkable selectivity displayed by our (NHC)Pt(dvtms) complexes prompted us to investigate their potential in the hydrosilylation of alkynes. From the onset, we focused on the use of low catalyst loading and well-defined molecular complexes, in an effort to provide an effective and economically viable method for the hydrosilylation of alkynes.

$$C_6H_{13}\text{—}\equiv\text{—} + Me_3Si\text{-O-Si(H)(Me)-O-}SiMe_3 \xrightarrow[\text{o-Xy, 80°C}]{5\times10^{-3}\ \text{mol\% Pt}} \underset{C_9H_{19}}{\text{vinylsilane β-(E)}} + \underset{}{C_6H_{13}\text{-α-vinylsilane}}$$

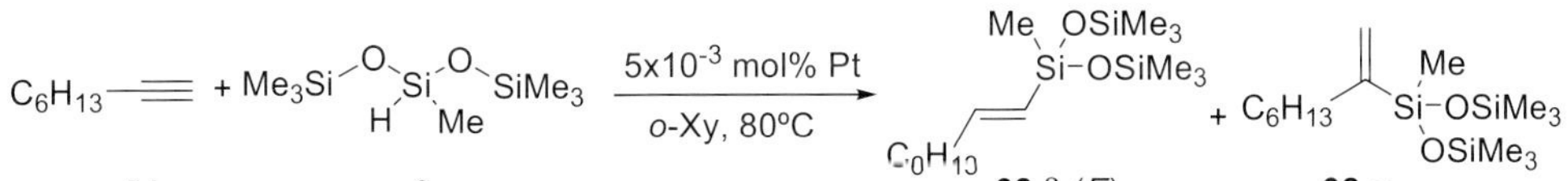

31   **9**   **32 β-(E)**   **32 α**

Scheme 6.14 Model hydrosilylation reaction of alkynes.

Initially, a family of (NHC)Pt(dvtms) complexes was screened in a model reaction involving the hydrosilylation of oct-1-yne (**31**) by bis(trimethylsilyloxy)methylsilane (**9**) (Scheme 6.14). The MD^{H}M siloxane represents a challenging case for the hydrosilylation of alkynes, since related dialkoxy silanes yield modest selectivities [79]. Vinylsilanes, bearing the bis(trimethylsiloxy)methyl substituent, and siloxanes in general have the major advantage of being particularly stable towards hydrolysis while still being amenable to cross-coupling reaction [18]. The conditions chosen for this model reaction were not optimized but served only as a guideline to investigate the effect of the NHC-ligand on the selectivity of the hydrosilylation.

Table 6.7 presents the results of this catalyst screening. As can be seen, the reaction times are rather long and can be in part explained by the very low catalyst loading employed (0.005 mol%). These stringent conditions allowed us to evaluate the relative efficacy of these catalysts. Thus, the complexes (IPr)Pt(dvtms) (**17b**) and (SIPr)Pt(dvtms) (**17d**), the most active and selective catalysts, provided a relatively high regiocontrol within a short reaction time (Table 6.7, entries 8 and 9). These transformations are rare examples in which enhanced reactivity accompanies high selectivity. Further examination of Table 6.7 reveals two distinct trends. The NHC ligands bearing alkyl substituents on both nitrogen atoms display very low regioselectivities and require prolonged reaction times (Table 6.7, entries 2–5). This feature appears to be independent of the steric bulk of the nitrogen substituents. For instance, there is no difference in regiocontrol when a adamantyl, *tert*-butyl or cyclohexyl group is present on the NHC ligand (Table 6.7, entries 3–5). Only a slight decline in selectivity is observed with the smallest NHC-ligand (Table 6.7, entry 2). Conversely, enhanced selectivity is observed when bulky aryl substituents are employed (Table 6.1, entries 6–9). This change in regioselectivity could be due either to a steric or to an electronic effect. To distinguish between these two possibilities, the p-tolyl substituted (NHC)Pt^0(dvtms) complex **17e** was synthesized. The low selectivity and mediocre reactivity observed when catalyst **17e** was employed suggests that the ortho,ortho'-substituents on the aryl group have a predominant effect on the regiocontrol of the hydrosilylation (Table 6.7, entry 1).

The intriguing break in the reactivity and selectivity pattern observed between the alkyl NHC and the hindered aryl NHC ligands prompted us to investigate further the structural parameters directing the selectivity of the addition. Previous examination of the X-ray crystal structures of complexes **14a–c**, **17a** and **17d** revealed that alkyl substituted NHCs (Me, Cy and ^{t}Bu) are always disposed perpendicularly to the plane formed by the trigonal planar arrangement of the dvtms ligand around the platinum center (Fig. 6.25A) [52]. This arrangement is maintained even in the case of the bulky I^{t}Bu-NHC substituent (**14c**) but results in a lengthening of the Pt–C$_{carbene}$ bond. This effect is seemingly due to the "spherical" type of steric hindrance induced by alkyl substituents. The ortho,ortho'-aryl substituted NHCs, however, appear to release their steric strain by tilting away, sometimes significantly, from the perpendicular arrangement, therefore considerably influencing the other ligands surrounding the platinum center. An interesting linear correlation can be observed between this tilt angle (θ) and the selectivity of the hydrosilylation reaction (Fig. 6.25B). It thus transpires that, by tilting away from the perpendicular arrangement, the bulky ortho,ortho'-aryl NHC ligands induce a complementary, gear-like distortion of the substituents in the platinum complex.

6.5 Hydrosilylation of Alkynes

Table 6.7 Catalyst screening for the hydrosilylation of oct-1-yne by MD^{H}M.

Entry[a]	Catalyst	β-(E)/α[b]	t (h)[c]	Entry[a]	Catalyst	β-(E)/α[b]	t (h)[c]
1	(IpTol)Pt(dvtms) (**17e**)	1.5	77	5	(ICy)Pt(dvtms) (**14b**)	2.8	150
2	(IMe)Pt(dvtms) (**14a**)	1.6	44	6	(IMes)Pt(dvtms) (**17a**)	5.8	49
3	(IAd)Pt(dvtms) (**14d**)	2.3	55	7	(SIMes)Pt(dvtms) (**17c**)	6.4	50
4	(I^{t}Bu)Pt(dvtms) (**14c**)	2.5	55	8	(IPr)Pt(dvtms) (**17b**)	10.6	6
				9	(SIPr)Pt(dvtms) (**17d**)	10.1	3

a) Reaction conditions: oct-1-yne (0.5 mol L^{-1}), MD^{H}M (0.5 mol L^{-1}), [Pt] (0.005 mol%), 80 °C, o-xylene. The results are the average of at least two runs.
b) Ratio determined by GC analysis.
c) Time to completion of reaction (>95% conversion).

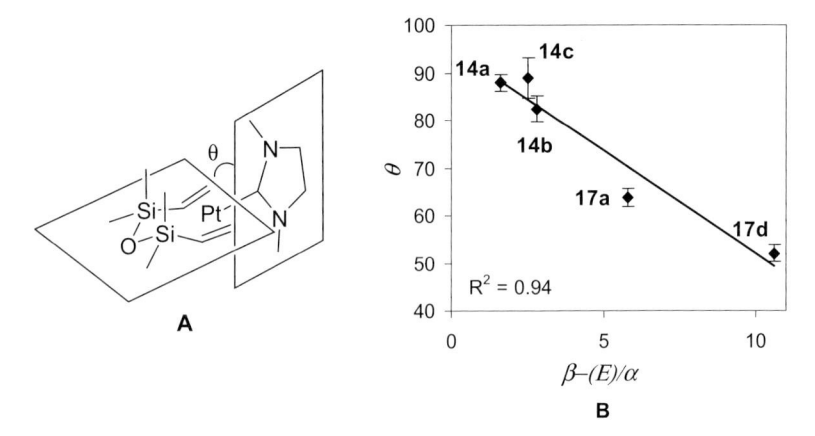

Fig. 6.25 Representation of the NHC tilt angle (θ) (A) and its correlation with the observed β-(E)/α selectivity in the model reaction (B).

Whilst the Tolman cone angle nicely correlates with the volume occupied by a phosphine ligand in a metal complex, no such generally applicable parameter is available for NHC ligands [97]. In an attempt to link reactivity with some structural features of NHC-containing organometallic derivatives, Nolan has introduced parameters quantifying the NHC steric hindrance: the A_H (Fig. 6.26) and A_L angles and the buried volume [54, 98]. To our knowledge, none of these parameters have been correlated successfully with observed reactivity and/or selectivity.

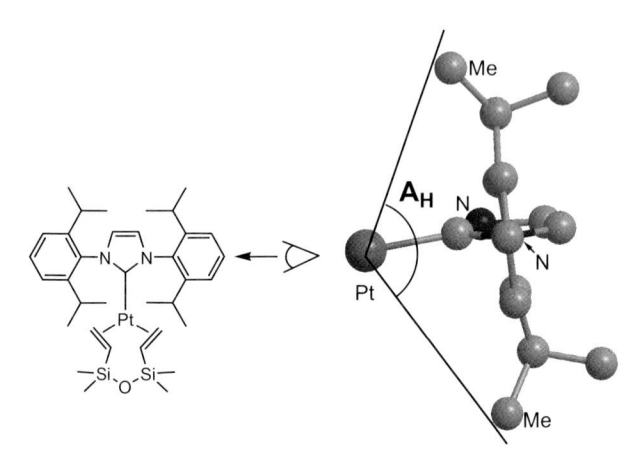

Fig. 6.26 Depiction of Nolan's A_H angle.

It appeared to us that the A_H angle might directly reflect the size of the NHC in a direction that incorporates the ortho,ortho'-substituents. Given our observation that these groups exert a major role on the selectivity of the hydrosilylation reaction, a correlation involving A_H and the β-(E)/α ratios was sought (Fig. 6.27).

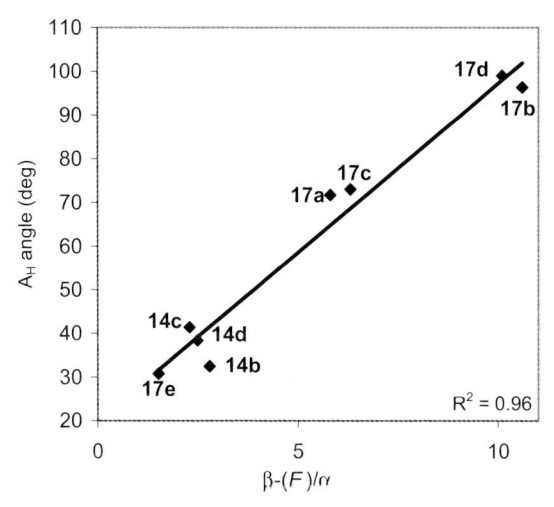

Fig. 6.27 Correlation between the A_H angle and β-(E)/α ratio for the hydrosilylation of oct-1-ene catalyzed by (NHC)Pt(dvtms) complexes.

As can be seen, an excellent correlation can be established between the A_H angle and the β-(E)/α ratio, indicating that, for a given alkyne, the selectivity in favor of the β-(E)-isomer increases as the steric hindrance around the NHC carbene ligand heightens. Such a correlation would possess a predictive value only if the electronic influence of the NHC ligand proved to be negligible. In this context, the cationic complex **34** was prepared by our standard procedure (Scheme 6.15) and the reactivity of three platinum complexes (**14a, 19a** and **34**), bearing NHCs with distinct electronic properties, was compared in the catalytic hydrosilylation of phenylacetylene by Et_3SiH.

Scheme 6.15 Synthesis of **34**.

Remarkably, the β-(F)/α ratios remained constant (ratio = 0.37), despite the large variation in the electronic effect of the NHC ligands. This observation further reinforces the validity of our proposed correlation. The overwhelming influence of the steric effect over the electronic properties of the NHC ligand, in the control of the regioselectivity of the hydrosilylation of alkynes, parallels similar observations reported previously for analogous bulky phosphine-Pd complexes [99].

The modulation of the steric and electronic nature of the silane was investigated next, using oct-1-yne as the substrate and complex **17b** as catalyst. These results are summarized in Table 6.8. Whilst the influence of the electronic nature of the silane is difficult to rationalize at this point, a steric hindrance effect can be noticed. In general, the greater the steric bulk of the silicon-hydride the higher the β-(E)/α ratio [100].

Table 6.8 Effect of the silane on the regioselectivity of the addition.

Entry[a]	Silane	β-(E)/α[b]	t (h)	Conv. (%)[c]
1	$^{t}BuMe_2SiH$	1(1)	92	85
2	$(EtO)_3SiH$	2	150	>99
3	$(Me_3SiO)Me_2SiH$	4.3 (3.22)	3	54
4	Et_3SiH	6.3 (2.77)	42	87
5	$(Me_3SiO)_2MeSiH$	10.6 (2.94)	6	>99
6	Me_2PhSiH	11.5 (5.26)	22	>99
7	Ph_3SiH	15.7 (20)	22	88

a) Reaction conditions: oct-1-yne (0.5 mol L^{-1}), silane (0.5 mol L^{-1}), 80 °C, o-xylene, (IPr)Pt(dvtms) (17b) (0.005 mol%).
b) Ratio determined by GC analysis; results obtained with PtCl$_2$(COD) are reported in brackets.
c) Conversion determined by GC.

Finally, the influence of the electronic nature of the alkyne was briefly studied (Table 6.9). In this context, phenylacetylene, bearing the electron-withdrawing phenyl substituent, was reacted with triethylsilane in the presence of complex 14a. The least hindered NHC ligand was selected to minimize the influence of steric effects. Interestingly, a 73:27 ratio of α and β-(E)-vinylsilanes was obtained, favoring the α-isomer (Table 6.9, entry 2). To the best of our knowledge, this is the first example of an inversion of regioselectivity in the hydrosilylation of alkynes catalyzed by platinum complexes.

Table 6.9 Alkyne effect on regioselectivity.

Entry[a]	R —≡	β-(E)/α[b]	% β-(E)	% α
1	n-C$_6$H$_{13}$	1.35	57	43
2	Ph	0.37	27	73

a) Reaction conditions: alkyne (0.5 mol L^{-1}), Et$_3$SiH (0.5 mol L^{-1}), (IMe)Pt(dvtms) (14a) (0.005 mol%), 80 °C, o-xylene.
b) Ratio determined by GC analysis.

6.5 Hydrosilylation of Alkynes

In terms of regiocontrol, the crucial step in the catalytic cycle is the migratory insertion, for which a rationalization is proposed below. When the alkyne coordinates to the platinum center, it can afford two possible, isomeric and most probably equilibrating, species **a** and **b** (Fig. 6.28). This coordination is non-symmetrical and can be qualitatively described by the strength of the orbital interactions between the platinum d orbitals and the alkyne π^*. According to Tsipis [92], when R' is an alkyl substituent the strongest interaction occurs between the Pt and the terminal carbon atom of the alkyne, favoring isomer **a**, which leads to the β-(E) product. Conversely, with an aryl-substituted alkyne the predominant interaction arises between the Pt center and the internal carbon of the acetylene derivative, encouraging the formation of **b** and providing the α isomer.

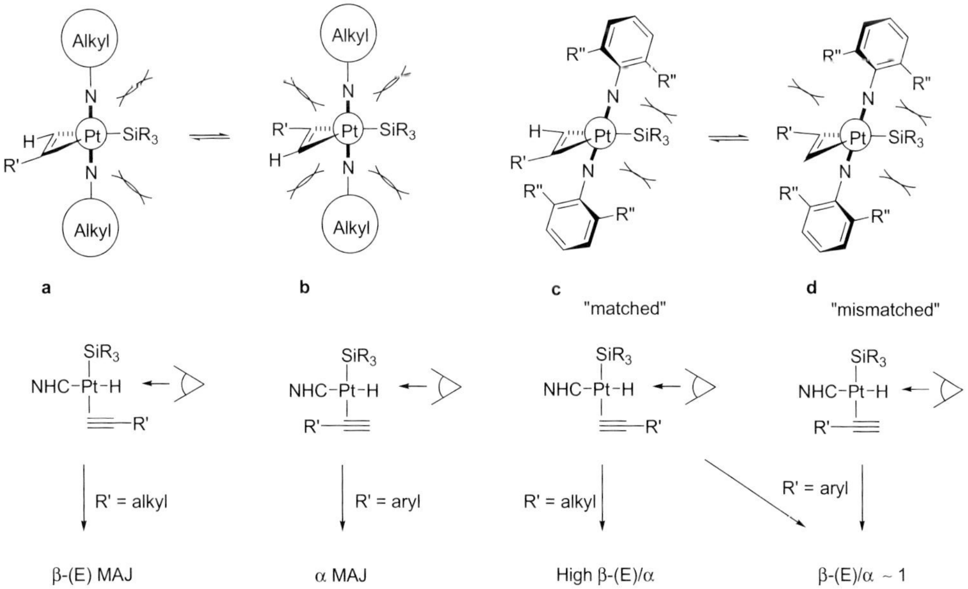

Fig. 6.28 Steric interactions involved in the hydrosilylation reaction.

These electronic influences can be enhanced or counterbalanced by the steric effects provided by the NHC ligands. A projection along the NHC-ligand plane of the four possible intermediates that can be formed upon coordination of an alkyne to a Pt(II)-silyl hydride complex is presented in Fig. 6.28. For alkyl-substituted NHC ligands, the steric interaction provided by the NHC moiety is small or negligible and the electronic nature of the alkyne will dictate the resulting β-(E)/α ratio (intermediates **a** and **b**).

The situation becomes more complex when aryl-substituted NHC ligands are employed. Thus, when an alkyl-substituted acetylene is used in conjunction with a bulky aromatic-containing NHC-Pt catalyst, both the steric and the electronic demands of the reaction partners are matched, leading to high regiocontrol (Fig.

6.28, intermediate **c**). However, when R' is an aryl substituent, complex **d**, which is electronically preferred, suffers now from a destabilizing steric repulsion between R' and R''. To alleviate these steric repulsions, the alkyne can flip around, leading to intermediate **c** (R' = alkyl). Unfortunately this coordination mode is opposite to the electronic demand of the acetylene derivative. Therefore, in this case, mismatched interactions between steric and electronic demands are present in both species **c** and **d**, and poor selectivities ensue. This model predicts that the combined use of an NHC ligand providing negligible steric hindrance, with a small silane, should lead to the preferential formation of the α-isomer, a prediction in full accord with the experimental results.

6.6
Summary

We have described a novel family of NHC-Pt^0 complexes. These reagents proved to be particularly active and selective in the hydrosilylation of alkenes and alkynes, providing the desired adducts in high yields and with excellent levels of regiocontrol, at very low catalyst loading (30 ppm Pt). An in depth kinetic study has revealed some salient features of the reactivity of these organometallic species and has shed some light on the mechanism followed by these catalysts. Particularly noteworthy are the influence of the olefinic ligands around the platinum on the rate of the reaction, the importance of the nature of the NHC substituents on the velocity and the regioselectivity of the hydrosilylation of alkynes, the unusual, σ-metathesis-like mechanism involved in the key-step of the catalytic cycle and the discovery that the (NHC)Pt(dvtms) complexes act essentially as reservoirs of the active NHC-Pt species. This last property is directly responsible for the high efficiency and selectivity displayed by our novel complexes, which afford, beside the desired adducts, only minute amounts of by-products. We believe that these results open up new vistas for the generation of even more active, selective and robust catalysts for the hydrosilylation of carbon–carbon double and triple bonds.

References

1 (a) Marciniec, B. *Comprehensive Handbook on Hydrosilylation*, Pergamon Press, Oxford, **1992**, p. 130. (b) Hiyama, T., Kusumoto, T. in *Comprehensive Organic Synthesis* ed. Trost, B. M., Fleming, I., Pergamon Press, Oxford, **1991**, vol. 8, p. 763. (c) Ojima, I., in *The Chemistry of Organic Silicon Compounds* ed. Patai, S., Rappoport, Z., John Wiley, Chichester, **1989**, p. 1479. (d) Reiche, J. A., Bery, D. H. *Adv. Organomet. Chem.* **1998**, *43*, 197. For a recent review on alkyne hydrosilylation see: Trost, B. M., Ball, Z. T. *Synthesis* **2005**, *6*, 853. For recent examples of platinum-catalyzed hydrosilylation of alkenes see: Yamamoto, Y., Ohno, T., Itoh, K. *Organometallics*, **2003**, *22*, 2267 and references cited therein.

2 Sheldon, R. *Pure Appl. Chem.* **2000**, *72*, 1233.

3 Trost, B. M. *Science* **1991**, *254*, 1471.

4 Rochow, E. G. *J. Am. Chem. Soc.* **1945**, *67*, 963.
5 Lewis, L. N., Stein, J., Gao, Y., Colborn, R. E., Hutchins, G. *Platinum Met. Rev.* **1997**, *41*, 66.
6 Blumenkopf, T. A., Overman, L. E. *Chem. Rev.* **1986**, *86*, 857.
7 Langkopf, E., Schinzer, D. *Chem. Rev.* **1995**, *95*, 1375.
8 Tamao, K., Ishida, N., Kumada, M. *Organometallics* **1983**, *2*, 2120..
9 Fleming, I., Henning, R., Plaut, H. *Chem. Commun.* **1984**, 29.
10 Jones, G. R., Landais, Y. *Tetrahedron* **1996**, *52*, 7599.
11 Tamao, K., Kumada, M., Maeda, K. *Tetrahedron Lett.* **1984**, *25*, 321.
12 Tamao, K., Akita, M., Maeda, K., Kumada, M. *J. Org. Chem.* **1987**, *52*, 1100.
13 Stamos, D. P., Taylor, A. G., Kishi, Y. *Tetrahedron Lett.* **1996**, *37*, 8647 and references cited therein.
14 Hatanaka, Y., Hiyama, T. *Synlett* **1991**, 845.
15 Hiyama, T., Shirakawa, E. *Top. Curr. Chem.* **2002**, *219*, 61.
16 Takahashi, K., Minami, T., Ohara, Y., Hiyama, T. *Tetrahedron Lett.* **1993**, *34*, 8263.
17 Lee, H. M., Nolan, S. P. *Org. Lett.* **2000**, *2*, 2053.
18 Denmark, S. E., Neuville, L. *Org. Lett.* **2000**, *2*, 3221.
19 Nakao, Y., Imanaka, H., Sahoo, A. K., Yada, A., Hiyama, T. *J. Am. Chem. Soc.* **2005**, *127*, 6952.
20 Denmark, S. E., Tymonko, S. A. *J. Am. Chem. Soc.* **2005**, *127*, 8004.
21 Armitage, D. A. in *Comprehensive Organometallic Chemistry* ed. Wilkinson, G., Stone, G. A. F., Abel, E. W., Pergamon, Oxford, **1982**, vol. 2, p. 117.
22 Grate, J. W. *Polym. News* **1999**, *24*, 149.
23 Karstedt, B. D., General Electric, US 3,715,334, **1973**.
24 Hitchcock, P. B., Lappert, M. F., Warhurst, N. J. W. *Angew. Chem. Int. Ed. Engl.* **1991**, *30*, 438.
25 Speier, J. L., Webster, J. A., Barnes, G. H. *J. Am. Chem. Soc.* **1956**, *79*, 974.
26 Lewis, L. N., Lewis, N. *J. Am. Chem. Soc.* **1986**, *108*, 7228.
27 Stein, J., Lewis, L. N., Gao, Y., Scott, R. A. *J. Am. Chem. Soc.* **1999**, *121*, 3693.
28 We observed that the formation of colloids was concommitant with the hydrogenation of the isomerised alkene.
29 Duin, M. A., Clement, N. D., Cavell, K. J., Elsevier, C. J. *Chem. Commun.* **2003**, 400–401.
30 Sprengers, J. W., Mars, M. J., Duin, M. A., Cavell, K. J., Elsevier, C. J. *J. Organomet. Chem.* **2003**, *679*, 149.
31 Duin, M. A., Lutz, M., Spek, A. L., Elsevier, C. J. *J. Organomet. Chem.* **2005**, 5804 and references cited therein.
32 Coqueret, X., Wegner, G. *Organometallics* **1991**, *10*, 3139.
33 Chandra, G., Lo, P. Y., Hitchcock, P. B., Lappert, M. F. *Organometallics* **1987**, *6*, 191.
34 Beuter, G., Heyke, O., Lorenz, I. P. *Z. Naturforsch., B, Chem.Sci.* **1991**, *46*, 1694.
35 Hitchcock, P. B., Lappert, M. F., MacBeath, C., Scott, F. P. A., Warhurst, N. J. W. *J. Organomet. Chem.* **1997**, *528*, 185.
36 Arduengo, A. J. J. I., Harlow, R. L., Kilne, M. *J. Am. Chem. Soc.* **1991**, *113*, 361.
37 Arduengo, A. J. J. I. *Acc. Chem. Res.* **1999**, *32*, 913.
38 Bourissou, D., Guerret, O., Gabbaï, F. P., Bertrand, G. *Chem. Rev.* **2000**, *100*, 39.
39 Arduengo, A. J. J. I., Camper, S. F., Calabrese, J. C., Davidson, F. *J. Am. Chem. Soc.* **1994**, *116*, 4391.
40 Arnold, P. L., Cloke, F. G. N., Geldbach, T., Hitchcock, P. B. *Organometallics* **1999**, *18*, 3228.
41 Lin, G., Jones, N. D., Robert A. Gossage, McDonald, R., Cavell, R. G. *Angew. Chem. Int. Ed.* **2003**, *42*, 4054.
42 Hasan, M., Kozhevnikov, I. V., Siddiqui, M. R. H., Femoni, C., Steiner, A., Winterton, N. *Inorg. Chem.* **2001**, *40*, 795.
43 Quezada, C. A., Garrison, J. C., Tessier, C. A., Youngs, W. J. *J. Organomet. Chem.* **2003**, *671*, 183.
44 Anthony, J., Arduengo, I., Dias, H. V. R., Calabrese, J. C., Davidson, F. *Organometallics* **1993**, *12*, 3405.
45 Arduengo, A. J., Jr. III, Krafczyk, R., Schmutzler, R., Craig, H. A., Goerlich, J. R., Marshall, W. J., Unverzagt, M. *Tetrahedron* **1999**, *55*, 14523.

46 Herrmann, W. A., Köcher, C., Goossen, J. L., Artus, G. R. J. *Chem. Eur. J.* **1996**, *2*, 1627.
47 Bassindale, A. R., Brown, S. S. D., Lo, P. *Organometallics* **1994**, *13*, 738.
48 Chianese, A. R., Li, X., Janzen, M. C., Faller, J. W., Crabtree, R. H. *Organometallics* **2003**, *22*, 1663.
49 Warhurst, N. J. W. Ph.D. Thesis, University of Sussex, **1990**.
50 Markó, I. E., Stérin, S., Buisine, O., Mignani, G., Branlard, P., Tinant, B., Declercq, J.-P. *Science* **2002**, *298*, 204.
51 Markó, I. E., Michaud, G., Berthon-Gelloz, G., Buisine, O., Stérin, S. *Adv. Synth. Catal.* **2004**, 1429.
52 Berthon-Gelloz, G., Brière, J. F., Buisine, O., Stérin, S., Michaud, G., Mignani, G., Tinant, B., Declercq, J.-P., Chapon, D., Markó, I. E. *J. Organomet. Chem.* **2005**, 6156.
53 Hartley, F. R. in *Comprehensive Organometallic Chemistry* ed. Abel, E. W., Stone, F. G. A., Wilkinson, G., Pergamon, Oxford, **1982**, vol. 6, p. 474.
54 Viciu, M. S., Navarro, O., Germaneau, R. F. III, R. A. K., Sommer, W., Marion, N., Stevens, E. D., Cavallo, L., Nolan, S. P. *Organometallics* **2004**, *23*, 1629.
55 Herrmann, W. A., Elison, M., Fisher, J., Köcher, C., Artus, G. R. J. *Angew. Chem. Int. Ed.* **1995**, *34*, 2371.
56 McGuinness, D. S., Cavell, K. J. *Organometallics* **1999**, *18*, 1596.
57 Vinyl silanes are excellent ligands for late, low-valent transition metals. It has been suggested that the coordinating ability of vinyl silanes is mostly due to the Si–C=C (dπ–pπ) interaction, which allows delocalization of the alkene's electron density onto the silicon atom via its d-orbitals. This results in enhanced back-bonding between the metal centre and the olefin delocalized π* orbital. Therefore, greater thermal stability of the vinylsilane-containing complexes relative to their carbon analogues is observed.
58 Chandra, G., Lo, P. Y. K. Dow Corning Corp.: US Patent 4,593,084, **1984**.
59 Steffanut, P., Osborn, J. A., DeCian, A., Fisher, J. *Chem. Eur. J.* **1998**, *4*, 2008.
60 Liedtke, J., Loss, S., Widauer, C., Grutzmacher, H. *Tetrahedron* **1999**, *56*, 143.
61 Liedtke, J., Loss, S., Alcaraz, G., Gramlich, V., Grutzmacher, H. *Angew. Chem. Int. Ed.* **1999**, *38*, 1623.
62 Under "normal" conditions, the Pt complexes **14a** to **14c** are added to a mixture of silane **9** and olefin **10**. The "inverse" protocol implies the addition of the silane **9** to a mixture of alkene **10** and Pt catalysts **14a** to **14c**. No colloidal Pt is formed upon increasing the amount of Pt-carbene catalysts from 30 up to 300 ppm. The rate increases concomitantly with the amount of catalyst, and the reaction can become highly exothermic.
63 The high selectivity observed with this "inverse" procedure stems from the kinetics of the hydrosilylation reaction using (NHC)Pt(dvtms) complexes, which differ significantly from the Karstedt-catalyzed system. To obtain a fair comparison between the two catalysts, both reactions have been performed under identical, nonoptimized conditions.
64 Hydrosilylation of alcohol-containing substrates using the Karstedt catalyst results in significant formation of the corresponding silyl ethers (up to 40%).
65 The high reactivity of the Karstedt catalyst precludes its use in the case of epoxide-bearing substrates. For example, more than 50% of vinyl cyclohexyl epoxide (Table 6.4, entry 5) is decomposed in the presence of this complex.
66 Whilst most hydrosilylations are complete within 2–4 h, the reaction of allylacetone with **14b**, under identical conditions, requires up to 24 h to reach 80% conversion (Table 6.4, entry 6). In the case of the platinum-carbene complex **14b**, the catalyst is still active and prolonged reaction time leads to full conversion.
67 All the hydrosilylations performed using the carbene complex **10b** afford crude products that are at least 95% pure and that typically contain less than 2% of isomerized alkene. With Karstedt catalyst **8**, yellow solutions, including colloidal platinum species, are generated. The purity of the crude reaction

mixtures varies between 72 and 92% and the amount of isomerized alkene oscillates between 5 and 11%. In the case of epoxide-containing substrates, extensive decomposition is observed.
68 It was necessary to heat for several hours at 60 °C complex **8** in the presence of the imidazolinium salt and KOtBu to obtain good yields of the desired complex.
69 Herrmann, W. A. *Angew. Chem. Int. Ed.* **2002**, *41*, 1290.
70 Kücükbay, H., Cetinkaya, B., Guesmi, S., Dixneuf, P. H. *Organometallics* **1996**, *15*, 2434.
71 Hahn, F. E., Langenhahn, V., Meier, N., Lügger, T., Fehlhammer, W. P. *Chem. Eur. J.* **2003**, *9*, 704.
72 Seo, H., Kim, B. Y., Lee, J. H., Park, H.-J., Son, S. U., Chung, Y. K. *Organometallics* **2003**, *22*, 4783.
73 Hahn, F. E., Wittenbecher, L., Boese, R., Bläser, D. *Chem. Eur. J.* **1999**, *5*, 1931.
74 O'Brien, C. J., Kantchev, E. A. B., Chass, G. A., Hadei, N., Hopkinson, A. C., Organ, M. G., Setiadi, D. H., Tang, T.-H., Fang, D.-C. *Tetrahedron* **2005**, *61*, 9723 and references cited therein.
75 Metallinos, C., Barrett, F. B., Chaytor, J. L., Heska, M. E. A. *Org. Lett.* **2004**, *6*, 3641.
76 G. Berthon-Gelloz, I. E. Marko and S. P. Nolan, unpublished results.
77 Ciriano, M., Green, M., Howard, J. A. K., Proud, J., Spencer, J. L., Stone, G. A. F., Tsipis, C. A. *J. Chem. Soc., Dalton Trans.* **1978**, 801.
78 Green, M., Spencer, J. L., Stone, G. A. F., Tsipis, C. A. *J. Chem. Soc., Dalton Trans.* **1977**, 1519.
79 Green, M., Spencer, J. L., Stone, G. A. F., Tsipis, C. A. *J. Chem. Soc., Dalton Trans.* **1977**, 1525.
80 Chalk, A. J., Harrod, J. F. *J. Am. Chem. Soc.* **1965**, *87*, 16.
81 Lewis, L. N., Colborn, R. E., Grade, H., Garold L. Bryant, J., Sumpter, C. A., Scott, R. A. *Organometallics* **1995**, *14*, 2202.
82 Lewis, L. N., Stein, J., Colborn, R. E., Gao, Y., Dong, J. *J. Organomet. Chem.* **1996**, *521*, 221.
83 Hartmann, R., Chen, P. *Adv. Synth. Catal.* **2003**, *345*, 1353.
84 Rothenberg, G., Cruz, S. C., Strijdonck, G. P. F. v., Hoefsloot, H. C. J. *Adv. Synth. Catal.* **2004**, *346*, 467.
85 Trost, B. M., Ball, Z. T. *J. Am. Chem. Soc.* **2001**, *123*, 12726.
86 Trost, B. M., Ball, Z. T. *J. Am. Chem. Soc.* **2005**, *127*, 17644.
87 Kawanami, Y., Sonoda, Y., Mori, T., Yamamoto, K. *Org. Lett.* **2002**, *4*, 2825.
88 Doyle, M. P., High, K. G., Nesloney, C. L., Clayton, J. W., Jr., Lin, J. *Organometallics* **1991**, *10*, 1225.
89 Ojima, I., Clos, N., Donovan, R. J., Ingallina, P. *Organometallics* **1990**, *9*, 3127.
90 Mori, A., Takahisa, E., Yamamura, Y., Kato, T., Mudalige, A. P., Kajiro, H., Hirabayashi, K., Nishihara, Y., Hiyama, T. *Organometallics* **2004**, *23*, 1755.
91 Nagao, M., Asano, K., Umeda, K., Katayama, H., Ozawa, F. *J. Org. Chem.* **2005**, *70*, 10511.
92 Tsipis, C. A. *J. Organomet. Chem.* **1980**, *187*, 427.
93 Murphy, P. J., Spencer, J. L., Procter, G. *Tetrahedron Lett.* **1990**, *31*, 1051.
94 Denmark, S. E., Wang, Z. *Org. Lett.* **2001**, *3*, 1073.
95 Itami, K., Mitsudo, K., Nishino, A., Yoshida, J.-i. *J. Org. Chem.* **2002**, *67*, 2645.
96 Aneetha, H., Wu, W., Verkade, J. G. *Organometallics* **2005**, *24*, 2590.
97 Tolman, C. A. *Chem. Rev.* **1977**, *77*, 313.
98 Huang, J., Schanz, H.-J., Stevens, E. D., Nolan, S. P. *Organometallics* **1999**, *18*, 2370.
99 Strieter, E. R., Blackmond, D. G., Buchwald, S. L. *J. Am. Chem. Soc.* **2003**, *125*, 13978.
100 The electronic effect of alkoxy groups on silicon does not correlate in a simple way with their electronegativity and their influence is difficult to rationalize.

7
Ni-NHC Mediated Catalysis

Janis Louie

7.1
Introduction

Ni complexes have played an important role in the field of catalysis, yet their ability to promote reactions has generally paled in comparison to other group 10 metals such as palladium. With the application of NHC ligands, the number of Ni-catalyzed reactions has grown considerably in the past decade. As observed in other transition metal based catalyst systems, the replacement of traditional amine or phosphine ligands with electron-rich NHC ligands has led to a substantial enhancement in catalytic activity. This chapter summarizes the recent impact that the use of NHC ligands has played in furthering the field of Ni-mediated catalysis.

7.2
Rearrangement Reactions

7.2.1
Rearrangement Reactions of Vinyl Cyclopropanes

The use of Ni as a catalyst for the rearrangement of vinylcyclopropanes (VCPs) to cyclopentenes was first reported in 1979 [1]. In combination with a phosphine ligand, activated VCPs could be converted into the corresponding cyclopentenes at elevated temperatures. Recently, it was shown that replacement of the phosphine ligand with a NHC ligand led to a dramatic increase in catalytic activity [2]. Sterically hindered N-aryl substituted NHCs, such as IPr or SIPr, gave greater yields of isomerized products at faster reaction rates than less bulky NHCs (such as ICy and IiPrim). VCPs possessing an electron-withdrawing group, heteroatom, or phenyl group on the cyclopropane ring underwent rapid isomerization and afforded high yields of the corresponding cyclopentene (entries 1 and 2, Table 7.1). Furthermore, VCP **5**, a substrate lacking any functionality that could promote

N-Heterocyclic Carbenes in Synthesis. Edited by Steven P. Nolan
Copyright © 2006 WILEY-VCH Verlag GmbH & Co. KGaA, Weinheim
ISBN: 3-527-31400-8

rearrangement, readily isomerized under ambient conditions. In contrast, no conversion was observed when phosphine ligands were employed even after prolonged periods at elevated temperatures. Although slightly elevated temperatures were necessary, simple trisubstituted olefins (**7**) were also successfully isomerized. In contrast, the isomerization of 1,2-disubstituted olefins, possessing either cis or trans olefinic geometries, was generally sluggish even under more forcing conditions (100 °C). Nevertheless, VCP **9** afforded cyclopentene **10** in excellent yield. Bicyclic cyclopentene **12**, which possesses an exocyclic, tetrasubstituted double bond containing a trimethylsilyl group, was easily obtained from the isomerization of **11**.

Table 7.1 Ni-catalyzed isomerization of VCPs[a)]

Entry	Substrate	Product	Temp (time)	Yield[b)] (%)	Entry	Substrate	Product	Temp (°C) (time)	Yield[b)] (%)
1	**1** Ph	**2** Ph	RT (1 h)	96	4	**7** $C_{11}H_{23}$	**8** $C_{11}H_{23}$	60 (12 h)	94
2	**3**	**4**	RT (2 h)	93	5	**9** Ph	**10** Ph	60 (12 h)	92
3	**5** Bn	**6** Bn	RT (12 h)	93	6	**11** TMS	**12** TMS	55 (4 h)	91

a) Performed with 1 mol% Ni(COD)$_2$, 2 mol% IPr, 0.10 M substrate.
b) Isolated yields (average of at least two runs).

7.2.2
Rearrangement Reactions of Cyclopropylen-Ynes

Ni/NHC-based systems also catalyze the rearrangement reaction of cyclopropylen-ynes to afford three different heterocyclic based structures, two of which are distinct from those obtained employing Rh- and Ru-based catalysts (Reaction 1) [3]. Although the combination of Ni(COD)$_2$ (COD = cyclooctadiene) and SIPr displayed the fastest reaction rate, the size of the substituent on the alkyne (R) had a significant effect on the nature of the heterocyclic product formed (Table 7.2). When R was small [e.g., R = Me (**13a**), entry 1], a cyclopentane product (**14a**) was formed exclusively. However, a mixture of rearrangement products was obtained from substrates **13b** and **13c** (entries 2 and 3) that included the expected cyclopentane (**14b** and **14c**) in addition to a bicyclic seven-membered ring (**15b** and **15c**). Furthermore, when R was large (e.g., R = t-Bu (**13d**) or TMS (**13e**), entries 4 and 5), isomerized seven-membered rings (**16d** and **16e**) were formed exclusively in good yields.

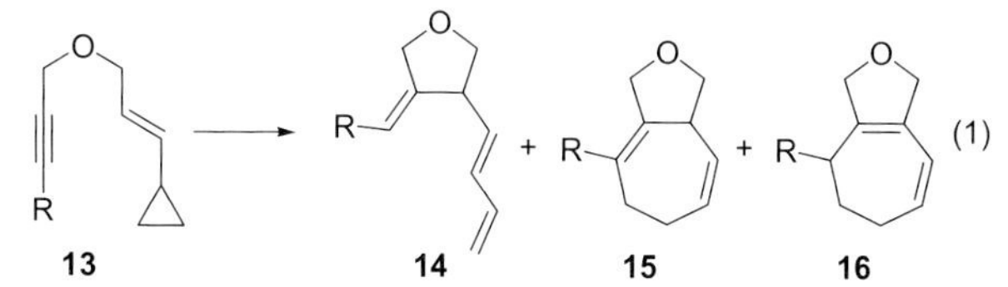

7.2 Rearrangement Reactions

Table 7.2 Product distribution in the Ni-catalyzed rearrangement of **13**[a)]

Entry	Substrate	14 : 15 : 16[b)]	Yield (%)[c)]
1	R = Me (**13a**)	1 : 0 : 0	54 (**14a**)
2	R = Et (**13b**)	3 : 2 : 0	34 (**14b**)
			27 (**15b**)
3	R = i-Pr (**13c**)	1 : 2 : 0	28 (**14c**)
			38 (**15c**)
4	R = t-Bu (**13d**)	0 : 0 : 1	82 (**16d**)
5	R = TMS (**13c**)	0 : 0 : 1	88 (**16e**)

a) Reaction conditions: 5 mol% Ni(COD)$_2$, 5 mol% SIPr, toluene, ambient temperature.
b) Determined by GC using naphthalene as an internal standard.
c) Isolated yield (average of two runs).

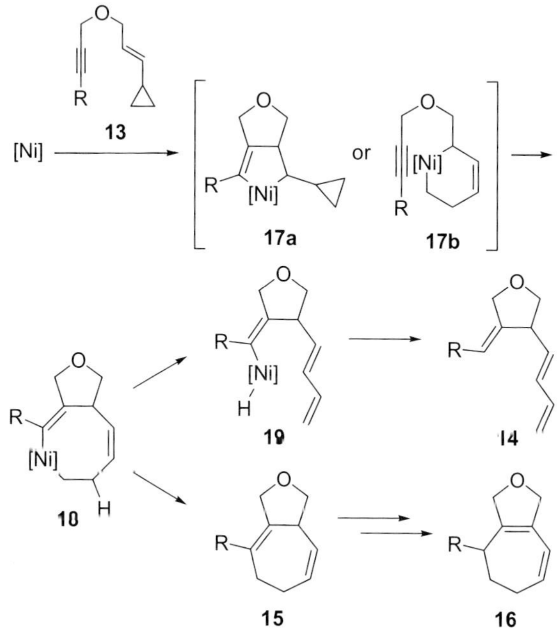

Scheme 7.1 Proposed mechanism for the rearrangement of **13**.

Scheme 7.1 shows a plausible mechanism that diverges at a common intermediate and may account for the product distributions. Reaction between the Ni catalyst and cyclopropylen-yne **13** would ultimately afford the eight-membered intermediate **18**, which could result from either initial oxidative coupling between an alkene and alkyne (**17a**) [4] or initial isomerization of the VCP (**17b**). Ultimately, β-hydride elimination from **18** and reductive elimination would afford cyclopentane product **14**. In contrast, if both the ligand and R are large, β-hydride elimination would be inhibited and direct reductive elimination would yield a seven-membered ring (**15**). Product **16** would arise from further isomerization of **15**.

By substituting SIPr for an N-alkyl substituted NHC ligand (I*t*Bu), the cyclopentane product (**14**) could be prepared selectively from cyclopropylen-yne substrates, regardless of substituent size (e.g., R), (Reaction 2) [5]. As shown in Table 7.3, cyclopentene products (**14a–d**, entries 1–4) were formed exclusively under mild conditions (room temperature, 2 h). In addition, the Ni/NHC catalyst system tolerated both ester (**20**) and amino (**21**) functionality (entries 5 and 6).

Table 7.3 Selective formation of cyclopentanes[a]

Entry	Cyclopropylen-yne	Product	Yield (%)[b]
1	X = O, R = Me (**13a**)	**14a**	54
2	X = O, R = Et (**13b**)	**14b**	52
3	X = O, R = i-Pr (**13c**)	**14c**	65
4	X = O, R = t-Bu (**13d**)	**14d**	79
5	X = C(CO$_2$Me)$_2$, R = Me (**20**)	**20a**	71[c]
6	X = NTs, R = Me (**21**)	**22**	86[c]
7	X = O, R = CH$_2$OMe (**23**)	**24**	64
8	**25**	**26**	61
9	**27**	**28**	55[c,d]
10	**29**	**30**	61

a) Reaction conditions: 5 mol% Ni(COD)$_2$, 5 mol% I*t*Bu, toluene, room temperature.
b) Isolated yield (average of two runs).
c) SIPr was used as the ligand instead of I*t*Bu.
d) Reaction was run at 40 °C.

7.3
Cycloaddition Reactions

An attractive method for the rapid construction of the heterocyclic core of numerous biologically active pharmacophores is the cycloaddition or rearrangement of unsaturated substrates. Considerable effort has focused on developing transition metal catalysts that mediate such transformations. Ultimately, reactions that require prohibitively harsh conditions (high temperatures, high pressures) may become practical (room temperature, atmospheric pressures) when a transition metal catalyst is employed. Of the transition metal based catalysts, Ni/NHC systems are some of the most versatile and have been used in the synthesis of both oxygen- and nitrogen-containing heterocycles.

7.3.1
Cycloaddition of Diynes and Carbon Dioxide

The Ni/phosphine-catalyzed coupling of two alkynes with CO_2 to afford pyrones (Reaction 3) was first discovered by Inoue and further developed by Tsuda [6, 7]. These reactions generally involve high pressures of CO_2 and elevated temperatures. In addition, only a limited number of diynes could be successfully converted into the corresponding pyrone. As with many cycloaddition reactions, oligomerization of the diyne was a major side reaction. These obstacles were overcome when IPr was used as the ligand in lieu of phosphines [8]. The steric bulk of this ligand helped to suppress oligomerization of the diyne. As a result, various bicyclic pyrones were obtained in high yields (Table 7.4). Notably, all pyrones were obtained using ambient pressures and relatively low reaction temperatures.

Ni/IPr serves as a general catalyst system for the coupling of diynes and CO_2. To date, this catalyst does not provide pyrones from either terminal diynes or sterically hindered diynes. Terminal diynes (i.e., substrate **31d**, where R = H) oligomerized at a faster rate than CO_2 incorporation. In contrast, sterically hindered diynes (R = *t*-Bu or TMS) did not react under any conditions (elevated temperature and pressure).

Asymmetrical diynes, including diynes possessing one sterically demanding substituent, underwent clean conversion into pyrones (Reaction 4) [9]. As shown in Table 7.5, when the steric difference between the two terminal substituents on the diyne is small (e.g., Me vs. Et), a nearly equal mixture of two pyrone regioisomers was obtained (entry 1). However, as the relative difference between the two terminal groups was increased, the regioselectivity of the reaction improved and one isomer was preferentially formed. Furthermore, the use of a diyne that contained a methyl group and a very bulky group (such as *t*-Bu or TMS) afforded only one regioisomer (entries 3 and 4) as determined by single-crystal X-ray analysis (Fig. 7.1).

7 Ni-NHC Mediated Catalysis

$$\text{diyne} + CO_2 \xrightarrow[\text{(IPr)}]{\substack{5\text{ mol\% Ni(COD)}_2 \\ 10\text{ mol\%}}} \text{pyranone} \quad (3)$$

CO$_2$ (1 atm)

Table 7.4 Ni-catalyzed cycloaddition of diynes and carbon dioxide[a]

Entry	Substrate	Product	Yield (%)[b]	Entry	Substrate	Product	Yield (%)[b]
1 a b c	31a R = Me 31b R = Et 31c R = iPr	32a 32b 32c	93 94 86	4	37	38	82
2 a b	33a R = Bn 33b R = TBDMS	34a 34b	93 92	5 a b	39a R = CO$_2$Et, R' = Me 39b R = H, R' = Et	40a 40b	97 75
3	35	36	96	6 a b	41a R = H 41b R = Me	42a 42b	41[c] 53[c]

a) Reaction conditions: 5 mol% Ni(COD)$_2$, 5 mol% IPr, 0.10 M substrate in toluene at 60 °C, 2 h.
b) Isolated yield (average of two runs).
c) Mixture of regioisomers.

$$31\text{e-h} \xrightarrow[\text{CO}_2\text{ (1atm), 60 °C}]{\substack{5\text{ mol\% Ni(COD)}_2 \\ 10\text{ mol\% IPr}}} 32\text{e-h} + 32'\text{e-h} \quad (4)$$

Table 7.5 Nickel- cycloaddition of CO$_2$ and asymmetrical diynes[a]

Entry	Substrate	R$_L$	Product	32:32'[b]	Yield (%)[c]
1	31e	Ethyl	32e	62:38	75
2	31f	i-Pr	32f	80:20	64
3	31g	t-butyl	32g	100:0	64
4	31h	TMS	32h	100:0	83

a) Reaction conditions: 5 mol% Ni(COD)$_2$, 10 mol% IPr, 0.10 M substrate in toluene at 60 °C, 30 min.
b) Determined by GC and ^{1}H NMR analysis.
c) Isolated yield (average of two runs).

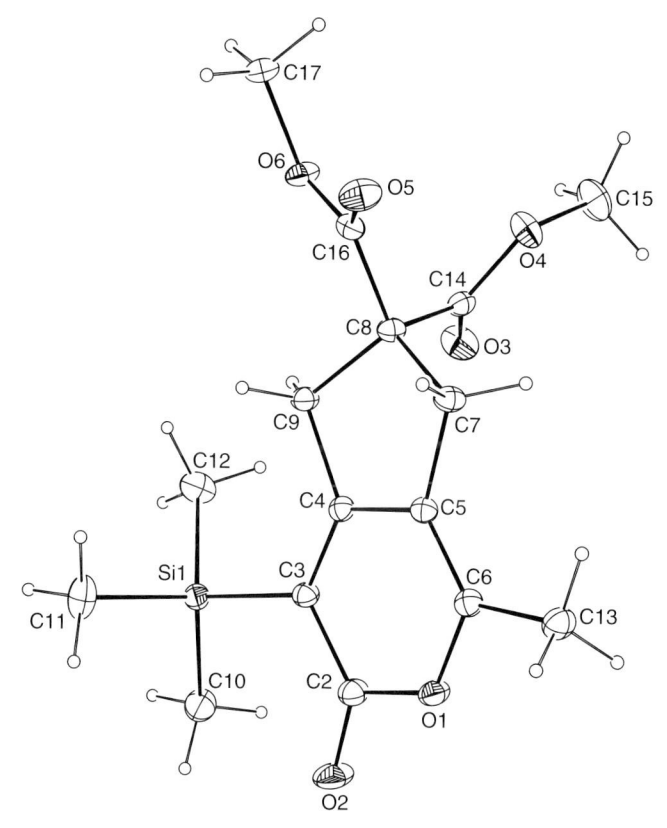

Fig. 7.1 X-ray structure of **32h**.

7.3.2
Cycloaddition of Unsaturated Hydrocarbons and Carbonyl Substrates

Pyrans constitute another class of oxygen-containing heterocycles that have been prepared from Ni-catalyzed cycloaddition reactions. The coupling of diynes and aldehydes could be mediated by the combination of a Ni(0) catalyst and a phosphine ligand; however, reaction temperatures exceeded 130 °C [10]. By replacing the phosphine ligand with SIPr, a striking increase in catalytic activity was observed and cycloadducts were obtained at room temperature (Reaction 5) [11]. Both aryl and aliphatic aldehydes were successfully incorporated into the dienones (Table 7.6). Furthermore, despite the depressed reactivity of unactivated ketones [12], the cyclization of cyclohexanone and **39a** proceeded smoothly and afforded pyran in good yield (Reaction 6). The increase in the overall catalytic activity probably most likely stems from the ability of the NHC ligand to enhance carbon–oxygen bond-forming reductive elimination.

Table 7.6 Ni-catalyzed cycloadditions of diynes **39** and aldehydes[a]

Entry	Diyne	Aldehyde	Dienone product	Yield (%)[b]
1	39a R$_1$ = Me	43a R$_2$ = C$_6$H$_5$	44a	78
2	39a	43b R$_2$ = 4-MeO-C$_6$H$_4$	44b	91
3	39a	43c R$_2$ = 4-CF$_3$-C$_6$H$_4$	44c	65[c]
4	39a	43d R$_2$ = i-Pr	44d	58[d,e]
5	39a	43e R$_2$ = n-Pr	44e	72[d,e]
6	39c R$_1$ = Et	43a	44f	80

a) Reaction conditions: 0.1 M diyne, 0.125 M aldehyde, 5 mol% Ni(COD)$_2$, 10 mol% SIPr, room temperature.
b) Isolated yields (average of two runs).
c) 10 mol% Ni(COD)$_2$, 20 mol% SIPr.
d) 10 mol% Ni(COD)$_2$, 10 mol% SIPr.
e) Product exists as an equilibrium mixture of dienone (major) and ether (minor).

7.3.3
Cycloaddition of Diynes and Isocyanates

Nitrogen-based heterocycles can also be prepared through Ni/NHC-catalyzed cycloaddition reactions. For example, Ni/SIPr catalyzed the cycloaddition of diynes with isocyanates under the mildest conditions to date [13]. In particular, excellent yields of pyridones are obtained from diynes and isocyanates at room temperature using only 3 mol% catalyst. As shown in Reaction 7 (Table 7.7), various diynes were subjected to these optimized conditions. Both aryl and alkyl isocyanates were readily converted into the respective 2-pyridone. Sterically hindered substrates appeared to have very little effect on the reaction as excellent product yields were obtained with bulky isocyanates and bulky diynes (entries 1e and 3). The catalyst system was unaffected by a diyne containing an internal amino group (**41c**) but a slightly lowered yield of product was observed with the analogous ether (**41d**) (entries 5a and 5b, respectively).

7.3 Cycloaddition Reactions

Table 7.7 Ni-catalyzed cycloaddition of diynes and isocyanates[a]

Entry	Diyne	Isocyanate	Product	Yield (%)[b]	Entry	Diyne	Isocyanate	Product	Yield (%)[b]
1	31a	NCO (aryl)			5	X	PhNCO		
a		45a R = H	46a	88	a	41c X = NTs	45a	50c	78
b		45b R = p-MeO	46b	82	b	41d X = O	45a	50d	42
c		45c R = p-CF$_3$[c]	46c	70					
d		45d R = o-CH$_3$	46d	83	6	31d	R-NCO		
e		45e R = 2,6-dimethyl	46e	79					
2	31a	CyNCO 45f	47	90	a		45a R = Ph	51a	74[d]
					b		45f R = Cy	51b	83[d]
					c		45h R = Bn	51c	87[d]
3	31c	BuNCO 45g	48	85	7	Et—≡—Et 52	45a	53	90
4					8	31h	45g	54	94
a	39b	45f	49a n = 2, R = Et, R' = Cy	96					
b	39c	45g	49b n = 3, R = Me, R' = Bu	38[c]					

a) Reaction conditions: 3 mol% Ni(COD)$_2$, 3 mol% SIPr, 0.10 M diyne in toluene at RT, 30 min.
b) Isolated yield (average of two runs).
c) Run at 80 °C.
d) 0.005 M in diyne.

The increased reactivity of isocyanates, relative to carbon dioxide, was reflected in the wider range of cycloaddition partners. For example, terminal diynes (**31d**) as well as non-tethered alkynes (e.g., hex-3-yne, **52**) were also successfully converted into 2-pyridones (Table 7.7, entries 6a and 7, respectively) rather than undergoing rapid telomerization to aromatic by-products. Importantly, the cycloaddition of an asymmetrical diyne (**31h**) afforded a pyridone with the larger substituent in the 3-position (**54**) (entry 8). Thus, the same regioselectivity with pyrone products was observed, indicating that a similar cycloaddition mechanism is most likely involved.

NHCs were found to react with isocyanates to afford isocyanurates (Reaction 8) [14, 15]. Although SIPr was an effective catalyst for isocyanurate formation (for a wide variety of isocyanates), no isocyanurate was observed in most Ni-catalyzed cycloaddition reactions of diynes and isocyanates (Reaction 9). Furthermore, isocyanurates were not formed reversibly during the reaction since no pyridones were obtained when isocyanurates were used as the sole source of isocyanate. This data further highlights the efficacy of the Ni/NHC catalyst system.

$$R-N=\cdot=O \xrightarrow[\text{THF, RT, 1h}]{0.1\ \text{mol\% SIPr}} \text{isocyanurate} \quad (8)$$

$$\text{diyne} \xrightarrow[\text{1 eq R'N}=\cdot=O]{\substack{3\ \text{mol\% Ni(COD)}_2 \\ 3\ \text{mol\% SIPr}}} \text{pyridone} \;+\; \cancel{\text{isocyanurate}} \quad (9)$$

7.3.4
Cycloaddition of Diynes and Nitriles

The combination of Ni(0) and a phosphine ligand had been used to catalyze the cycloaddition of diynes with CO_2 [7], aldehydes [10], and isocyanates [16]. The corresponding cycloaddition with nitriles, however, had not been demonstrated. The absence of this cycloaddition reaction may be due to the inability of Ni/PR$_3$ systems to facilitate the hetero-oxidative coupling of an alkyne and a nitrile [17]. By employing an NHC ligand, the nucleophilicity of the Ni catalyst was enhanced, which led to a greater interaction with the nitriles. Thus, diynes and nitriles could be converted into pyridines, and under ambient conditions (Reaction 10, Table 7.8) [18]. In general, both aryl and alkyl nitriles were readily converted into the respective pyridine, although alkyl nitriles gave slightly diminished yields. Both aryl nitriles bearing either an electron-donating group (p-OMe, **55b**) or an electron-withdrawing group (p-CF$_3$, **55c**) readily cyclized (entries 1b and 1c, respectively). This is in contrast to the Ni/NHC-catalyzed route to pyridones, where electron-deficient isocyanates required slightly elevated temperatures for complete cycloaddition [13]. Notably, sterically hindered nitriles [such as o-tolunitrile (**55d**), tert-butyl nitrile (**55g**), and naphthalene-1-carbonitrile (**55h**)] delivered the desired pyridines (entries 1d, 2c, and 5, respectively). Pyridine yield was unaffected when degassed, but not dried, acetonitrile was employed (entry 2a). Diynes devoid of internal substitution, such as dodeca-3,9-diyne (**39b**), and diynes containing either an internal amino group (**41c**) or the analogous ether (**41d**) also coupled with aryl nitriles to give pyridines (entries 3b and 4a–c). Heteroaryl nitriles were readily converted into pyridines in high yields (entry 6). In accordance with previous

7.3 Cycloaddition Reactions | 173

cycloadditions of diynes and heterocumulene-type substrates, the coupling of diyne **31h** and acetonitrile (**55e**) afforded pyridine **62** as a single regioisomer in 58% yield (entry 8). Initial hetero-oxidative coupling of the TMS-terminated alkyne and nitrile followed by insertion of the methyl-terminated alkyne explains the observed regioselectivity.

(10)

Table 7.8 Ni-catalyzed cycloaddition of diynes and nitriles[a]

Entry	Diyne	Nitrile	Product	Yield (%)[b]	Entry	Diyne	Nitrile	Product	Yield (%)[b]
1	31a	—CN / Aryl	56a–d		5	31a	naphthyl-CN (55h)	59	91
a		55a R = H	56a	86					
b		55b R = p-MeO	56b	64					
c		55c R = p-CF$_3$	56c	94					
d		55d R = o-Me	56d	81					
					6	31a	pyrrolyl-CN (55i)	60	97
2	31a	Alkyl-CN	56e–g						
a		55e MeCN	56e	69 (69)[c]					
b		55f i-BuCN	56f	72					
c		55g t-BuCN	56g	56	7	Et—≡—Et (52)	55a	61	82
3			57a–c		8	E,E-diyne (31h) with TMS	55e	62	58[d]
a	39b	55a	57a n = 2, R = Et, R' = Ph	92					
b	39b	55e	57b n = 2, R = Et, R' = Me	46					
c	39c	55e	57c n = 3, R = Me, R' = Me	29					
4									
a	41c X = NTs	55a	58a X = NTs, R = Ph	78					
b	41d X = O	55a	58b X = O, R = Ph	93					
c	41d X = O	55e	58c X = O, R = Me	37					

a) Reaction conditions: 3 mol% Ni(COD)$_2$, 6 mol% SIPr, 0.10 M diyne in toluene at RT, 30 min.
b) Isolated yield (average of two runs).
c) Isolated yield of reaction with MeCN that was degassed, but not dried.
d) Run at 80 °C.

7.4
Reductive Coupling Reactions

7.4.1
Reductive Coupling Reactions: No added Reductant

Murakami and coworkers recently reported that cyclobutanones can be coupled with alkynes to afford ring-expanded cyclohexenones (such as **67**, Reaction 11) [19]. While phosphine ligands were generally employed to facilitate the reaction, the authors also demonstrated that IPr was an effective ligand. In reactions involving asymmetrically substituted alkynes, such as 1-phenyl-1-propyne (Scheme 7.2, R_3 = Ph, R_4 = Me), the methyl group was located α to the carbonyl group in the major product. The observed regioselectivity can be explained in terms of a favorable electronic interaction when the aryl substituent (R_3) is located on the α carbon in nickelacycle **65**. A similar phenomenon has been observed in other nickel promoted coupling reactions involving alkynes [20].

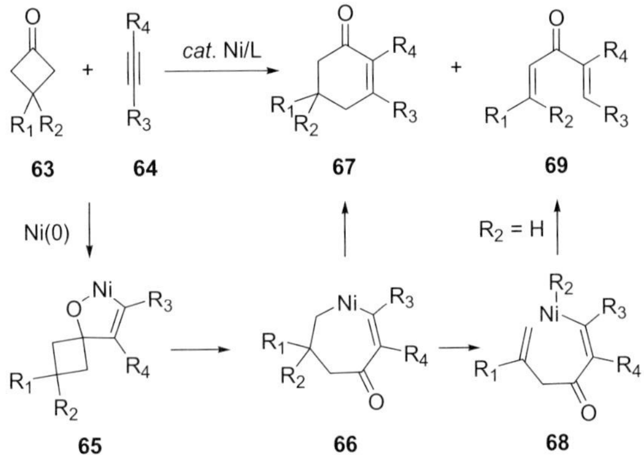

Scheme 7.2

In analogy to cycloadditions described above, the first step of these ring expansion reactions involves the initial oxidative coupling between the carbonyl and the alkyne to afford a nickelapentenacycle (**65**, Scheme 7.2) [21]. Subsequent β-carbon elimination relieves ring strain and affords the seven-membered nickelacycle **66** that reductively eliminates the cyclohexenone product and regenerates the catalyst. When a β-hydrogen is available (i.e., R_2 = H), β-H elimination becomes competitive with reductive elimination and acyclic products (**69**) are seen in appreciable amounts. However, replacement of the phosphine ligand with an NHC ligand such as IPr appeared to suppress this side reaction and afforded good yields of the desired cyclohexenone (Table 7.9).

7.4 Reductive Coupling Reactions

[Reaction scheme: cyclobutanone + alkyne with cat. Ni/L → products 67 and 69 (11)]

Table 7.9 Ligand effects.

Entry	Mol% Ni(COD)$_2$	L (mol%)	67 (%)	69 (%)
1	10	P(c-Hex)$_3$ (20)	41	54
2	10	PPh$_3$ (20)	37	26
3	10	I^{i}Pr (10)	61	
4	20	I^{i}Pr (20)	79	

7.4.2
Reductive Coupling Reactions in the Presence of a Reductant

Nickel-catalyzed cyclizations, couplings, and cycloadditions involving three reactive components have been actively researched over the past decade [22]. Central to these reactions is the involvement of a low-valent nickel capable of facilitating oxidative coupling of an unsaturated hydrocarbon (such as an alkyne, allene, or alkene) and a carbonyl substrate (such as an aldehyde or ketone). The use of NHC as ligands has been evaluated for couplings of aldehydes. Such reactions typically afford O-protected allylic alcohols in good yields.

In 2001, Mori and coworkers showed that the use of NHC ligands can dramatically influence the olefinic geometry in the Ni-catalyzed coupling reaction of 1,3-dienes and aldehydes [23]. Specifically, when Ni/PPh$_3$ is used as the catalyst, homoallylic silyl alcohol products were obtained in the E configuration. However, when PPh$_3$ was replaced with I^{i}Pr, homoallylic alcohol products were obtained in the Z configuration. The reaction of diene **70** with a handful of aryl aldehydes was investigated. Electron-withdrawing substituents on the aldehydes seemed to somewhat retard the reaction. Yields were generally good and ranged from 54 to 95% (Reaction 12).

[Reaction scheme 12: MOMO-substituted diene 70 + R-ArCHO, 5–10 mol% NiCl$_2$, 5–10 mol% (I^{i}Pr·HCl), BuLi/THF, Ni(0)/NHC cat, Et$_3$SiH, THF, 50 °C → product 71 (OSiEt$_3$), 54–95%]

This paper was one of the first to demonstrate the generation of a Ni(0)/NHC catalyst *in situ* from air-stable Ni(II) precursors and an NHC·HCl salt. It was known that the addition of base to NHC·HCl generates the free NHC carbene ligand, and that Ni(II) can be reduced by organolithium reagents. Mori combined these protocols by using BuLi to reduce the Ni(II) starting material and to deprotonate I^{i}Pr·HCl. Although Grignard reagents were also evaluated, no Ni(0)/NHC species were observed by ^{13}C NMR.

A few years later, Mori found that a silyl diene could serve as a substrate in Ni-catalyzed coupling reactions with aryl aldehydes (Scheme 7.3) [24]. No comment on the ability to use more substituted diene partners was made. IMes proved to be a superior ligand to I^{i}Pr, in contrast to reactions of aryl dienes described above. However, in analogy to the reactions of aryl dienes, reactions run with PPh$_3$ and reactions run with IMes displayed differences in product distribution. That is, reactions run with PPh$_3$ gave E products whereas those run with IMes gave Z products (Scheme 7.3). Interestingly, higher yields were obtained when an equivalent of PPh$_3$ was added to reactions. Possibly, the added PPh$_3$ serves to stabilize the coordinatively unsaturated Ni-NHC complex, thereby increasing the lifetime of the catalyst. In all cases, reaction times were consistently longer in reactions run with IMes than with PPh$_3$.

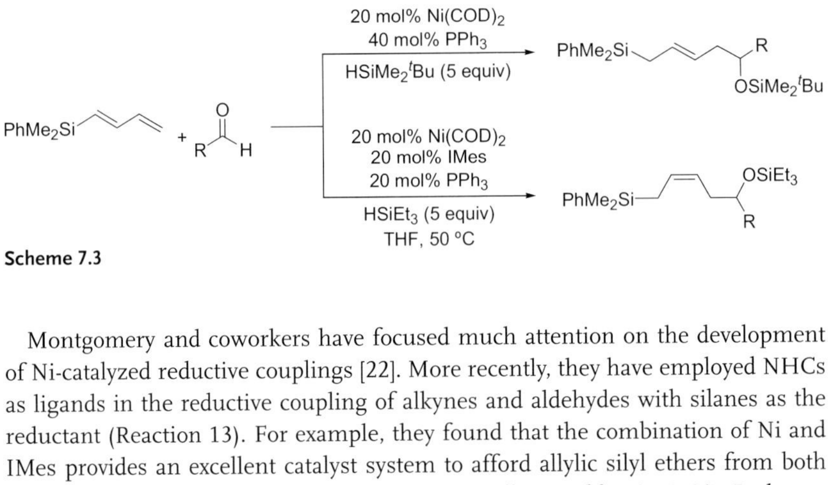

Scheme 7.3

Montgomery and coworkers have focused much attention on the development of Ni-catalyzed reductive couplings [22]. More recently, they have employed NHCs as ligands in the reductive coupling of alkynes and aldehydes with silanes as the reductant (Reaction 13). For example, they found that the combination of Ni and IMes provides an excellent catalyst system to afford allylic silyl ethers from both aromatic and aliphatic aldehydes in good to excellent yields (56–84%). Both aromatic and aliphatic aldehydes, including electron-rich aromatic aldehydes and sterically demanding aliphatic aldehydes, were used as coupling partners. The alkyne may be internal or terminal, with aromatic or aliphatic substitution patterns being tolerated in both cases. In almost all cases, good regioselectivity was observed (generally 98:2), except with an internal aliphatic alkyne (1.3:1).

Interestingly, reactions run with the NHC ligand displayed different reactivity than their original Ni(COD)$_2$/PBu$_3$ system. It appears that the two catalyst systems may proceed through different mechanisms (Scheme 7.4). Crossover experiments revealed that significant crossover occurred in reactions run with PBu$_3$

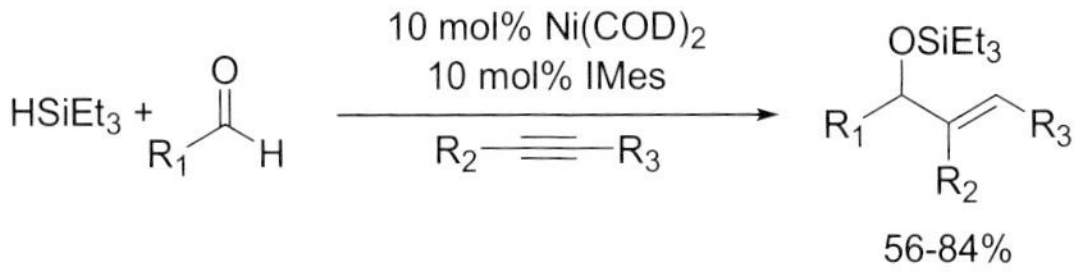

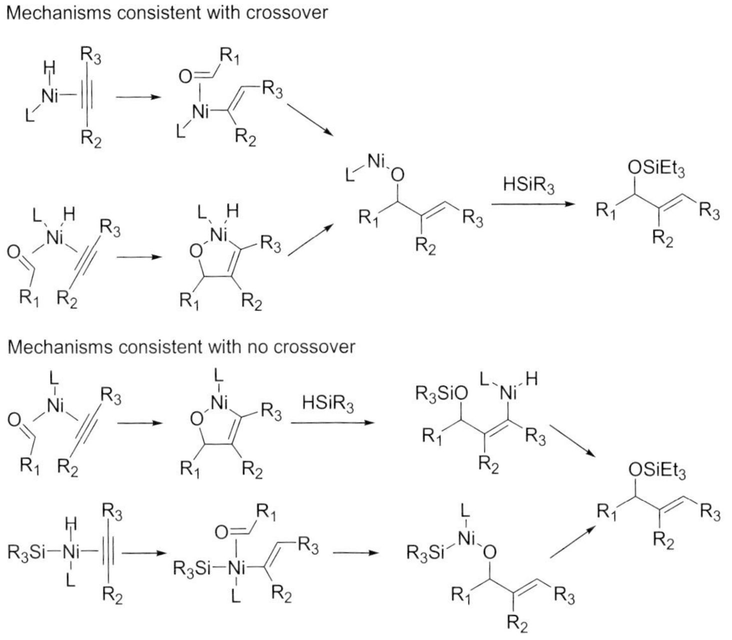

Scheme 7.4 Possible mechanisms.

whereas negligible crossover was observed in those run with IMes. Although the actual mechanism, and the reason for the difference in crossover between the two reactions, is still not fully understood, two distinct mechanisms are clearly involved. In reactions run with PBu$_3$, the authors suggest that either a nickel hydride or nickel silyl species, but not both, is involved. In contrast, the lack of crossover seen with IMes suggests that oxidative coupling of the aldehyde and alkyne and subsequent reaction of the silane may be operative. Alternatively, the Ni/IMes catalyst may oxidatively add the silane, undergo successive alkyne and aldehyde insertions, and ultimately reductively eliminate the product.

Montgomery and coworkers later used this procedure for the macrocyclization of ynals (Scheme 7.5) [25]. Macrocyclic rings ranging in size (e.g., 14- to 22-membered rings) and all possessing endocyclic (E)-olefins were obtained from terminal ynals in good yields (62–70%) using a catalyst derived from Ni(COD)$_2$, KO^{t}Bu, and IMes·HCl. Internal alkynes were also examined. Phenyl-substituted alkynes afforded macrocycles possessing an exocyclic olefin selectively, regardless of

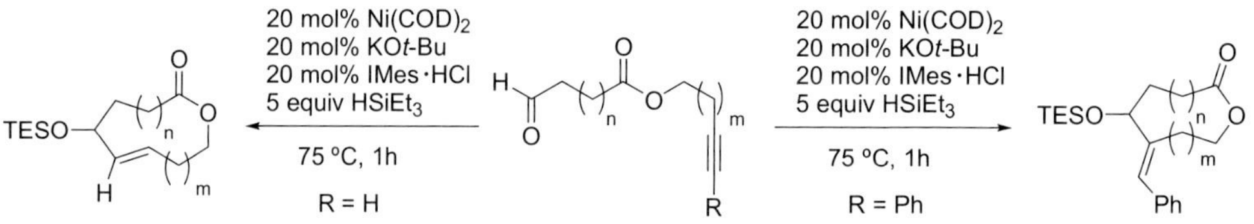

Scheme 7.5

ligand employed (phosphine or NHC). However, the selectivity for exocyclic versus endocyclic olefins diminished with methyl-substituted alkynes.

Jamison and coworkers have used a similar approach for the coupling of allenes, aldehydes, and silanes (Reaction 14) [26]. They first explored the use of phosphines such as P(Cyp)$_3$. They observed an excellent ratio of allylic and homoallylic products (>95:5) but significant erosion of enantiomeric purity (95 to 62%) occurred. The use of IPr solved this problem and a range of internal allenes were converted into the corresponding allylic silyl ethers in yields ranging from 40 to 80%, although it appears this reaction is limited to aryl aldehydes. Aryl aldehydes bearing either electron-donating or -withdrawing groups showed similar reactivity. In all cases, the Z geometry corresponded to attachment of the aldehyde to the more hindered face of the allene.

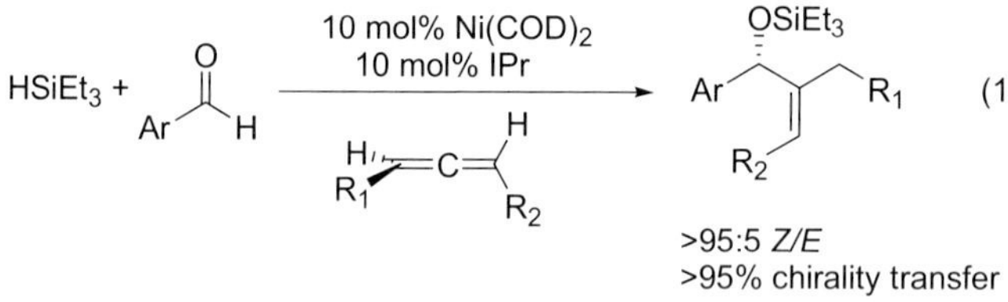

7.5
Oligomerization and Polymerization

When Ni(II)-NHC complexes contain an alkyl, aryl, or acyl group, reductive elimination can occur, affording Ni(0) compounds and 2-mediated organoimidazolium salts (Reaction 15). This pathway results in catalyst decomposition for reactions by Ni-NHC systems [27]. In Ni-NHC-catalyzed olefin dimerization, Cavell and Wasserscheid showed that this decomposition is inhibited when reactions are run in ionic liquids rather than more classical solvents such as toluene [28].

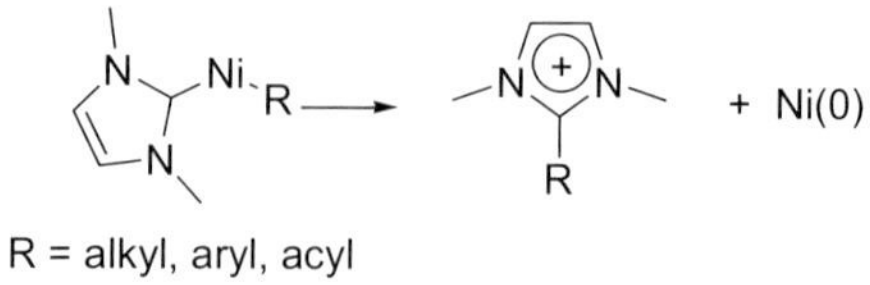

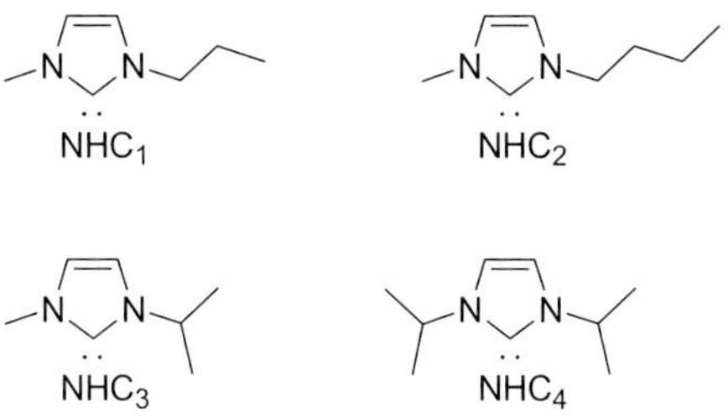

Fig. 7.2 NHC ligands.

A series of Ni(NHC)$_2$I$_2$ complexes were prepared and evaluated as catalysts in both toluene and ionic liquids (NHC ligands shown in Fig. 7.2). In toluene, no butene oligomers were formed at 20 °C. Instead, butene was incorporated into the imidazolylidene cation in the 2-position (Scheme 7.6). These results suggest that although Ni hydrides and alkyls were being formed, these species reductively eliminated too rapidly for chain growth to occur. In contrast, all reactions run in a buffered melt (composed of a mixture of 1-butyl-3-methylimidazolium chloride, AlCl$_3$, and N-methylpyrrole) showed complete conversion into butene dimers. Interestingly, greater turnover numbers were observed in reactions catalyzed by Ni-NHC complexes versus NiCl$_2$(PCy$_3$)$_2$ (Table 7.10). In addition, Ni(NHC$_1$)$_2$I$_2$ displayed different selectivity than NiCl$_2$(PCy$_3$)$_2$ toward different isomers in the dimerization of propene. Desirable highly branched propene dimers were obtained in higher ratios with NiCl$_2$(PCy$_3$)$_2$. Changes in the organic side-chain of the carbene did not lead to an increase in branching, which may suggest the formation of a common active species resulting from incorporation of the imidazolium cation onto the Ni complex. Also, in ionic liquids, phosphine dissociation may not play a significant role.

Scheme 7.6

Table 7.10 1-Butene dimerization in IL.

Entry	Catalyst	Yield (%)	TON	TOF (h^{-1})
1	NiI$_2$(NHC$_1$)$_2$	56.3	2815	5630
2	NiI$_2$(NHC$_2$)$_2$	70.2	3510	7020
3	NiI$_2$(NHC$_3$)$_2$	38.2	1910	3820
4	NiI$_2$(NHC$_4$)$_2$	50.7	2535	5070
5	NiCl$_2$(PCy$_3$)$_2$	29.5	1475	2950

7 Ni-NHC Mediated Catalysis

By changing the NHC ligands to NHCs possessing a hemilabile pyridine linkage, Jin and coworkers were able to use Ni(II)-NHC complexes as catalysts for the polymerization of norbornene and ethylene in the presence of methylaluminoxane (MAO) as a cocatalyst [29]. The Ni complexes were prepared as shown in Scheme 7.7. Although the free carbenes of **72** could not be generated successfully, the desired Ni compounds (**73**) could be prepared via the preparation of Ag-NHCs and subsequent reaction with Ni(PPh$_3$)$_2$Cl$_2$. X-Ray analysis revealed that both compounds possess essentially square-planar geometries and the two chelates adopt a cis arrangement around the nickel atom.

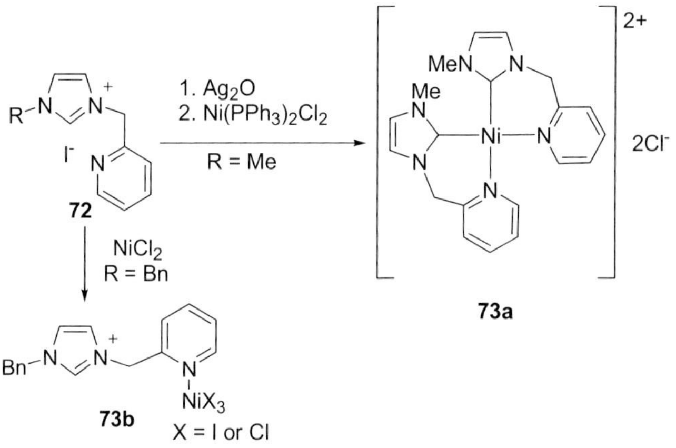

Scheme 7.7

The catalytic activity for olefin polymerization was evaluated for complex **73a**. High molecular weight addition-type polynorbornene with a moderate molecular weight distribution ($M_w = 10^6$, $M_w/M_n = 2.3–3.5$) was obtained when **73a** was activated with MAO. The activity was highest at 80 °C [10^7 g of PNB (mol of Ni) h^{-1}], resulting from an increase in the concentration of active catalyst centers at that temperature. However, further increases in temperature lead to catalyst decomposition rather than higher turnover numbers.

Complex **73a** displayed moderate catalytic activity towards the polymerization of ethylene (3.3×10^5 g mol^{-1} h^{-1}). In addition, higher molecular weight distributions were observed ($M_w/M_n = 12.8$). ^{13}C NMR analysis of the polyethylene showed that methyl branches predominate (with ca. 3.4 methyl branches per 1000 carbon atoms), suggesting that chain walking does not effect polymerization to a high degree. When only the pyridine moiety (and not the imidazolium salt) is ligated (**73b**) [30], ethylene polymerization occurs twice as effectively [6×10^5 g-PE (mol of Ni) h^{-1}] under similar conditions (only 30 rather than 60 min).

7.6 Hydrogenation

Fort and Schneider showed recently that the combination of Ni(0) and IMes catalyzed the transfer hydrogenation of imines to the corresponding amines in the presence of NaOCHEt$_2$ (Reaction 16) [31]. Various aldimines and ketimines were reduced in good to excellent yields under mild conditions. A range of NHC ligands were explored, including ones possessing pendant, hemilabile pyridines. However, only IMes was effective (97%). Surprisingly, under identical conditions, IPr showed only 4% yield. Although ICyPic gave respectable results (86%), IMes-Pic showed marginal activity (15%). Clearly, no correlation between catalyst activity and NHC ligand could be rationalized. Nevertheless, the authors used the combination of Ni(0) and IMes to catalyze the transfer hydrogenation of various aldimines and ketimines. Amines possessing a wide range of functionalities were obtained in yields ranging from 65 to 99%. No hydrogenation was observed with cyano- and pyridine-substituted imines, which is likely due to ligand displacement and subsequent catalyst deactivation.

$$R_1R_2C=NR_3 \xrightarrow[\substack{5 \text{ mol\% IMes HCl} \\ 50 \text{ mmol NaOCHEt}_2 \\ \text{dioxane, 100 °C}}]{5 \text{ mol\% Ni(COD)}_2} R_1R_2CH-NHR_3 \quad 65\text{-}99\% \quad (16)$$

7.7 Conclusions

The number of synthetic methods catalyzed by Ni complexes has risen since the discovery of NHCs as ligands for transition metals. As seen with most of the reactions outlined above, sterically hindered NHC ligands have played the greatest role in the advancement of nickel catalysis. The increased steric bulk in conjunction with the enhanced donating ability of these ligands helps to generate and stabilize the highly active nickel species. Clearly, the onset of these NHCs has opened new vistas in the ever-growing field of Ni-mediated catalysis.

References

1. (a) T. Hudlicky, J. W. Reed in *Comprehensive Organic Synthesis* ed. B. Trost, I. Fleming, L. A. Paquette, Pergamon, Oxford, **1992**, vol. 5, pp. 899–970. (b) M. Murakami, S. Nishida, *Chem. Lett.* **1979**, 927.
2. G. Zuo, J. Louie *Angew. Chem. Int. Ed.* **2004**, *43*, 2777.
3. For Rh: (a) P. A. Wender, C. M. Barzilay, A. J. Dyckman *J. Am. Chem. Soc.* **2001**, *123*, 179. (b) P. A. Wender, D. A. Sperandio *J. Org. Chem.* **1998**, *63*, 4164. (c) P. A. Wender, H. Takahashi, B. Witulski *J. Am. Chem. Soc.* **1995**, *117*, 4720. For Ru: (d) B. M. Trost, F. D. Toste, H. Shen *J. Am. Chem. Soc.* **2000**, *122*, 2379.
4. (a) M. Zhang, S. L. Buchwald *J. Org. Chem.* **1996**, *61*, 4498. (b) P. A. Wender, T. E. Smith *J. Org. Chem.* **1995**, *60*, 2962. (c) K. Tamao, K. Kobayashi, Y. Ito *J. Am. Chem. Soc.* **1988**, *110*, 1286.
5. G. Zuo, J. Louie *J. Am. Chem. Soc.* **2005**, *127*, 5798.
6. (a) Y. Inoue, Y. Itoh, H. Hashimoto *Chem. Lett.* **1978**, 633. (b) Y. Inoue, Y. Itoh, H. Hashimoto *Chem. Lett.* **1977**, 855. (c) Y. Inoue, Y. Itoh, H. Kazama, H. Hashimoto *Bull. Chem. Soc. Jpn.* **1980**, *53*, 3329.
7. (a) T. Tsuda, R. Sumiya, T. Saegusa *Synth. Commun.* **1987**, *17*, 147. (b) T. Tsuda, S. Morikawa, R. Sumiya, T. Saegusa *J. Org. Chem.* **1988**, *53*, 3140. (c) T. Tsuda, S. Morikawa, T. Saegusa *J. Chem. Soc., Chem. Commun.* **1989**, 9. (d) T. Tsuda, N. Hasegawa, T. Saegusa *J. Chem. Soc., Chem. Commun.* **1990**, 945. (e) T. Tsuda, S. Morikawa, N. Hasegawa, T. Saegusa *J. Org. Chem.* **1990**, *55*, 2978. (f) T. Tsuda, K. Maruta, K. Kitaike *J. Am. Chem. Soc.* **1992**, *114*, 1498.
8. J. Louie, J. E. Gibby, M. V. Farnworth, T. N. Tekavec *J. Am. Chem. Soc.* **2002**, *124*, 15188.
9. T. N. Tekavec, A. Arif, J. Louie *Tetrahedron* **2004**, *60*, 7431.
10. T. Tsuda, T. Kiyoi, T. Miyane, T. Saegusa *J. Am. Chem. Soc.* **1988**, *110*, 8570.
11. T. N. Tekavec, J. Louie *Org. Lett.* **2005**, *7*, 4037.
12. K. Miller, T. F. Jamison *Org. Lett.* **2005**, *7*, 3077.
13. H. A. Duong, M. J. Cross, J. Louie *J. Am. Chem. Soc.* **2004**, *126*, 11438.
14. W. Schössler, M. Regitz *Chem. Ber.* **1974**, *107*, 1931.
15. H. A. Duong, M. J. Cross, J. Louie *Org. Lett.* **2004**, *6*, 4679..
16. (a) P. Hong, H. Yamazaki *Tetrahedron Lett.* **1977**, 1333. (b) H. Hoberg, B. W. Oster *Synthesis* **1982**, 324. (c) R. A. Earl, K. P. C. Vollhardt *J. Org. Chem.* **1984**, *49*, 4786.
17. J. J. Eisch, X. Ma, K. I. Han, J. N. Gitua, C. Krüger *Eur. J. Inorg. Chem.* **2001**, 77.
18. M. M. McCormick, H. A. Duong, G. Zuo, J. Louie *J. Am. Chem. Soc.* **2005**, *127*, 5030.
19. M. Murakami, S. Ashida, T. Matsuda *J. Am. Chem. Soc.* **2005**, *127*, 6932.
20. (a) J. Eisch, G. A. Damasevitz *J. Organomet. Chem.* **1975**, *96*, C19. (b) H. A. Duong, J. Louie *J. Organomet. Chem.* **2005**, *690*, 5098.
21. (a) J. Chan, T. F. Jamison *J. Am. Chem. Soc.* **2004**, *126*, 10682. (b) K. M. Miller, T. F. Jamison *Org. Lett.* **2005**, *7*, 3077.
22. (a) J. Montgomery, *Acc. Chem. Res.* **2000**, *33*, 467. (b) J. Montgomery *Angew. Chem., Int. Ed.* **2004**, *43*, 3890.
23. Y. Sato, R. Sawaki, M. Mori *Organometallics* **2001**, *20*, 5510.
24. R. Sawaki, Y. Sato, M. Mori *Org. Lett.* **2004**, *6*, 1131.
25. B. Knapp-Reed, G. M. Mahandru, J. Montgomery *J. Am. Chem. Soc.* **2005**, *127*, 13156.
26. S.-S. Ng, T. F. Jamison *J. Am. Chem. Soc.* **2005**, *127*, 7320.
27. K. J. Cavell, D. S. McGuinness *Coord. Chem. Rev.* **2004**, *248*, 671.
28. D. S. McGuinness, W. Mueller, P. Wasserscheid, K. J. Cavell, B. W. Skelton, A. H. White, U. Englert *Organometallics* **2002**, *21*, 175.
29. X. Wang, S. Liu, G.-X. Jin *Organometallics* **2004**, *23*, 6002.
30. X. Wang, S. Liu, L. Weng, G.-X. Jin *J. Organometal. Chem.* **2005**, *690*, 2934.
31. S. Kuhl, R. Schneider, Y. Fort *Organometallics* **2003**, *22*, 4184.

8
Asymmetric Catalysis with Metal N-Heterocyclic Carbene Complexes

Marc Mauduit and Hervé Clavier

8.1
Introduction

Since the first characterization of metal complexes containing N-heterocyclic carbenes (NHC) ligands by Öfele [1] and Wanzlick [2] in 1968 and the seminal contribution of the group of Arduengo, who isolated the first stable free NHC **1** in 1991 (Fig. 8.1) [3], NHCs have become, in the past decade, the focus of intense study through their application as ancillary ligands for catalytically active transition-metal complexes.

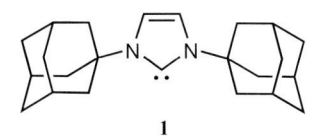

Fig. 8.1 First stable free NHC.

NHCs show many interesting properties that make them valuable as ligands in catalysis. A combination of their powerful s-donating and weak π-accepting character allows for the generation of a stronger bond to the metal than their phosphines homologues and leads to the formation of an interestingly robust electron-rich metal complex. Consequently, metal-NHC complexes tend to be air-stable, easy to handle and highly active in several catalytic transformations where harsh conditions are often required. In recent years, an exceptionally larger number of NHC-complexes have emerged and have been used successfully in many organometallic transformations, notably the C–C and C–N cross-coupling reactions [4], as well as the extremely popular metathesis reaction [5].

Development of chiral versions of these valuable ligands for enantioselective catalysis is a logical extension in this field. Historically, the first chiral NHC-complexes were reported by the group of Lappert in 1983 [6], before the isolation of the first stable NHC (**1**) by Arduengo. At this time, free NHC species were believed to be extremely unstable and therefore impossible to isolate. The strategy

N-Heterocyclic Carbenes in Synthesis. Edited by Steven P. Nolan
Copyright © 2006 WILEY-VCH Verlag GmbH & Co. KGaA, Weinheim
ISBN: 3-527-31400-8

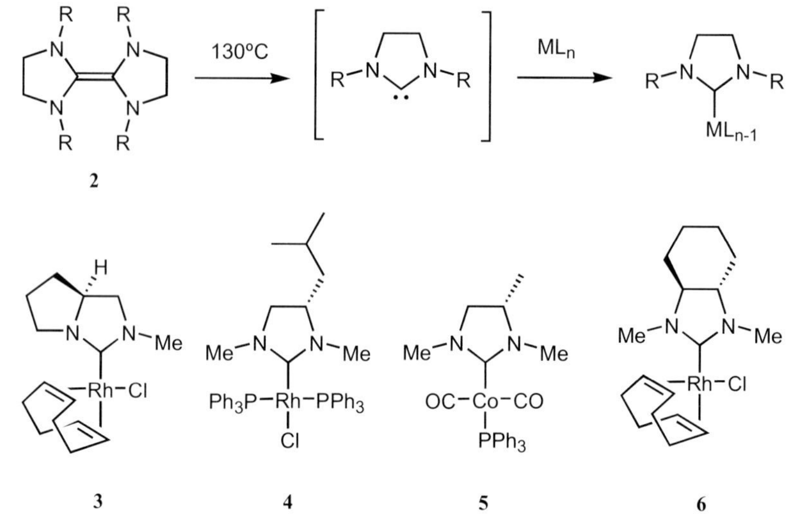

Scheme 8.1

employed by Lappert was to generate the desired chiral NHC complex via an *in situ* formation of the free NHC ligand. This was achieved by heating electron-rich tetramine **2** and allowing the spontaneous dimerization of the unstable free NHC in the presence of the metallic precursor (Scheme 8.1). Starting from natural amino acids, terpene derivatives or C_2-symmetric diamines, several chiral tetraaminoethenes were synthesized and condensed with various rhodium or cobalt complexes to lead to the corresponding NHC-complexes **3–6**. Whereas these original complexes were fully characterized by NMR and X-ray studies [7], none of them were evaluated in stereoselective reactions.

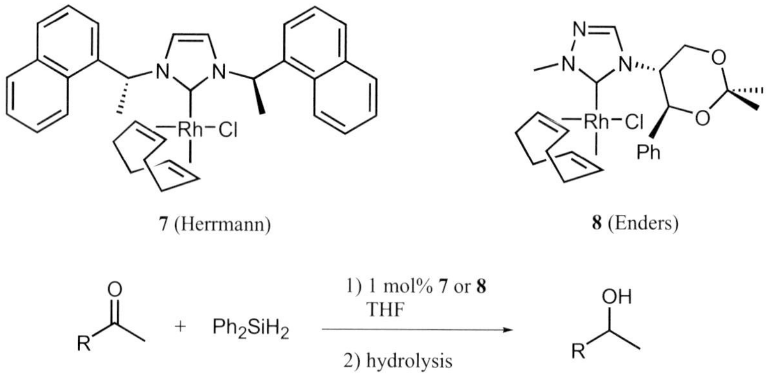

7 (Herrmann) **8 (Enders)**

Scheme 8.2

7: R= Ph, 90% yield, 32% ee
8: R= Cy, 75% yield, 44% ee

Asymmetric catalysis based on chiral NHC-complexes truly began in the mid-1990s with the original work of Herrmann [8] and Enders [9], who reported the stereoselective hydrosilylation of, respectively, acetophenone and cyclohexyl methyl ketone catalyzed by the Rh-NHC complexes **7** and **8** with moderate enantioselectivities (Scheme 8.2). Complexes of various transition metals containing a chiral NHC unit have been synthesized since then and used in asymmetric catalysis, affording moderate to high enantiomeric excess [10, 11]. This chapter summarizes the state of the art in this area through an overview of applications in asymmetric catalysis.

8.2
Concept, Design and Synthesis of Chiral NHC Complexes

8.2.1
Synthesis of Ligand Precursors

The principal classes of electron-rich N-heterocyclic carbenes usually employed in catalysis are represented in the Fig. 8.2. The growing and tremendous interest in chiral transition metal complexes of NHCs is closely related to the nature of the NHC ligand precursors, i.e., azolium salts, which are easily accessible by classical synthetic methods.

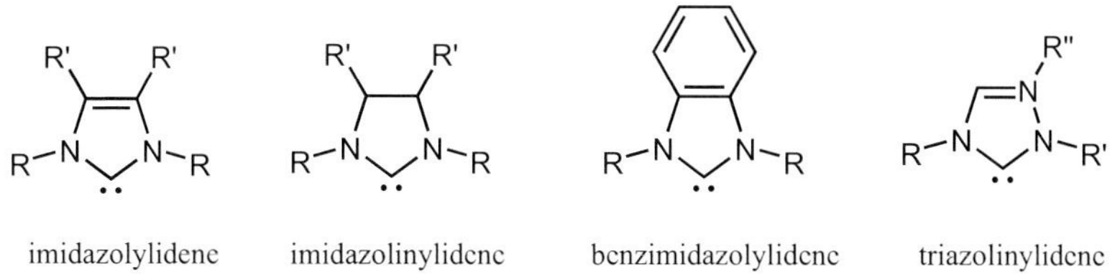

imidazolylidene   imidazolinylidene   benzimidazolylidene   triazolinylidene

Fig. 8.2 Principal class of NHCs used in catalysis.

Most often based on imidazole and imidazoline structures, these salts can be synthesized by various and complementary synthetic routes (Scheme 8.3): (1) the one-pot condensation of glyoxal, paraformaldehyde and primary amines to form symmetrical imidazolium salts (unsymmetrical salts can also be made by a similar route), (2) alkylation of substituted imidazoles (or imidazolines) with an appropriate alkyl-/aryl halide, and (3) the orthoformate route to convert vicinal N,N-disubstituted-1,2-diamines (symmetric or not) into the corresponding imidazolinium salts under acidic conditions.

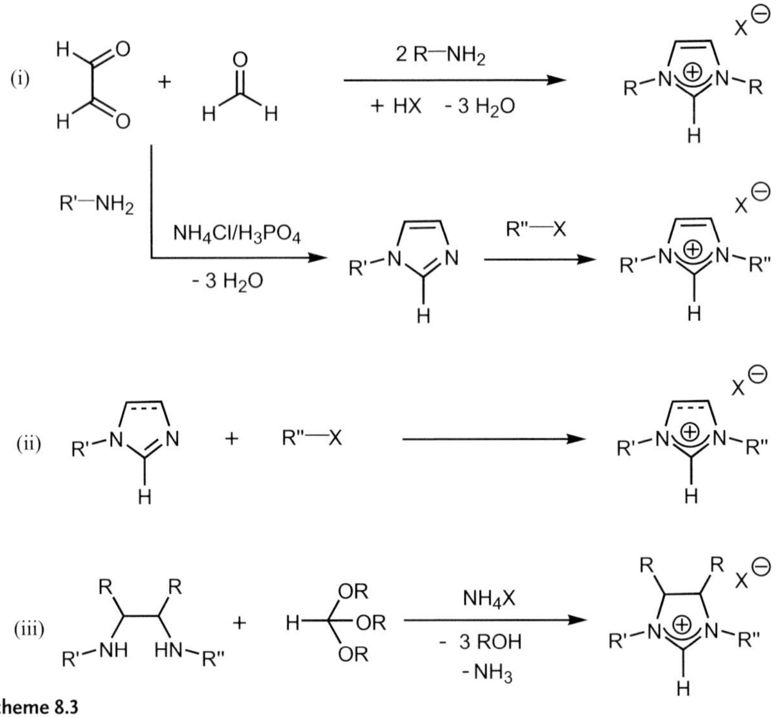

Scheme 8.3

8.2.2
Synthesis of NHC Complexes

The formation of NHC-metal complexes is mainly based on three principal routes (Scheme 8.4): (1) thermal cleavage of electron-rich alkenes (as mentioned above with the Lappert work), (2) complexation of the free stable NHC, and (3) *in situ* deprotonation of the corresponding azolium salts. Complexation of the free carbene is the best method, leading to NHC-complexes in high yield. However, as the original framework of the azolium salt leads to unstable carbene species in many cases, the *in situ* formation of free NHC in the presence of a complex precursor is often preferred, thus avoiding the formation of undesired dimerization by-products.

Lin's group have also developed a convenient method [12], using a pre-isolated silver-NHC complex (as a protected free NHC form) in a metal-exchange reaction to form several metal-carbene complexes in high yields (Scheme 8.5). This last method is particularly attractive when the deprotonation route fails due to the presence of proton in the azolium NHC precursor that has a comparable acidity to the azolinium C–H.

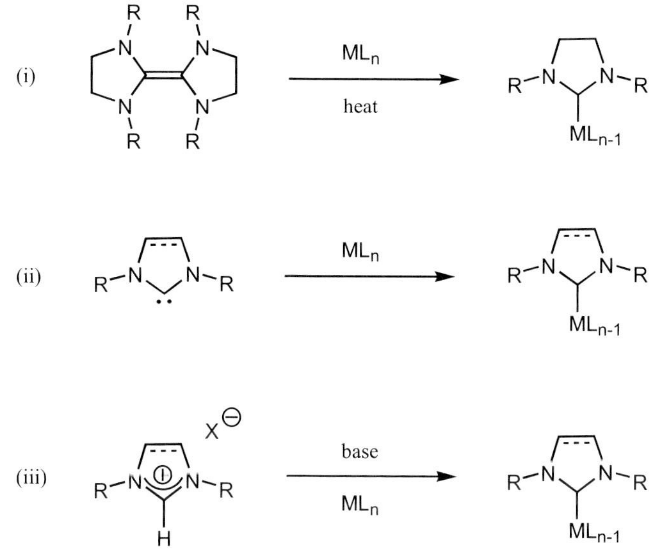

Scheme 8.4 Principal routes in the formation of NHC complexes.

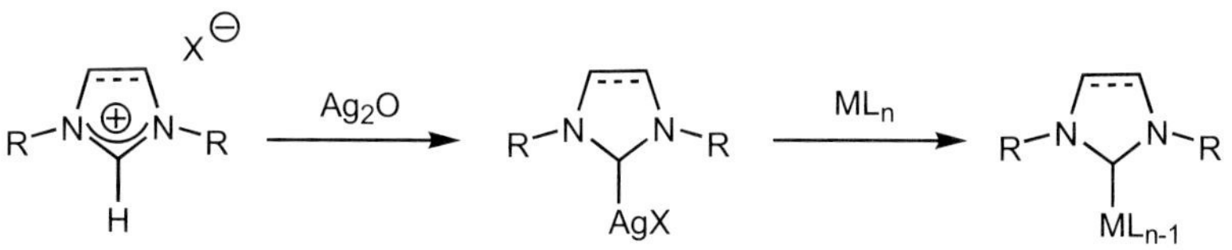

Scheme 8.5

8.2.3
Concept and Design of Chiral NHCs

The development and design of useful chiral ligands is generally guided by a few concepts. For example, the presence of a C_2-symmetric axis within the ligand architecture can serve to reduce significantly the number of possible competing diastereomeric transition states that play a crucial role in the stereoselective induction [13]. This concept was successfully reported through the development of chiral diphosphines (such as DIOP [14] and BINAP [15]), the bisoxazolines [16] and salen [17] derivatives, where high enantioselectivities are obtained in many asymmetric reactions (Fig. 8.3).

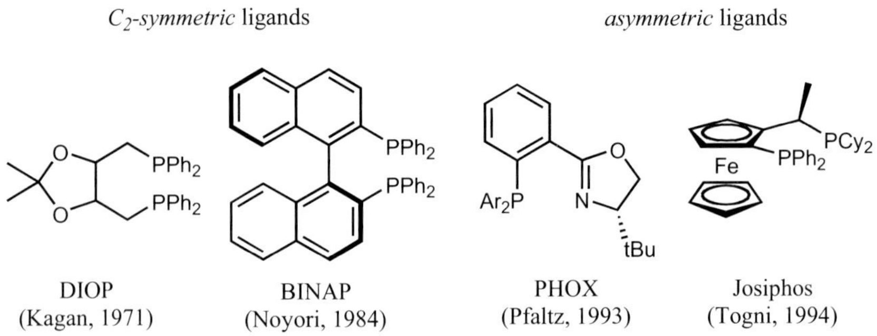

8 Asymmetric Catalysis with Metal N-Heterocyclic Carbene Complexes

C_2-*symmetric* ligands   *asymmetric* ligands

DIOP (Kagan, 1971)   BINAP (Noyori, 1984)   PHOX (Pfaltz, 1993)   Josiphos (Togni, 1994)

Fig. 8.3 Useful chiral ligands in asymmetric catalysis.

However, establishing a relationship between the ligand structure and the enantiocontrol of the catalytic transformation by postulating rules is not straightforward, and it is even more difficult to anticipate or predict the degree of stereoselectivity. If, as mentioned above, the C_2-symmetric concept is largely vindicated in asymmetric catalysis, *asymmetric* ligands such as those of Pfaltz [18] or Togni [19], which are phosphine-based ligands possessing a high degree of dissymmetry, have also proven to be excellent stereo-directing ligands. Consequently, the design of an efficient ligand is often guided by chemical intuition and followed, in most of cases, by the systematic screening of small ligand libraries to understand and to improve stereoselective induction [20].

The design of chiral NHC ligands has, logically, been inspired by work previously performed on chiral phosphine homologues that possess similar electronic donor properties. However, NHCs and phosphines architectures have quite different topologies. As emphasized by Burgess et al. in the first review on chiral-NHC complexes [20], the way to design chiral diphosphines is not directly applicable to the NHC unit. For example, the common topological feature of the aromatic substituents of chiral diarylphosphines is as an "edge to face" privileged orientation [21], which is impossible with the molecular architecture of an NHC as it has a planar chelation site. Consequently, replacement of the phosphorus unit by NHC in well-defined useful phosphino-derivative ligands is not necessarily fruitful. As a direct illustration (Fig. 8.4), the NHC-containing Josiphos analog **9** [22] was far less stereoselective in Rh-catalyzed hydrogenation of dimethylitaconate than the original Josiphos while the NHC-JM-Phos analog **10** [23] developed by Burgess and coworkers gave similar high enantioselectivities to the original JM-Phos [24] in the Ir-catalyzed hydrogenation of arenes.

8.2 Concept, Design and Synthesis of Chiral NHC Complexes

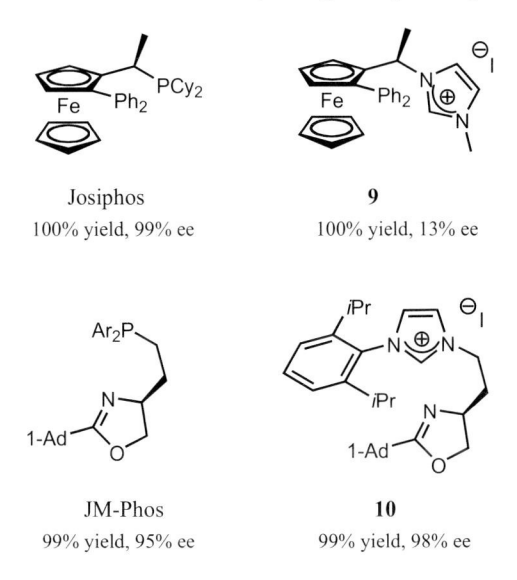

Josiphos
100% yield, 99% ee

9
100% yield, 13% ee

JM-Phos
99% yield, 95% ee

10
99% yield, 98% ee

Fig. 8.4 Phosphino mimics strategy.

Therefore, the key structural features in the NHC unit have to be further investigated and be clearly identified. Since the pioneering work of Herrmann and Enders in 1996, our understanding of stereocontrol in catalysis has continually progressed through the emergence of a growing number of chiral NHC complexes. Depending on the nature and the relative position of the chiral motif in the carbene framework, it is possible to distinguish four different classes of chiral NHC ligands used in asymmetric catalysis: (a) monodentate and bidentate NHCs possessing stereogenic centers on the nitrogen substituents, (b) monodentate and bidentate NHCs possessing stereogenic centers within the N-heterocycle, (c) monodentate and bidentate NHCs containing an element of planar chirality, and finally (d) bidentate NHCs containing an element of axial chirality. Figures 8.5 to 8.11 summarize these different classes of chiral NHCs through their corresponding azolinium salts precursors.

Application of chiral NHC ligands in asymmetric catalysis is continually under progress and, for instance, several metal-catalyzed transformations have been revisited: (a) asymmetric hydrogenation, (b) asymmetric 1,4-addition, (c) asymmetric 1,2 addition, (d) asymmetric hydrosilylation, (e) asymmetric olefins metathesis, (f) asymmetric allylic substitution, (g) asymmetric amide α-arylation and finally (h) catalytic kinetic resolution. Although most chiral NHC complexes show excellent catalytic activity, only a few have displayed high enantiocontrol. The following sections overview these applications in asymmetric catalysis.

Herrmann (1996) [8]

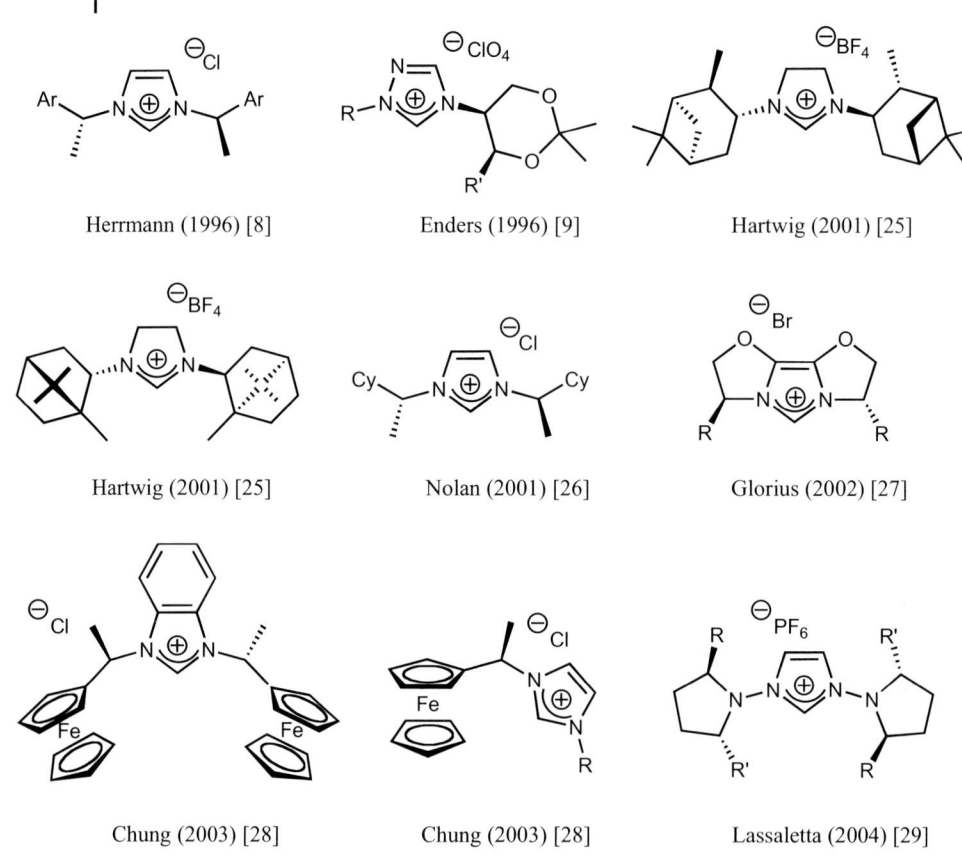

Enders (1996) [9]

Hartwig (2001) [25]

Hartwig (2001) [25]

Nolan (2001) [26]

Glorius (2002) [27]

Chung (2003) [28]

Chung (2003) [28]

Lassaletta (2004) [29]

Fig. 8.5 Monodentate NHC ligand precursors possessing stereogenic centers on the nitrogen substituents.

8.2 Concept, Design and Synthesis of Chiral NHC Complexes

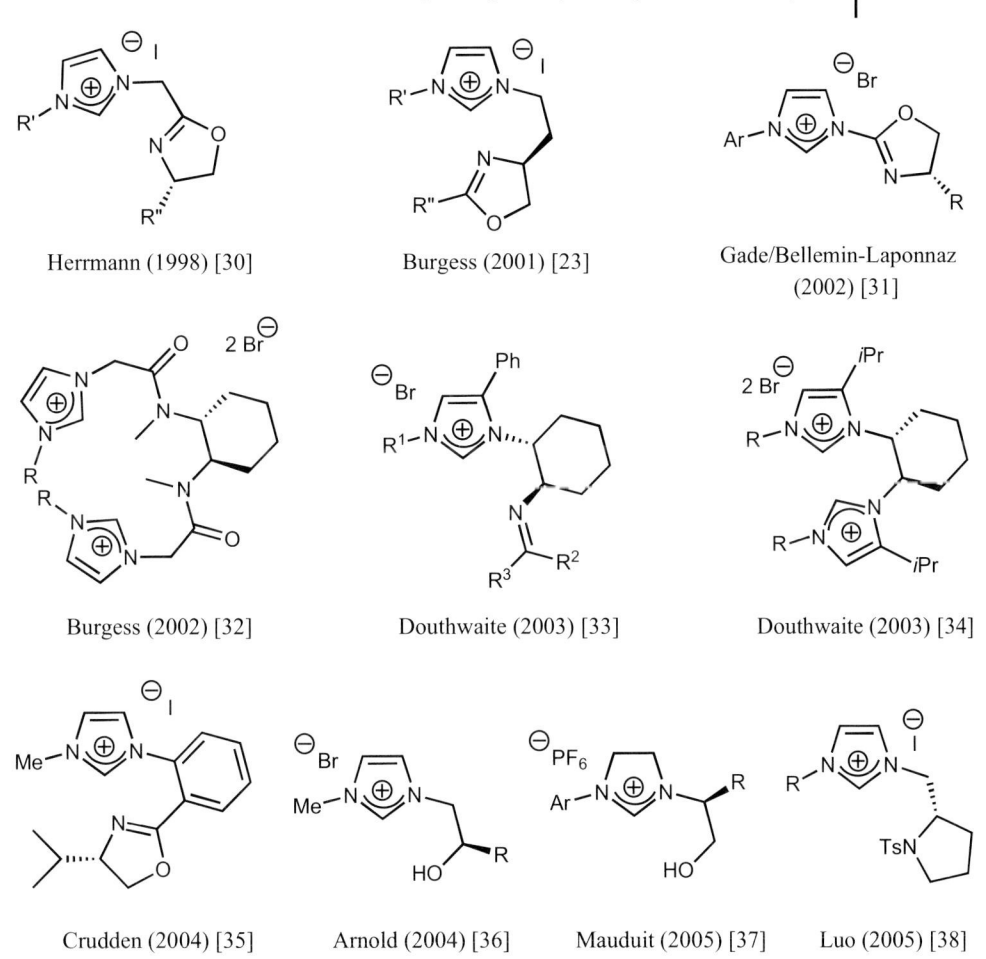

Herrmann (1998) [30]

Burgess (2001) [23]

Gade/Bellemin-Laponnaz (2002) [31]

Burgess (2002) [32]

Douthwaite (2003) [33]

Douthwaite (2003) [34]

Crudden (2004) [35]

Arnold (2004) [36]

Mauduit (2005) [37]

Luo (2005) [38]

Fig. 8.6 Bidentate NHC ligand precursors possessing stereogenic centers on the nitrogen substituents.

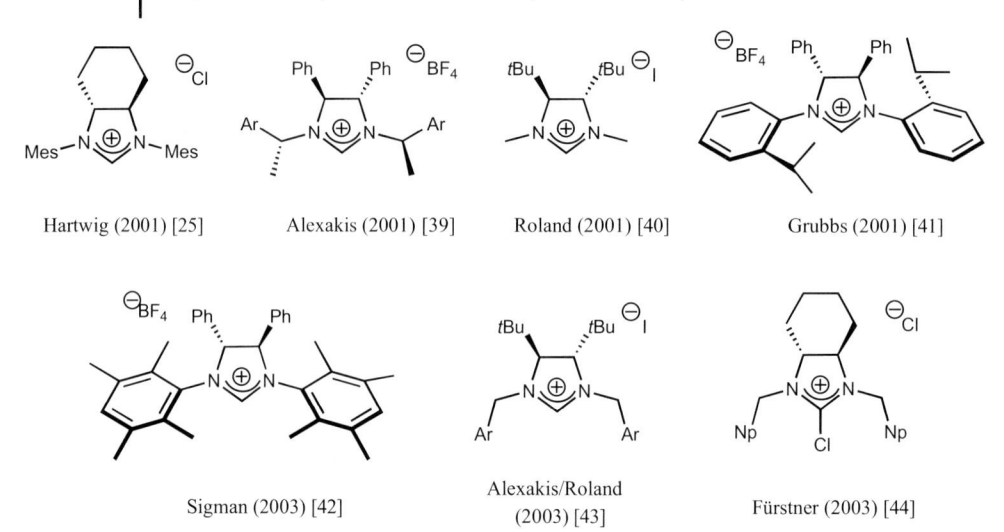

8 Asymmetric Catalysis with Metal N-Heterocyclic Carbene Complexes

Hartwig (2001) [25]   Alexakis (2001) [39]   Roland (2001) [40]   Grubbs (2001) [41]

Sigman (2003) [42]   Alexakis/Roland (2003) [43]   Fürstner (2003) [44]

Fig. 8.7 Monodentate NHC ligand precursors possessing stereogenic centers within the N-heterocycle.

Helmchen (2004) [45]   Hoveyda (2005) [46]

Fig. 8.8 Bidentate NHC ligand precursors possessing stereogenic centers within the N-heterocycle.

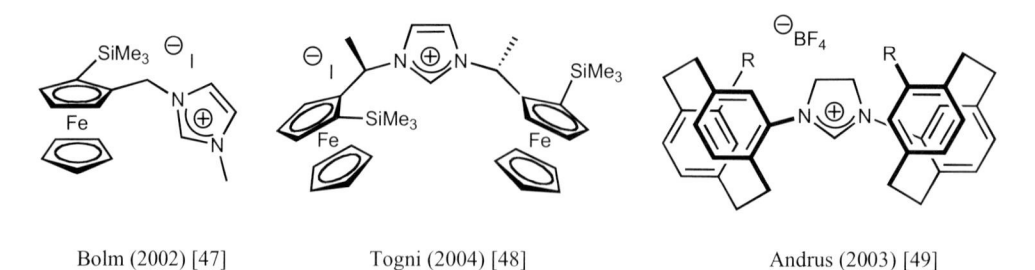

Bolm (2002) [47]   Togni (2004) [48]   Andrus (2003) [49]

Fig. 8.9 Monodentate NHC ligand precursors containing an element of planar chirality.

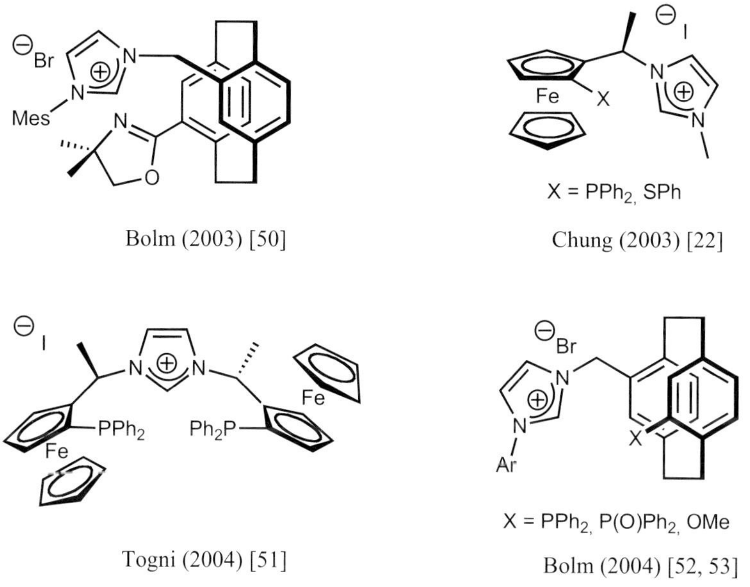

Bolm (2003) [50]

Chung (2003) [22]

X = PPh$_2$, SPh

Togni (2004) [51]

Bolm (2004) [52, 53]

X = PPh$_2$, P(O)Ph$_2$, OMe

Fig. 8.10 Bidentate NHC ligand precursors containing an element of planar chirality.

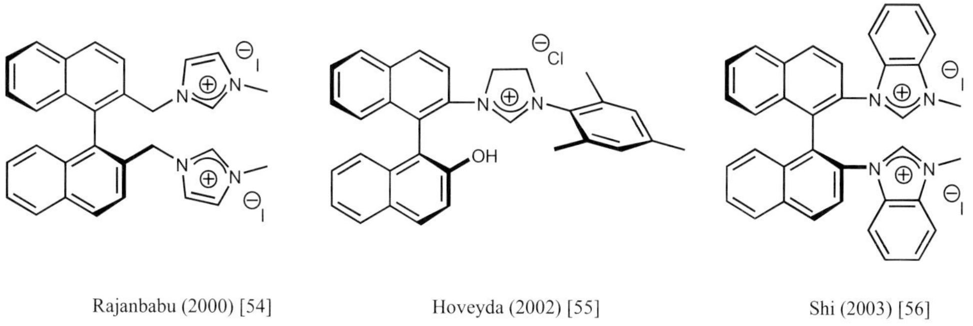

Rajanbabu (2000) [54]

Hoveyda (2002) [55]

Shi (2003) [56]

Fig. 8.11 Bidentate NHC ligand precursors containing an element of axial chirality.

8.3
Asymmetric Hydrogenation

Asymmetric hydrogenation of olefins catalyzed by chiral organometallic complexes is a well-established and widely used method to easily generate stereogenic centers with a very high degree of enantiocontrol. The first to develop this catalytic transformation using chiral NHC ligand was Burgess in 2001 [23] with the bidentate oxazoline-NHC iridium complex **11** (Scheme 8.6). Initially inspired by the Pfaltz/Hemlchen/Williams *asymmetric* phosphino-oxazoline (PHOX) ligands [18] that had proven to be useful in the iridium-catalyzed hydrogenation of unfunc-

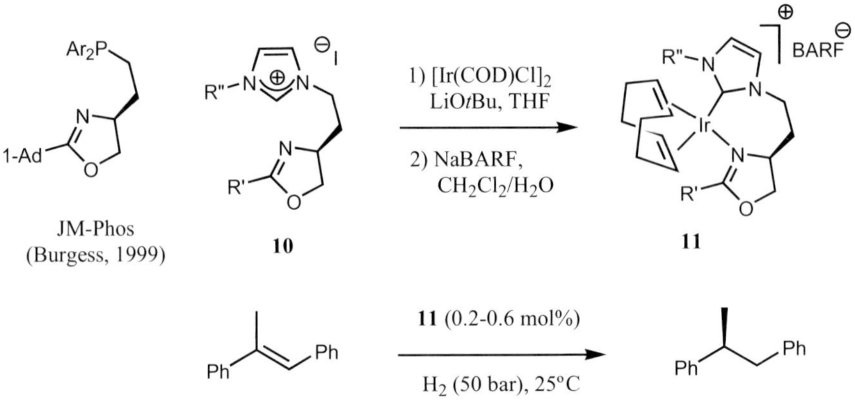

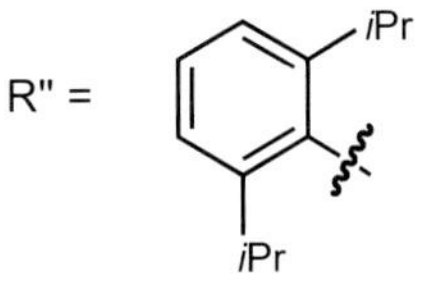

8 Asymmetric Catalysis with Metal N-Heterocyclic Carbene Complexes

Scheme 8.6

	R'	yield (%)	ee (%)
11a	Ph	25	13
11b	tBu	81	50
11c	1-Ad	99	98
11d	CHPh$_2$	12	25

tionalized alkenes, the design of the carbene ligand **10** bearing a chiral oxazoline unit is directly connected to their previous study concerning the JM-phos analogue [24]. Replacement of the phosphine unit by a NHC motif leads to a new and efficient class of bidentate ligand for iridium-catalyzed hydrogenation. Moreover, the easily and modular access of the NHC-oxazoline ligands allowed the screening of a large library through variation of the substituents at the N-heterocyclic carbene and at the 2-position of the oxazoline ring. Scheme 8.6 illustrates the screening of iridium complexes **11** in the enantioselective hydrogenation of (E)-1,2-diphenylpropene, showing clearly the importance of ligand design. The iridium complex **11c**, bearing an adamentyl oxazoline group and a 2,6-diisopropylphenyl as N-bulky substituent is the most efficient catalyst, reaching 98% e.e. and a yield of 99%.

This new class of NHC-oxazoline iridium complex (**11c**) enables, in a major improvement, the asymmetric hydrogenation at room temperature and ambient pressure of H$_2$ with high enantioselectivities (Scheme 8.7).

Recently, Burgess and coworkers [57] have extended their studies of the best NHC-oxazoline iridium complex **11c** and investigated applications in the hydrogenation of aryl-substituted dienes, which are relatively difficult substrates to hydrogenate with high stereocontrol. Good e.e.s (up to 99%) can be obtained

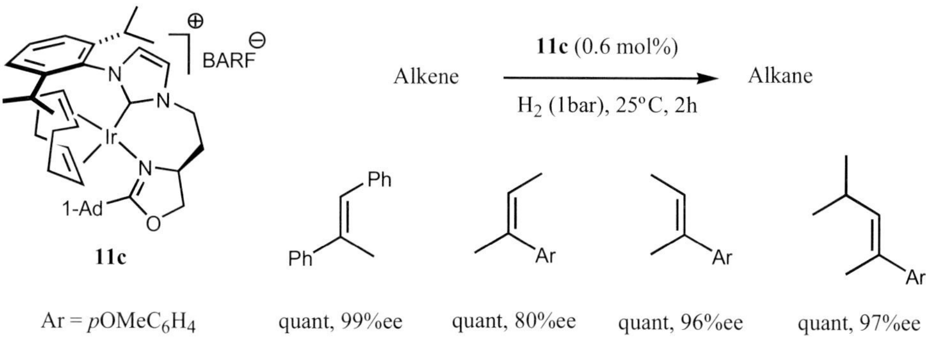

8.3 Asymmetric Hydrogenation

Scheme 8.7

Ar = pOMeC$_6$H$_4$

Alkene → Alkane, 11c (0.6 mol%), H$_2$ (1 bar), 25°C, 2h

- Ph/Ph: quant, 99%ee
- Ar (trisubstituted with methyls): quant, 80%ee
- Ar: quant, 96%ee
- Ar: quant, 97%ee

for three different classes of dienes (A–C) and, remarkably, excellent ent:meso diastereoselectivities (up to 20:1.0) were observed with diene C (Scheme 8.8). Importantly, the iridium complex **11c** shows better activity than Crabtree's Ir(py)(Pcy$_3$)(COD)PF$_6$ complex [58], the most widely recognized homogeneous catalyst for hydrogenation of unfunctionalized and highly hindered alkenes.

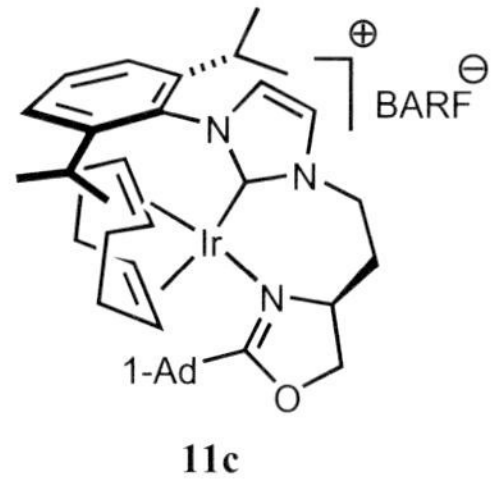

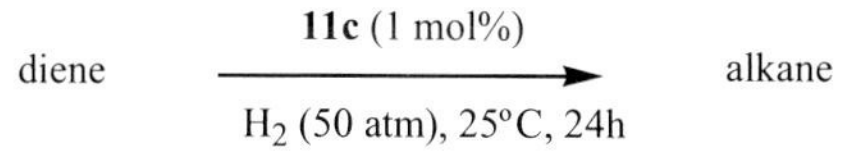

Scheme 8.8

diene → alkane, 11c (1 mol%), H$_2$ (50 atm), 25°C, 24h

diene	Conv. (%)	Yield (%)	ent:meso	ee (%)
A	100	96	1.0:2.9	87
B	100	69	1.3:1.0	98
C	100	100	20:1.0	99

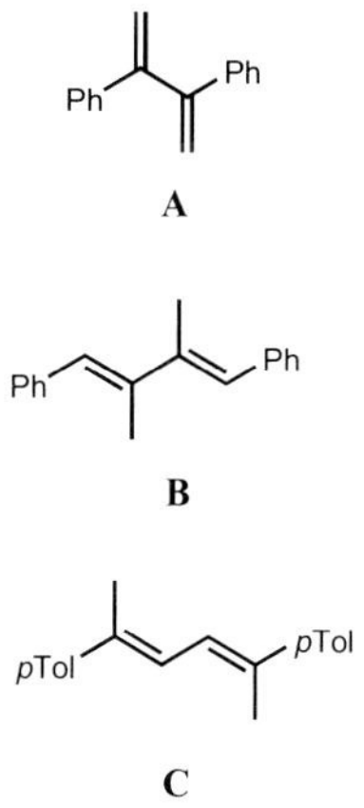

A: Ph/Ph diene
B: Ph/Ph diene
C: pTol/pTol diene

Chung and coworkers reported in 2003 the asymmetric hydrogenation of dimethylitaconate catalyzed by chiral bidentate ferrocenyl-NHC iridium and rhodium complexes **16–18** (Scheme 8.9) [28]. Directly inspired by the original Togni Josiphos ligand [19], the NHC ligands possess two different elements of symmetry, a planar chirality in the ferrocenyl motif possessing a chelating function (as thioether or phosphinane) and a chiral center at the carbon atom anchoring the NHC and the ferrocene. Their rhodium and iridium complexes **12–14** were characterized by X-ray diffraction, showing that bidentate S,C$_{carb}$ and P,C$_{carb}$ coordination is only effective for the rhodium transition metal. Owing to the presence of two NHC units coordinated to the metal, complexes **13** and **14** were inactive in the enantioselective hydrogenation while rhodium complex **12** reached 18% e.e. with a moderate conversion. The activity was improved by the *in situ* formation of

196 | 8 Asymmetric Catalysis with Metal N-Heterocyclic Carbene Complexes

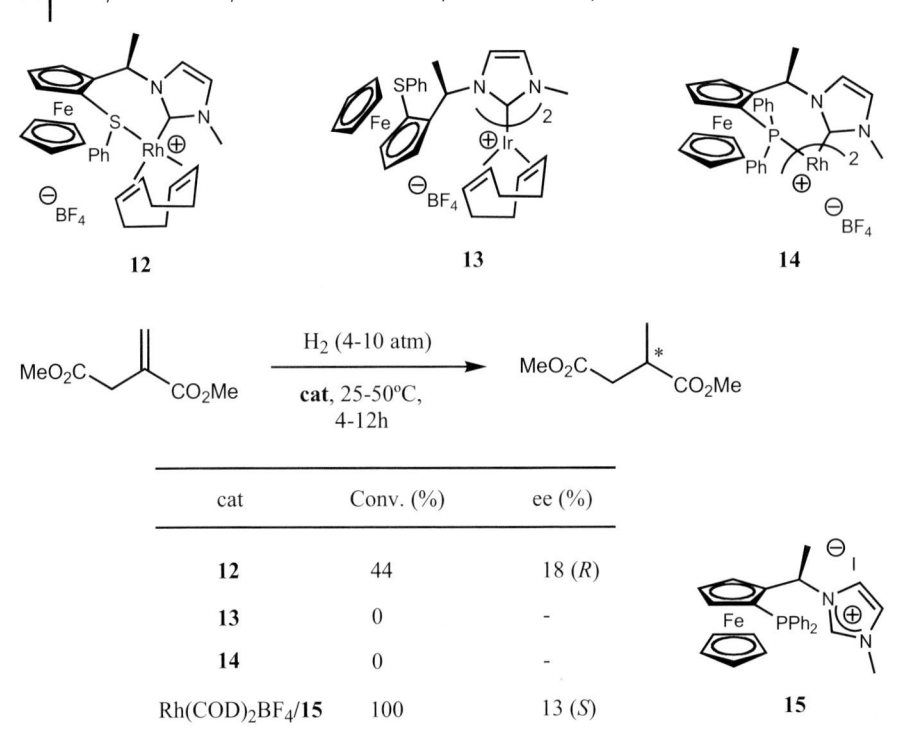

Scheme 8.9

cat	Conv. (%)	ee (%)
12	44	18 (R)
13	0	-
14	0	-
Rh(COD)$_2$BF$_4$/15	100	13 (S)

P,C$_{carb}$ chelated Rh-COD catalyst from the imidazolium salt **15**, with a complete conversion but a lower stereoselectivity.

Chung's group also investigated the monodentate version of these ferrocenyl based-NHC ligands in the catalytic transfer hydrogenation of 4'-methylacetophenone [28]. However, their corresponding rhodium complexes **16** and **17** gave very low stereoselectivities while a 52% e.e. is reached with iridium-catalyst **18**, which contains a chiral C_2-symmetric N-heterocyclic carbene (Scheme 8.10).

Other NHC-iridium complexes (**19**), based on the planar chiral pseudo-orthodisubstituted paracyclophane skeleton bearing an oxazoline moiety and imidazolylidene unit, have been reported by Bolm and coworkers [50]. Application of these catalysts to the hydrogenation of functionalized and simple olefins gave low to moderate enantioselectivity. The highest selectivity is obtained with dimethylitaconate (46% e.e.) with catalyst **19c** (Scheme 8.11). Interestingly, steric properties of the substituents on both chelating units play a major role in the activity of catalyst. Indeed, a sterically bulky group on the ligand (a mesityl unit to the imidazole ring (complex **19b**) or a tert-butyl group on the oxazoline (complex **19c**) is crucial to obtain an acceptable activity of the iridium catalyst. Both activity and selectivity should therefore be improved with the combination of an N-mesityl on the imidazolylidene and a tert-butyl group on the oxazoline moiety but, unfortunately, this ideal NHC could not be synthesized.

8.3 Asymmetric Hydrogenation | 197

cat	Yield (%)	ee (%)
16	>99	6.4 (R)
17	>99	6.7 (R)
18	98	52.6 (S)

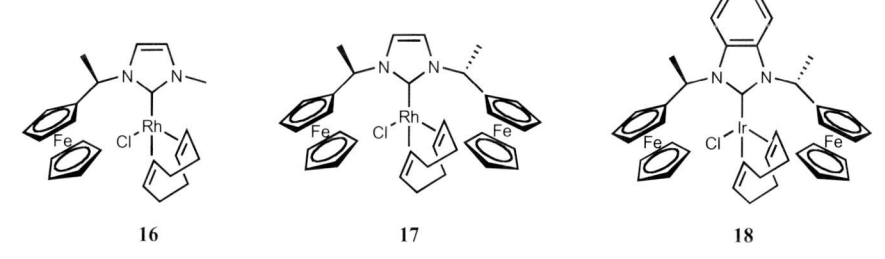

Scheme 8.10

(S_p)-19a R'=R"=R'''= Me
(S_p)-19b R'=R"=Me, R'''= Mes
(S,S_p)-19c R'=tBu, R"=H, R'''= Me

alkene	cat	p(H$_2$) (bar)	T (°C)	Conv. (%)	ee (%)
A	19a	50	50	40	0
	19b	50	50	76	15 (R)
	19c	50	50	61	4 (R)
B	19b	50	50	100	4
C	19b	50	25	100	9 (R)
C	19b	1	25	46	46 (R)

Scheme 8.11

Since the combination of oxazoline and chiral planar paracyclophane appears fruitless, Bolm and coworkers have extended these studies through the preparation of a novel planar chiral paracyclophane imidazolylidene bearing a phosphino substituent instead of the oxazoline unit [52]. Containing only one element of symmetry as the planar chirality, iridium complexes **20** have been applied successfully in the catalyzed hydrogenation of several alkenes, giving products in higher enantioselectivities (Scheme 8.12). For unfunctionalized olefins, the most selective catalyst was **20a**, bearing a phenyl substituent at the imidazole ring (up to 82% e.e. for (*E*)-1,2-diphenylpropene), while catalyst **20b**, containing a more bulky substituent, gave lower enantioselectivities. Inversely, in the hydrogenation of functionalized alkenes, the best e.e.s are obtained with the highly hindered catalyst **20b**, reaching 89% with dimethylitaconate. This is another example in which the imidazole substituent appears to play an important role in controlling the activity and the stereoselective induction of the chiral NHC-complexes.

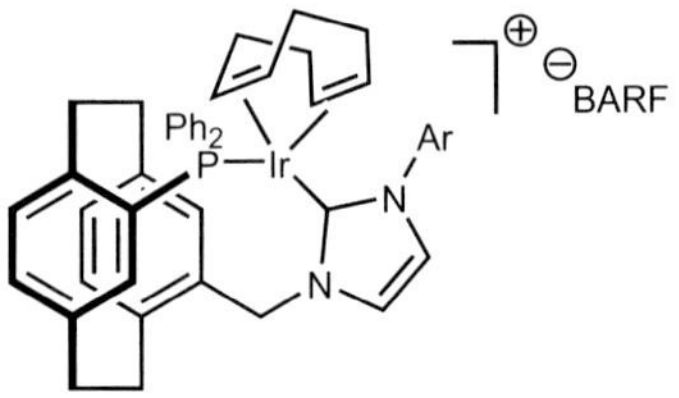

alkene	cat	Conv. (%)	ee (%)
A	20a	89	82 (*R*)
	20b	41	22 (*S*)
B	20a	100	79 (*R*)
	20b	99	37 (*R*)
C	20a	100	3 (*S*)
	20b	100	89 (*R*)

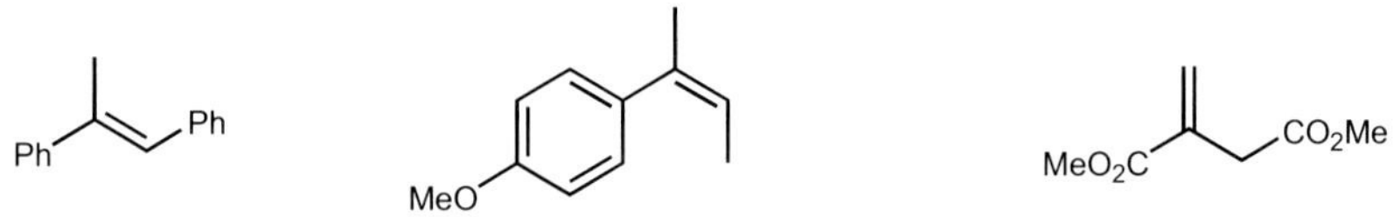

alkene → alkane; **20a-b** (1 mol%), H$_2$ (50 bar), 25°C

(*R*$_p$)-**20a** Ar = Ph
(*R*$_p$)-**20b** Ar = 2,6-*i*Pr$_2$Ph

Scheme 8.12

Finally, the novel NHC-rhodium complex **21**, in which the bidentate diphenylphosphino-imidazolylidene ligand contains the 1,2-diphenylethylendiamine skeleton in combination with ortho-monosubstituted N-aryl substituents in the imidazoline ring, has been recently studied by Helmchen and coworkers (Scheme 8.13) [45]. The stereo-directing ligand is mainly guided by steric repulsion between the o-aryl substituents and the backbone phenyl groups in which their mutual anti-conformation is preferred, permitting therefore efficient transmission of the

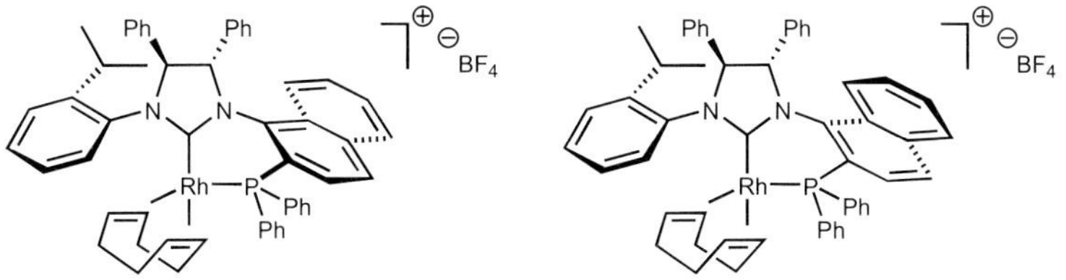

Scheme 8.13

(S,S)-21 (1:2 mixture of atropoisomers)

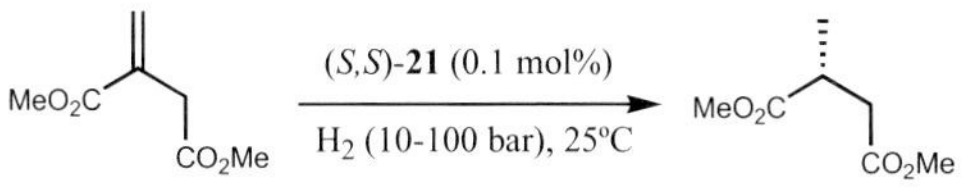

H_2 (bar)	Conv. (%)	ee (%)
10	19	89 (R)
20	>99	98 (R)
50	79	97 (R)
100	49	63 (R)

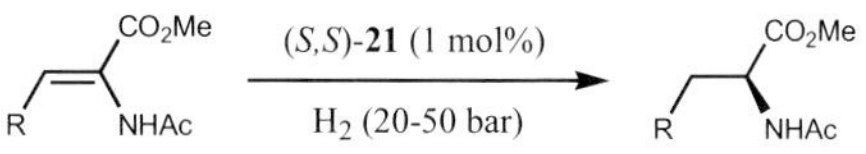

R	H_2 (bar)	T (°C)	Conv. (%)	ee (%)
H	20	25	50	98 (S)
H	20	70	100	99 (S)
H	50	70	100	99 (S)
Ph	30	70	97	97 (S)
Ph	50	25	72	99 (S)
Ph	50	70	100	97 (S)

chiral information to the active site of catalyst. NHC-rhodium complex **21**, where two atropisomers co-exist in a ratio of 1:2, was evaluated in the catalytic hydrogenation of functionalized olefins. For dimethylitaconate, the hydrogen pressure influences both the activity and the selectivity of the reaction – optimum yields and 98% e.e. were observed with a catalyst loading of 0.1 mol% at ambient temperature under a H_2 pressure of 20 bar. For methyl (Z)-acetamidoacrylate, 1 mol% catalyst loading is required and total conversion with 99% enantioselectivity could be obtained at 70 °C and 50 bar of H_2. Concerning methyl (Z)-acetamidocinnamate, the selectivity was practically independent of temperature and pressure.

8.4 Asymmetric 1,4-Addition

1,4-Conjugate addition of nucleophilic reagents to α,β-unsaturated compounds is one of the most powerful methods for the formation of carbon–carbon bonds in organic synthesis [59]. Owing to the importance of the metal-catalyzed conjugate addition as a versatile synthetic tool, considering the wide variety of acceptors and donors that can be used, attempts to realize a catalytic asymmetric version of this reaction were, unsurprisingly, made at a very early stage in organometallic chemistry.

8.4.1
Copper-NHC Complexes

In the 1990s, increasing interest in the catalytic asymmetric 1,4-addition led to the development of the enantioselective copper-catalyzed 1,4-addition of dialkylzinc reagents to enones [60]. Since the original experiment reported by the group of Alexakis in 1993 (32% e.e. using an ephedrine-based phosphorus ligand) [61], various successful catalytic systems, often based on phosphorus ligands, have been developed [62]. Arduengo and coworkers initially isolated stable copper complexes bearing an NHC unit in 1993 but their catalytic properties were not investigated [63]. An important contribution in the use of NHC ligands in copper-catalyzed conjugate addition was made by Woodward and coworkers in 2001 [64]. They effectively demonstrated that a copper complex containing an Arduengo carbene strongly accelerates the addition of diethylzinc to cyclohexenone. An asymmetric version of this addition using chiral NHC ligands was published independently by Alexakis [39] and Roland [40] in 2001. Based on the C_2-symmetry element containing the 1,2-di-*tert*-butylethylendiamine skeletons, the silver diaminocarbenes **22** developed by Roland catalyses efficiently the addition of diethylzinc to cyclohexenone in the presence of the Cu(OTf)$_2$, but the enantioselectivity only reached 23% e.e. (Scheme 8.14). Similar activity and selectivity (27% e.e.) were observed by Alexakis with the imidazolinium salt **23**, which possess phenyl groups instead of the *tert*-butyl in the backbone, despite the presence of a supplementary chiral N-substituents. Interestingly, the selectivity increased when the addition was performed at lower temperature with **23**, reaching a maximum of 50% e.e.

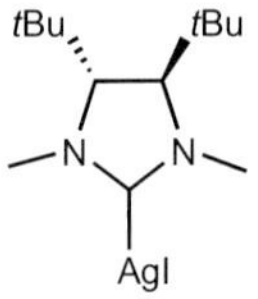

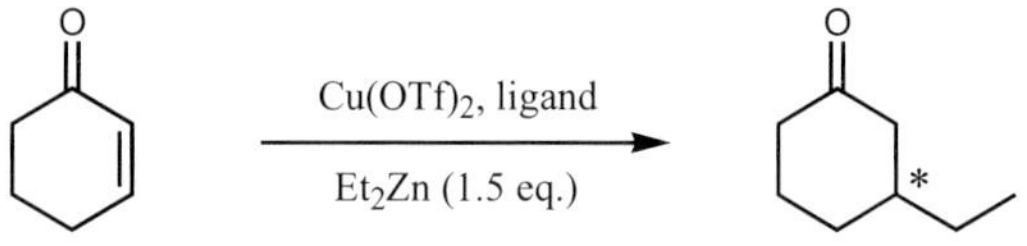

Ligand	T (°C)	Time (h)	Yield (%)	ee (%)
22	0	0.15	98	23 (S)
23	-20	2	90	27 (S)
23	-78	16	91	50 (S)

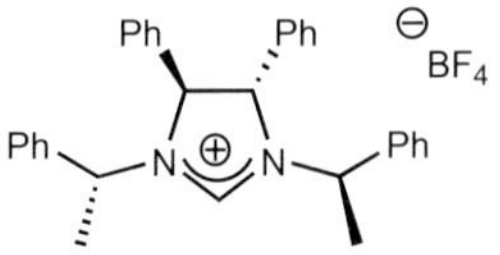

Scheme 8.14

The groups of Alexakis and Roland have jointly extended their studies of monodentate NHC by several slight modifications of both the experimental conditions and the ligand structure [43]. Firstly, use of silver carbene **24a** instead of its imidazolium salts carbene precursors gave increased selectivity of up 62% e.e. (while only 54% e.e. is reached with the corresponding salt). This improved selectivity could be explained by the incomplete *in situ* formation of the azolium salt to the corresponding carbene ligand. Moreover, there were some advantages in using the silver NHC ligands: they are stable, readily available in high yields and easy to handle. Use of the more bulky chiral silver carbene **25** derived from the original Herrmann imidazolium salt did not improve selectivity (59% e.e.). Introduction of N-benzyl substituents instead of the methyl groups in the backbone of the ligand **22** increased the stereoselectivity for cyclohexenone. A maximum of 69% e.e. is reached with the silver carbene **26b**, which bears methoxy substituents in the meta positions of the phenyl rings. Here, steric repulsions between the N-substituents and the backbone *tert*-butyl groups permit a C_2-symmetric privileged orientation of the benzylic moieties and, therefore, lead to an efficient chirality transfer from the backbone to the active metal site. Various other substrates were systematically screened with these silver diaminocarbenes. The highest selectivity was found with cyclohepten-2-one (93% e.e.) while the addition to acyclic Michael acceptors gave moderate selectivity, ranging from 42 to 75% e.e. (Scheme 8.15).

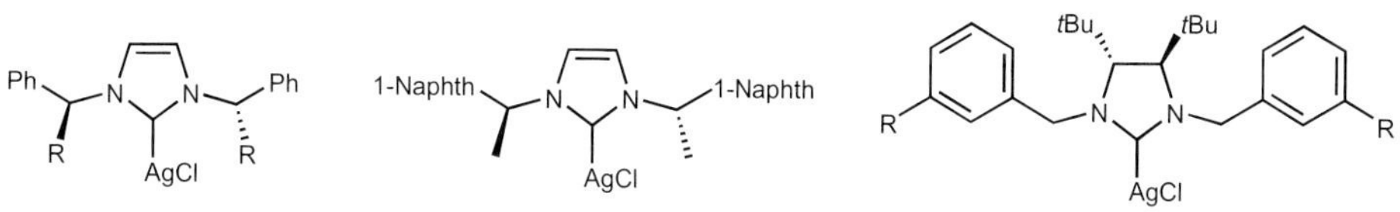

Enone	Ligand	Cu salt	Yield (%)	ee (%)
A	24a	CuTC	99	62 (S)
A	24b	Cu(OAc)$_2$	92	55 (S)
A	25	Cu(OAc)$_2$	87	59 (S)
A	26a	CuTC	100	58 (S)
A	26b	CuTC	99	69 (S)
B	25	Cu(OAc)$_2$	95	93 (S)
C	24a	CuTC	67	42 (S)
D	24a	Cu(OAc)$_2$	100	75 (R)

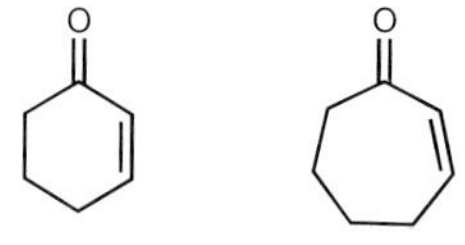

Scheme 8.15

In 2004, Arnold and coworkers described the first chiral bidentate alkoxy-imidazolydene-copper(II) complex, which contains a stereogenic center directly bound to the chelating alkoxy side chain [36]. Bearing a *tert*-butyl as sterically hindered group, the copper complex **27** catalyzed very efficiency the addition of diethylzinc to cyclohex-2-enone within 2 h at −30 °C while only 51% enantioselectivity is reached (Scheme 8.16).

Scheme 8.16

Recently, Mauduit and coworkers have also developed a new class of bidentate alkoxy-imidazolylidene ligand (**28**) [37], which differ from the Arnold NHC ligand by the position of the stereogenic center in the ethyloxy side chain. Unexpectedly, the selectivity of the addition is improved when the chiral center is linked to the nitrogen of the imidazoline ring. Whereas the nature of the alkyl group on the chiral sp³ carbon induced a slight variation of stereoselectivity for cyclohexenone, a close relationship between temperature and selectivity was found and, curiously, optimal e.e.s were observed when the conjugate addition is performed at ambient temperature, with e.e.s ranging from 85 to 87% (Scheme 8.17). More interestingly, the best enantioselectivity (89% e.e.) is reached without loss of activity by the use of inexpensive and easier to handle organic bases such as DBU instead of nBuLi (with which most organic solvents are incompatible) [65].

A wide range of enones and dialkylzinc reagents were also tested [65], showing interesting enantioselectivities of up to 93% with cyclic enones (Scheme 8.18). However, unfortunately, the addition to acyclic Michael acceptors using these alkoxy-NHCs was unfruitful, giving poor enantioselectivities (up to 37%).

28a R= Ph   **28b** R= tBu   **28c** R= iBu

Ligand	Base	Solvent	Yield (%)	ee (%)
28a	nBuLi	Et$_2$O	>99	86 (S)
28b	nBuLi	Et$_2$O	>99	87 (S)
28c	nBuLi	Et$_2$O	>99	86 (S)
28c	DBU	Et$_2$O	>99	87 (S)
28c	DBU	Toluene	>99	**89 (S)**
28c	DBU	THF	>99	**89 (S)**
28c	DBU	CH$_2$Cl$_2$	>99	82 (S)
28c	DBU	AcOEt	>99	85 (S)

Relationship between the temperature and the selectivity
catalyst: Cu(OTf)$_2$/ **28c**

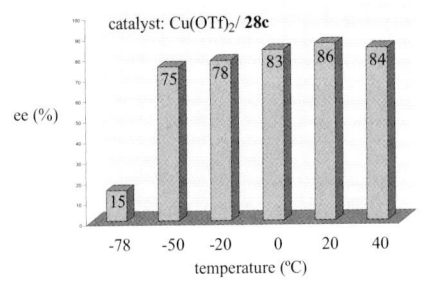

Scheme 8.17

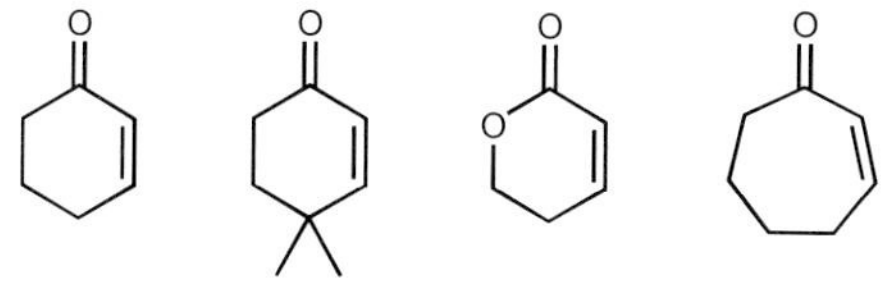

Scheme 8.18

Enone	Alkylzinc	Time (h)	Yield (%)	ee (%)
A	Me$_2$Zn	3	>99	88 (S)
A	iPr$_2$Zn	1	>99	79 (S)
B	Et$_2$Zn	12	>99	93 (S)
C	Et$_2$Zn	1	>99	72 (S)
D	Et$_2$Zn	0.5	>99	90 (S)

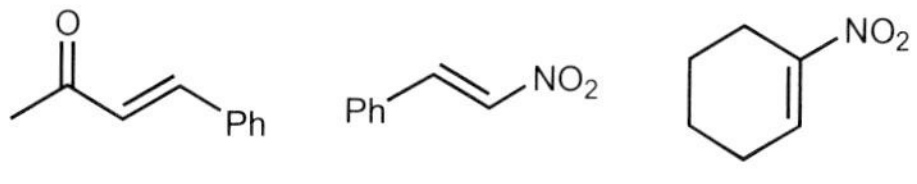

Enone	Alkylzinc	Time (h)	Yield (%)	ee (%)
E	Et$_2$Zn	1	>99	34 (S)
F	Et$_2$Zn	1	>99	21 (S)
G	Et$_2$Zn	1	>99	37 (S)

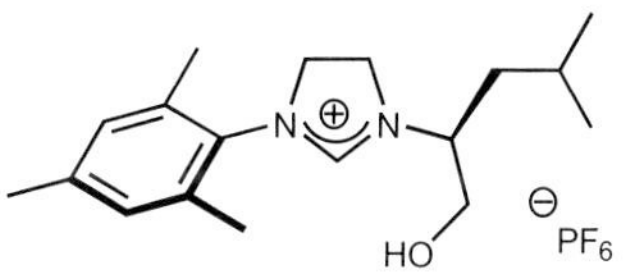

8.4.2
Rhodium-NHC Complexes

One limitation of the copper-catalyzed asymmetric conjugate addition of alkylzinc reagents is that only aliphatic organozinc reagents can be successfully used. The introduction of aryl or alkenyl groups has proved difficult. The rhodium-BINAP catalyzed conjugate addition of organoboronic acids to enones, developed by Miyaura and Hayashi in 1998, is a remarkable alternative in the introduction of these groups, with high yields and excellent selectivities [66]. At the end of 2003, Andrus and coworkers reported the fruitful application of chiral monodentate NHC ligands in this area [49]. Using novel imidazolinium salts **29**, containing paracyclophanes as chiral bulky N-substituents, in the presence of [Rh(acac)(C$_2$H$_4$)$_2$] enables efficient catalysis of the reaction of phenyl boronic acid and cyclohex-2-enone at 60 °C in THF/water in high yields; higher temperatures are required with the original [Rh(acac)(C$_2$H$_4$)$_2$]/BINAP catalytic system (Scheme 8.19). This improvement in activity can be linked directly to a previous communication by Fürstner's group in 2001 of the use of achiral NHC ligands in the 1,2-addition of boronic acid to aldehyde, showing the beneficial effect of bulky N-substituents [67]. Concerning the stereoselectivity of the reaction, the best e.e. was found with the rhodium catalyst derived from ligand **29d**, bearing anisyl units on the paracyclophane motif (98% e.e.).

A wide variety of arylboronic acids and potassium trifluoroborates have been screened in the addition to cyclic enones with this optimized catalyst Rh/**29d**, and excellent yields and selectivities were generally observed (Scheme 8.20). Extension of this procedure to acyclic enones was also explored, giving high yields; however, the selectivity was significantly lower (e.e.s ranging from 73 to 91%).

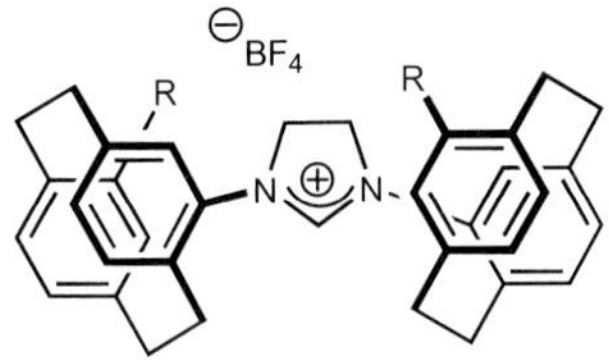

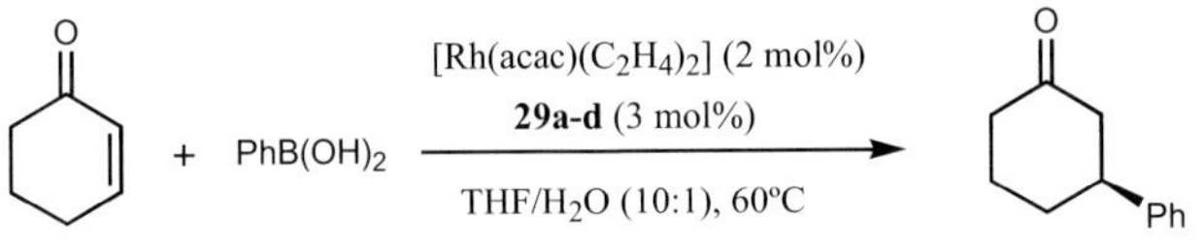

29a R= H
29b R= Ph
29c R= Cy
29d R= o-anisyl

Ligand	Yield (%)	ee (%)
29a	69	61 (S)
29b	86	91 (S)
29c	89	97 (S)
29d	96	98 (S)

Scheme 8.19

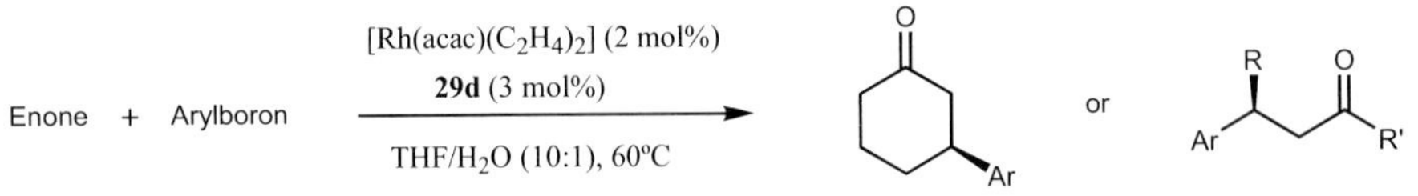

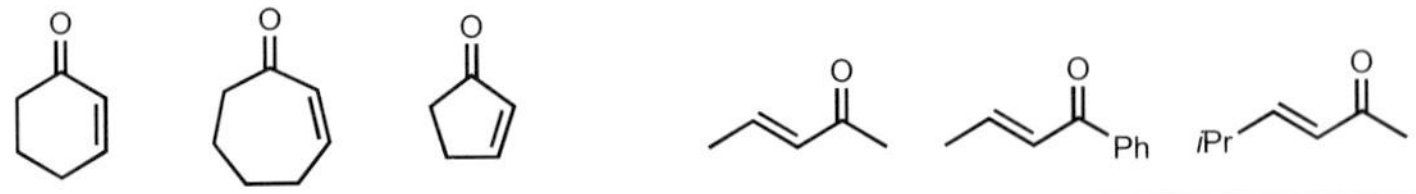

Enone + Arylboron → [Rh(acac)(C$_2$H$_4$)$_2$] (2 mol%), **29d** (3 mol%), THF/H$_2$O (10:1), 60°C

		cyclic enone			acyclic enone		
Arylboron		cyclohexenone	cycloheptenone	cyclopentenone	Me-enone	Ph-enone	iPr-enone
PhB(OH)$_2$	time	3h	3.5h	3.5h	2.5h	2.5h	2.5h
	Yield (%)	96	94	97	95	92	83
	ee (%)	98	93	92	81	78	91
PhBF$_3$K	time	1h	1h	1h	45 min.	45 min.	45 min.
	Yield (%)	96	93	95	96	95	94
	ee (%)	91	85	90	80	73	81
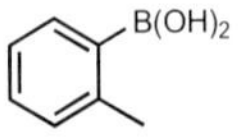 o-tolyl-B(OH)$_2$	time	3h	3.5h				
	Yield (%)	90	90				
	ee (%)	93	85				
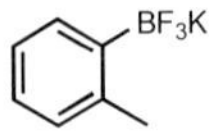 o-tolyl-BF$_3$K	time	1h	45 min.				
	Yield (%)	95	94				
	ee (%)	90	87				

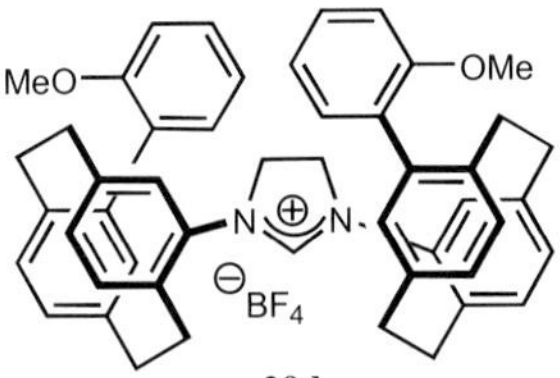

29d

Scheme 8.20

8.4.3
Palladium-NHC Complexes

Based on a previous study on the triphosphine Pigiphos [68], Togni and coworkers have synthesized the first C_2-symmetric tridentate PCP ligand in which two different elements of symmetry co-exist, a planar chirality in the ferrocenyl motif possessing a chelating phosphine function and a chiral center directly anchored to the nitrogen of the carbene unit [48]. Their palladium and ruthenium complexes were isolated and then characterized by X-ray diffraction. Only the palladium complex **30** was tested in the aza-Michael reaction [69] of morpholine onto methylacrylonitrile, giving the enantio-enriched amination product in high yield but with a modest selectivity (37% e.e.) (Scheme 8.21).

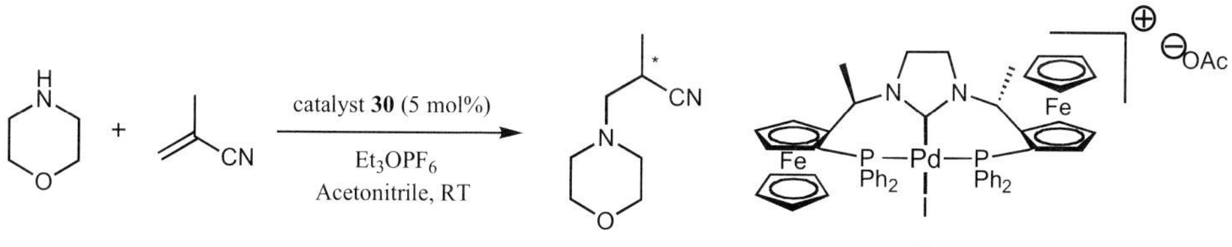

Scheme 8.21

8.5
Asymmetric 1,2-Addition

The addition of organometallic reagents to aldehydes is one of the general methods for the synthesis of secondary alcohols and particularly diarylmethanols, a common structural core unit of a wide variety of pharmacologically active compounds. Usually, organolithium and organomagnesium are used; however, their large-scale application is difficult due to their high sensitivity to moisture. Miyaura reported progress in this area in 1998 through the rhodium-catalyzed addition of arylboronic acids to aldehydes, giving diarylmethanol compounds in excellent yields [70]. Nevertheless, based on chiral phosphine ligands, the selectivity of the 1,2-addition (41% e.e.) is much lower than with the 1,4-conjugate addition (Section 8.4). In 2005, based on a recent report from Fürstner's group relating that NHCs are useful ligands for the rhodium-catalyzed addition of arylboronic acids to aldehydes [67], an asymmetric version was developed by Bolm and coworkers using the new chiral paracyclophane NHC ligands **31** [53]. Albeit the aryl transfer reaction to aromatic aldehydes gave the corresponding diarylmethanol adducts in good to excellent yields, the selectivity remains modest with these planar chiral NHC ligands (up to 38% e.e.) (Scheme 8.22).

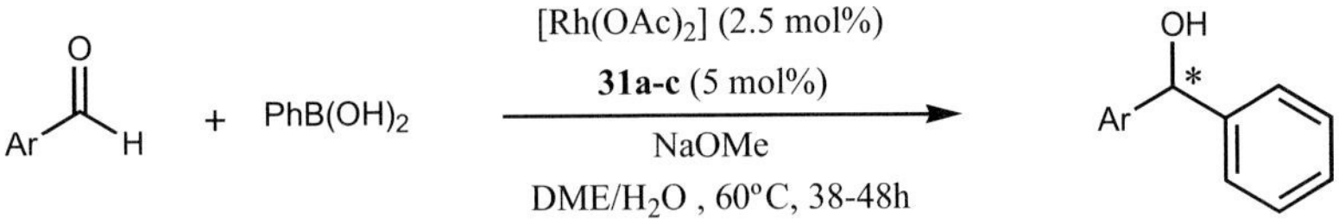

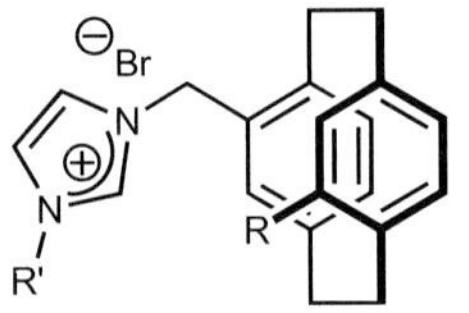

Scheme 8.22

Ar-CHO + PhB(OH)$_2$ → Ar-CH(OH)-Ph

[Rh(OAc)$_2$] (2.5 mol%), 31a-c (5 mol%), NaOMe, DME/H$_2$O, 60°C, 38-48h

31a R= P(O)Ph$_2$; R'= Me
31b R= P(O)Ph$_2$; R'= (S)-CHMePh
31c R= OMe; R'= (R)-CHMePh

Ligand	Ar	Yield (%)	ee (%)
(R_p)-31a	4-ClC$_6$H$_4$	72	20 (S)
(S,R_p)-31b	4-ClC$_6$H$_4$	84	29 (S)
(R,R_p)-31c	4-ClC$_6$H$_4$	57	10 (R)
(S,R_p)-31b	4-biphenyl	90	30 (S)
(S,R_p)-31b	1-naphthyl	85	38 (S)

At the same time, Luo and coworkers reported the use of bidentate imidazolium salts **32**, derived from L-proline amino acid, in the rhodium-catalyzed addition of phenylboronic acid to 4-chlorobenzaldehyde (Scheme 8.23) [38]. The highest yield was found with the salt **32c**, which contains a mesityl unit, showing the major role played by bulky substituents in the activity. The pre-synthesized NHC-Rh complex **33** did not allow improvements in either the activity or selectivity of the addition. The low enantioselectivity observed (around 20% e.e.) strongly shows that asymmetric rhodium-catalyzed 1,2-addition remains a challenge and that further structural modifications on the NHC will be necessary for improvements.

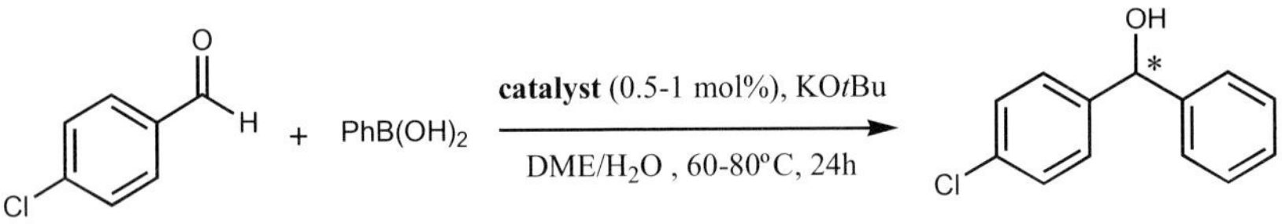

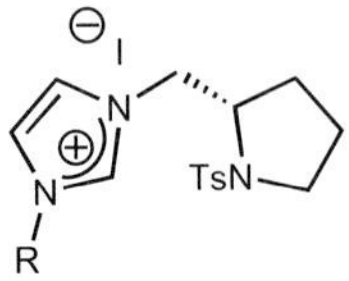

Scheme 8.23

4-Cl-C$_6$H$_4$-CHO + PhB(OH)$_2$ → 4-Cl-C$_6$H$_4$-CH(OH)-Ph

catalyst (0.5-1 mol%), KOtBu, DME/H$_2$O, 60-80°C, 24h

32a R= Me
32b R= Ph
32c R= Mes

catalyst	Yield (%)	ee (%)
[Rh(COD)Cl$_2$]/32a	68	-
[Rh(COD)Cl$_2$]32b	82	-
[Rh(COD)Cl$_2$]32c	95	17 (R)
33	96	21 (R)

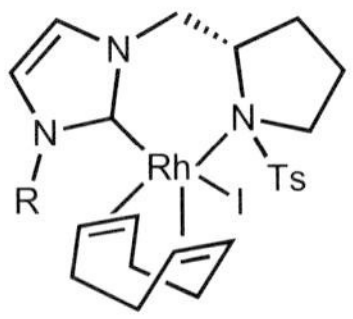

8.6
Asymmetric Hydrosilylation

Within the field of catalytic asymmetric synthesis, hydrosilylations of carbon–carbon and carbon–heteroatom double bonds are valuable alternatives to asymmetric hydrogenation. Indeed, the silylated product can be hydrolyzed easily under acidic or basic conditions to give the reduced final product. Several transition metal complexes, such as Ti, Rh, Ru, or Ir, have displayed high catalytic activity in the hydrosilylation of carbonyl compounds. Although these protocols are effective for the reduction of various ketones and aldehydes, the reported catalysts are quite inactive toward sterically hindered substrates. Furthermore, the available stoichiometric procedures for the reduction of highly hindered carbonyl compounds require up to 40 equiv. of the reducing agent. The efficiency of the metal NHC complexes represents an alternative to solve these drawbacks [71]. N-Heterocyclic carbenes have been used in hydrosilylation for many years. In 1977, Nile and Hill reported the hydrosilylation of alkynes, alkenes and ketones catalyzed by rhodium carbene complexes [72]. Unsurprisingly, the first application of chiral NHC was in asymmetric hydrosilylation. Over the last decade, a few groups have shown interest in this reaction and different classes of chiral carbenes have been tested with rhodium and, more recently, ruthenium complexes.

8.6.1
Rhodium-NHC Complexes

Herrmann's group have been at the forefront of investigations into stereoselective catalysis with chiral NHCs [8, 73]. The rhodium catalyst **7**, containing a C_2-symmetric imidazolydene, was used for the hydrosilylation of acetophenone (Scheme 8.24). The first enantiomeric excesses reported were low to modest (between 10 and 32% e.e.) but significant enough to demonstrate the ability of chiral NHC to generate an enantiomeric control of the reaction. The selectivity was subsequently improved with modified carbene **34**, bearing a chelating oxazoline unit, reaching 70% e.e. [30].

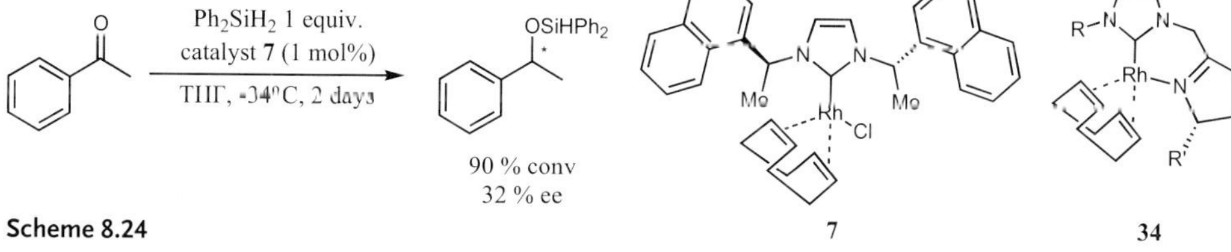

Scheme 8.24

Enders's group has made an important contribution to the synthesis and applications of new chiral NHCs. Soon after Herrmann's initial report, Ender and co-workers described the preparation of C_1-symmetrical (triazolinylidene)rhodium-(COD) complexes **8** and examined their ability to catalyze enantioselective hydrosilylation reactions (Scheme 8.25) [9, 74, 75]. Slightly better selectivities, up to just over 40% e.e., were reached with different aryl- or cyclohexyl- methyl ketones. Modification of the NHC structure (**8a-b**, methyl and phenyl groups instead of a *tert*-butyl group) did not increase the stereocontrol of the reduction.

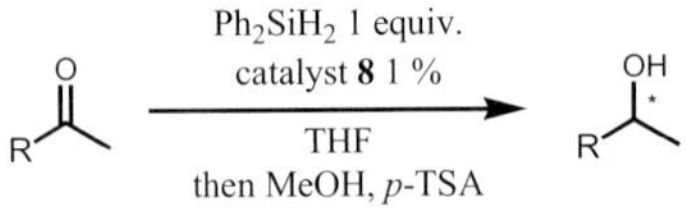

entry	catalyst	ketone, R =	temperature (°C)	time	yield (%)	ee (%)
1	8a	Ph	22	4h	90	20 (S)
2	8a	1-Np	42	4h	80	37 (R)
3	8a	2-Np	2	5d	90	19 (S)
4	8a	Cy	-10	6d	75	44 (S)
5	8b	Cy	2	4d	80	43 (S)
6	8c	Cy	22	3d	70	43 (R)

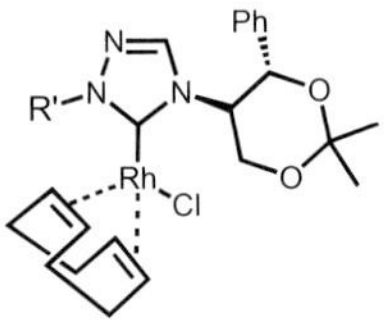

8a R' = Me
8b R' = Ph
8c R' = tBu

Scheme 8.25

The rhodium complex **35**, containing a planar chiral stable carbene (Fig. 8.12), with a ferrocenyl pattern developed by Bolm and coworkers, has been used for the hydrosilylation of acetophenone and propiophenone with diphenylsilane as reducing agent [47]. After 24 h at –10 °C followed by hydrolysis, only ca. 50% conversion was observed and the formed product was racemic.

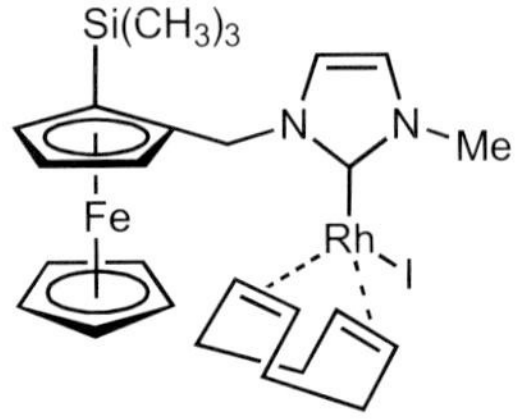

Fig. 8.12 Planar chiral NHC complex developed by Bolm.

Selectivity was greatly improved by the group of Shi using a NHC based on a 1,1′-binaphthalenyl unit [56]. This novel axially orientated chiral biscarbene (**36**) can be considered as a NHC analogue of the more useful BINAP chiral diphosphine ligand. The proficiency of this ligand in the catalytic enantioselective hydrosilylation of acetophenone process is excellent (Scheme 8.26). Enantiomeric excesses are up to 92% for aryl methyl ketones and bulky alkyl methyl ketones but are slightly lower for less bulky alkyl methyl ketones. These results are very encouraging and illustrate the potential of this carbene in other asymmetric reactions, following the example of BINAP.

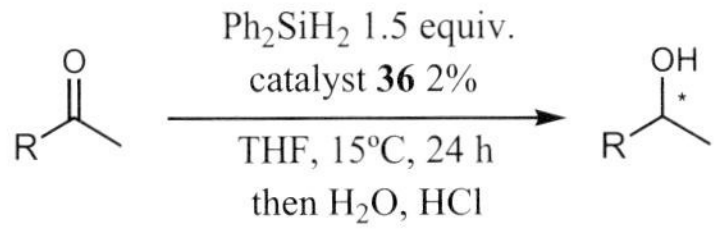

Scheme 8.26

Ph$_2$SiH$_2$ 1.5 equiv.
catalyst **36** 2%
THF, 15°C, 24 h
then H$_2$O, HCl

entry	ketone R =	yield (%)	ee (%)
1	Ph	87	98
2	2-Np	91	96
3	p-Tolyl	93	98
4	Ad	96	96
5	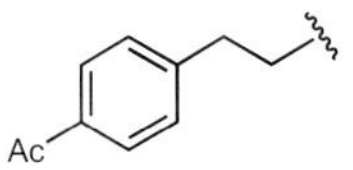 Ac-C$_6$H$_4$-CH$_2$CH$_2$-	87	71
6	C$_7$H$_{15}$	86	67

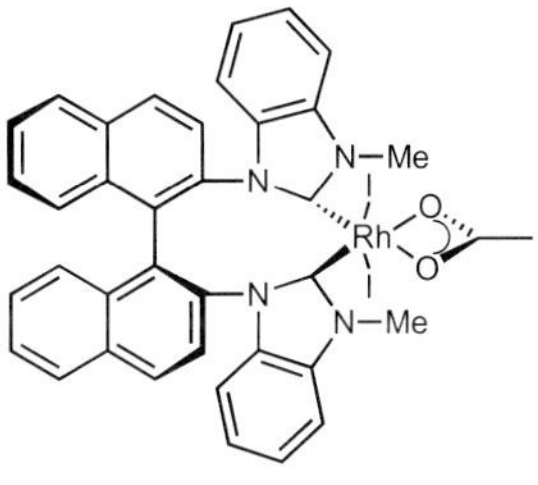

36

In ligand design, the chiral oxazoline is a useful structural pattern for asymmetric catalysis [76] – so its use to form new functionalized NHC ligands was evident. In 2002, Gade, Bellemin-Laponnaz and coworkers synthesized oxazolinyl-imidazolium salts by a direct and facile linkage between an imidazole and an oxazoline heterocycle [31]. These salts reacted with [{Rh(μ-OtBu)(nbd)}$_2$] to generate *in situ* the desired NHC-rhodium complexes **37** in high yields. These rhodium catalysts were then evaluated in the hydrosilylation of acetophenone, showing efficient activity. The best enantioselectivity was given by catalyst **37b**, where a more bulky alkyl group on the oxazoline unit (*tert*-butyl) was combined with moderate steric hindrance of the N-arylic substituent (mesityl) of the imidazolydene (Scheme 8.27).

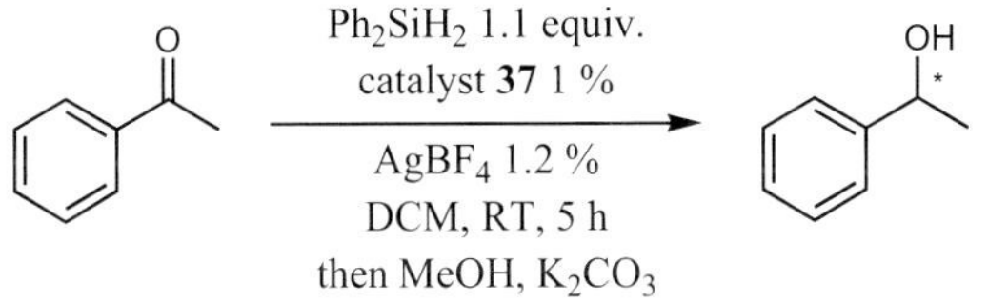

Ph$_2$SiH$_2$ 1.1 equiv.
catalyst **37** 1 %
AgBF$_4$ 1.2 %
DCM, RT, 5 h
then MeOH, K$_2$CO$_3$

entry	catalyst		ee (%)
1	**37a**	Ar = Mes, R = iPr	27 (S)
2	**37b**	Ar = Mes, R = tBu	65 (S)
3	**37c**	Ar = Ph, R = tBu	0
4	**37d**	Ar = 2,6-(iPr)$_2$Ph, R = tBu	20 (S)
5	**37e**	Ar = CH(Ph)$_2$, R = tBu	54 (R)
6	**37f**	Ar = CH(1-Np)$_2$, R = tBu	56 (S)

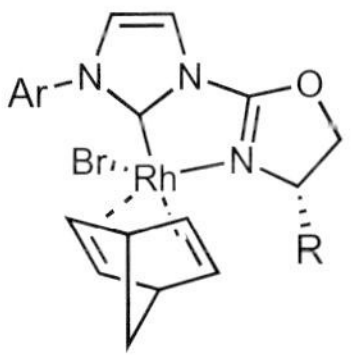

37

Scheme 8.27

Both the yield and the selectivity were greatly improved by adding a catalytic amount of silver salt and performing the reaction at −60 °C. Under these optimized conditions, good to excellent e.e.s (up to 99% e.e.) were obtained with a wide variety of ketones, including non-bulky alkyl methyl ketones, which are relatively difficult substrates to hydrosilylate with high stereocontrol (Scheme 8.28).

In 2004, Crudden's group synthesized another type of oxazolinyl-carbene rhodium complex (**38**), in which an aromatic phenyl ring was used as a rigidified linker/spacer of both chelating functions [35]. Unfortunately, the selectivity of the hydrosilylation of acetophenone was disappointing, reaching only 10% e.e. (Scheme 8.29). The poor enantiocontrol of the reaction can be explained by the presence the unhindered methyl group, which is unable to block the coordination site cis to the carbene.

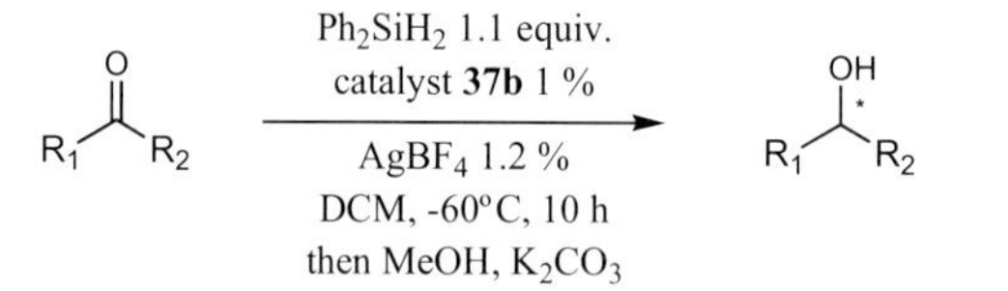

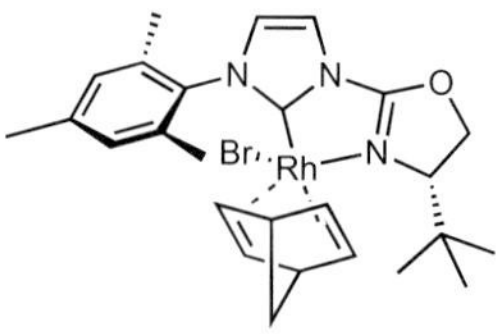

entry	ketone	yield (%)	ee (%)
1	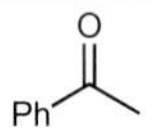 Ph-CO-Me	90	92
2	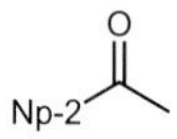 Np-2-CO-Me	91	99
3	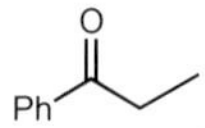 Ph-CO-Et	70	94
4	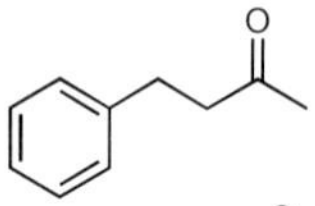 Ph-CH2CH2-CO-Me	77	97
5	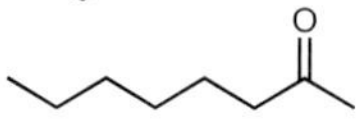 n-alkyl-CO-Me	79	95

Scheme 8.28

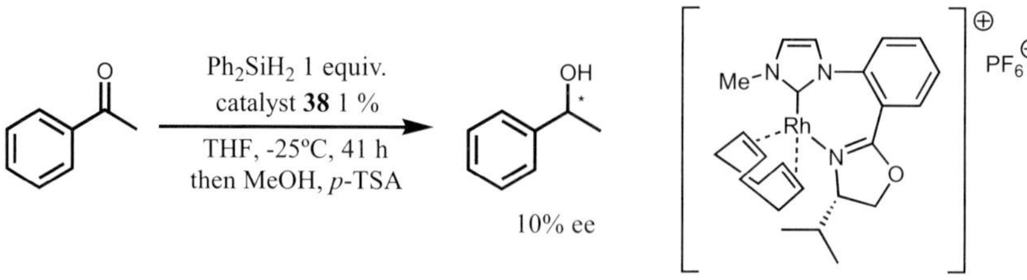

Scheme 8.29

8.6.2
Ruthenium-NHC Complexes

After the success of chiral NHC-rhodium complexes in the asymmetric conjugate addition based on the planar chirality of the paracyclophane-NHC ligands **29** (Section 8.4.2), Andrus and coworkers reported a new application in the ruthenium-catalyzed hydrosilylation of ketones [49]. *In situ* formation of the ruthenium catalyst from the bis-paracyclophane imidazolinium precursor **29** in the presence of [RuCl$_2$(PPh$_3$)$_3$] led to high yields and excellent selectivity in the case of acetophenone (Scheme 8.30). The presence of an anisyl group on the paracyclophane pattern is not essential while the removal of the phenyl unit could give rise to a significant loss in enantiomeric excess. Optimized catalyst Ru/**29d** afforded moderate to excellent selectivities in the asymmetric hydrosilylation of a large range of aryl ketones (e.e.s ranging from 58 to 97%).

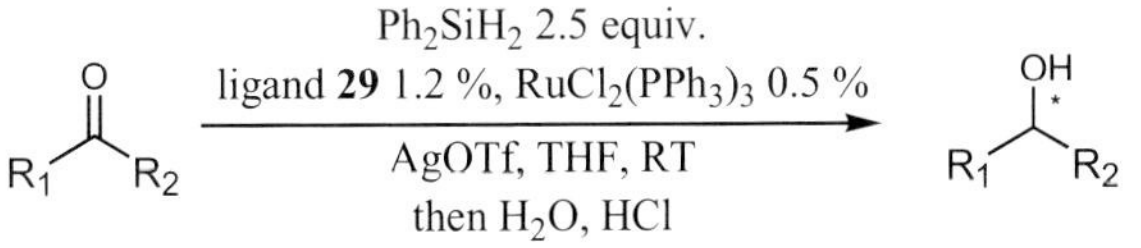

entry	catalyst	ketone	yield (%)	ee (%)
1	29a	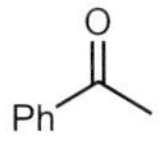 Ph-CO-Me	98	90
2	29b	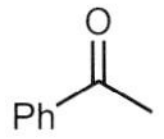 Ph-CO-Me	98	97
3	29d	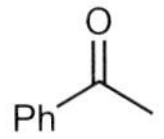 Ph-CO-Me	98	97
4	29d	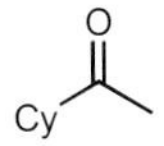 Cy-CO-Me	93	58
5	29d	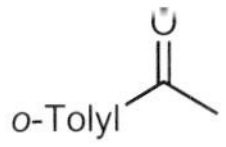 o-Tolyl-CO-Me	96	94
6	29d	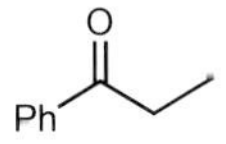 Ph-CO-Et	96	92

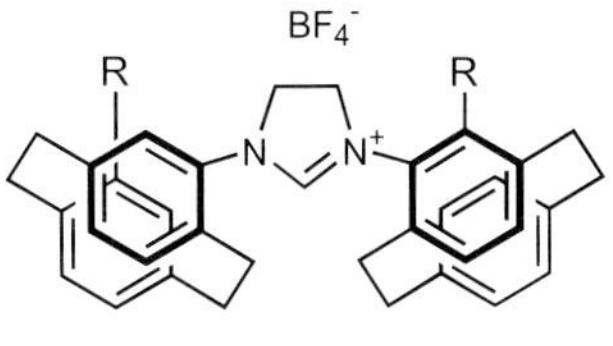

29a R = H
29b R = Ph
29d R = o-anisyl

Scheme 8.30

8.7
Asymmetric Olefin Metathesis

During the past few decades, catalytic olefin metathesis has gained considerable attention in the field of organic synthesis, allowing easy access to a wide range of small, medium and large ring carbo- and heterocycles as well as acyclic, unsaturated molecules [77, 78]. A major contribution in this area was the development of

8 Asymmetric Catalysis with Metal N-Heterocyclic Carbene Complexes

well-defined homogeneous Grubbs complexes [5a], which are easy to handle in air and are more tolerant towards a large variety of functional groups than the original Schrock molybdenum-based complexes [79]. Second-generation ruthenium catalysts possessing an NHC unit have considerably better stability and activity over a larger range of substrates. However, the first enantiomerically pure chiral metathesis reaction was performed by molybdenum-based systems developed by Schrock and Hoveyda in 1998 [80]. The development of chiral ruthenium-based complexes containing a chiral NHC ligand has therefore become a real challenge.

8.7.1
Asymmetric Ring-closing Metathesis

In 2001, Grubbs and coworkers reported the synthesis of a set of chiral NHC ruthenium complexes (**39**) based on C_2-symmetry strategy [41]. These catalysts were employed in the desymmetrization of triolefins to form chiral dihydrofurans by asymmetric ring-closing metathesis (ARCM). Several factors strongly influence enantiomeric control of the reaction. Firstly, the 1,2-diphenylethylenediamine backbone leads to higher enantioselectivities than the 1,2-diaminocyclohexane skeleton. Secondly, replacement of the mesityl group with an ortho-monosubstituted N-aryl substituent significantly improves selectivity. Finally, substitution of the halide ligand of the ruthenium complex from Cl^- to the more bulky I^- further improves the enantioselectivity, albeit a slight reduction in activity occurs. The combination of all these parameters allows the highest selectivity (90% e.e.) with catalyst **39c** with the most favorable triene (Scheme 8.31).

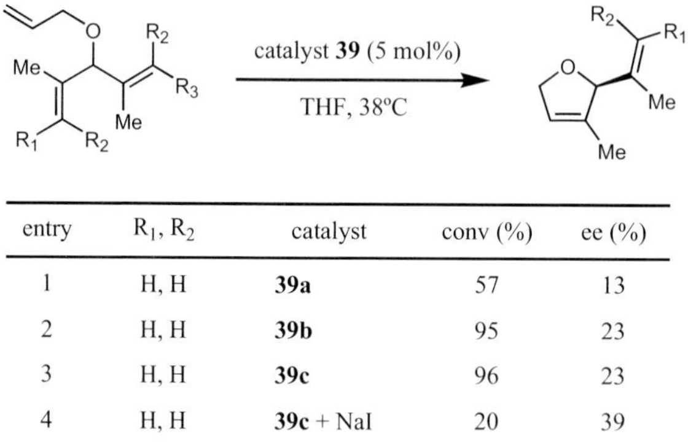

entry	R_1, R_2	catalyst	conv (%)	ee (%)
1	H, H	39a	57	13
2	H, H	39b	95	23
3	H, H	39c	96	23
4	H, H	39c + NaI	20	39
5	Me, H	39c + NaI	90	35
6	H, Me	39b + NaI	91	**85**
7	H, Me	39c + NaI	82	**90**

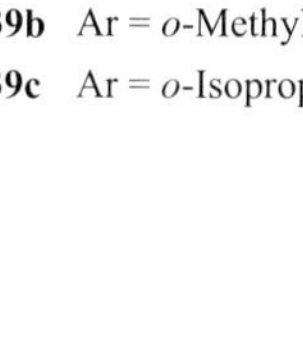

39a Ar = Mes
39b Ar = o-Methylphenyl
39c Ar = o-Isopropylphenyl

Scheme 8.31

Hoveyda and coworkers reported the second example of ARCM catalyzed by chiral NHC-complexes [81, 82]. Using a novel chiral bidentate anionic N-heterocyclic carbene ligand, containing an element of axial chirality from the binaphthalene pattern, Hoveyda's group isolated in high yield the chiral ruthenium catalyst **40a**. Reactions of triene desymmetrization can be promoted in the presence of 10 mol% of **40a** with excellent yields and good selectivities of up to 76% (Scheme 8.32).

However, the chiral NHC-Ru catalysts **39** and **40a** gave significantly lower enantioselectivities than the original Schrock–Hoveyda Mo-based catalysts [83].

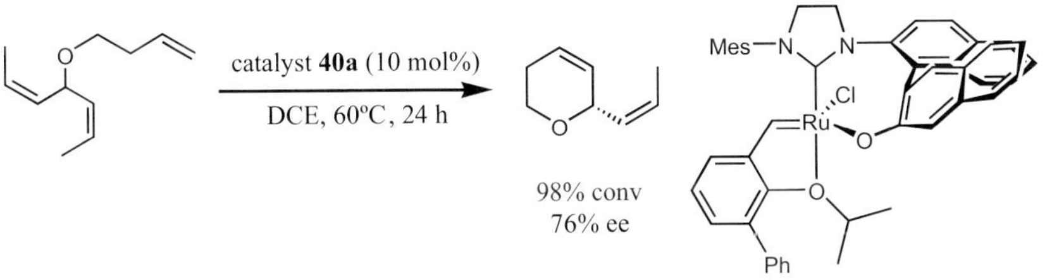

Scheme 8.32

8.7.2
Asymmetric Ring-opening Metathesis/Cross Metathesis

The AROM/CM reaction is a metal-catalyzed transformation in which chiral Ru-based complexes give similar enantioselectivities to chiral Mo-based catalysts. The structure of chiral NHC ligands based on the axially chiral binaphthalene skeleton (previously used in the ARCM) is very useful for this type of transformation [55, 82, 84]. Indeed, **40a-b** catalyzed efficiently the AROM/CM on a wide variety of substrates, giving almost exclusively the (E)-product (>98/2) in high yield with enantioselectivities ranging from 70 to 98% e.e. (Scheme 8.33). The highest stereocontrol of the asymmetric metathesis was found with complex **40b**, which contains an iodide instead of a chloride ligand.

In 2005, Hoveyda and coworkers extended their study on the AROM/CM reaction with the development of novel and readily available chiral bidentate NHC ligand using optically pure 1,2-diphenylethylendiamine skeleton as the source of chirality [46]. Here, the stereo-directing ligand is based on steric repulsions between backbone phenyl groups and the biphenol N-substituent, permitting an anti privileged orientation of the biphenol and leading, therefore, to the formation of a single atropisomer complex (**41**) after metal coordination. The Ru-catalysts **41a-b** showed similar high activity and selectivity to the first-generation chiral Ru-NHC complexes **40a-b** (Scheme 8.33, entries 6 and 7, 10 and 11).

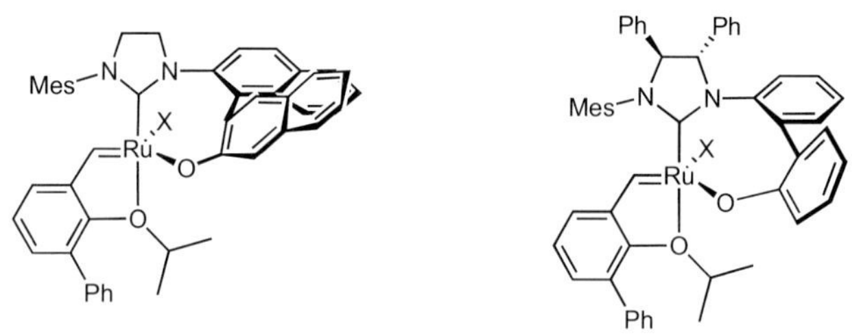

40a X = Cl
40b X = I

41a X = Cl
41b X = I

entry	substrates	product	catalyst	time (h), conv (%)	yield (%)	ee (%)
1	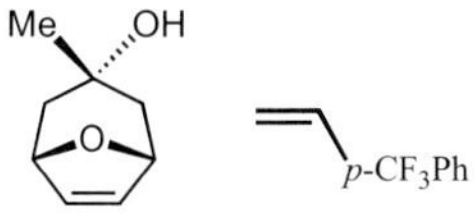	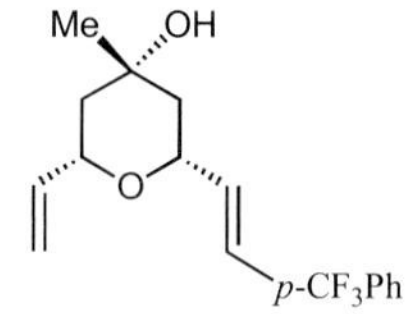	40a (2 %)	5, >98	75	74
2			40b (2 %)	0.1, >98	60	92
3	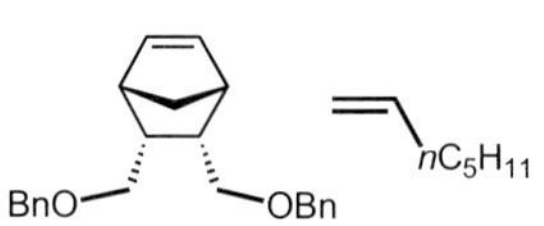	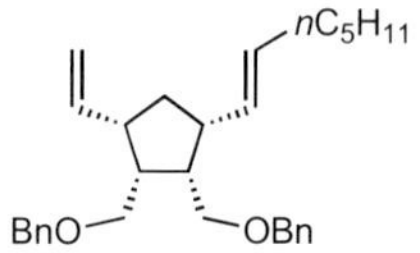	40a (5 %)	0.1, >98	65	91
4	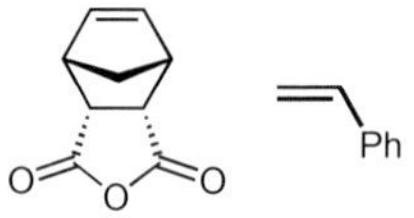	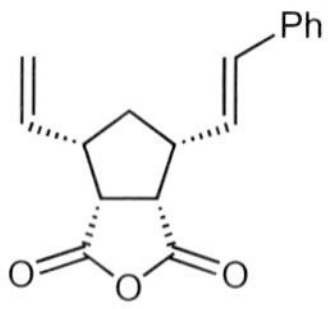	40b (5 %)	4, >98	78	>98
5			40a (5 %)	0.3, >98	60	70
6			41a (5 %)	0.1, >98	59	84
7			41b (5 %)	2, >98	50	90
8	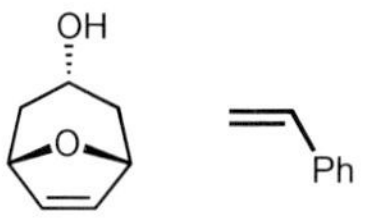	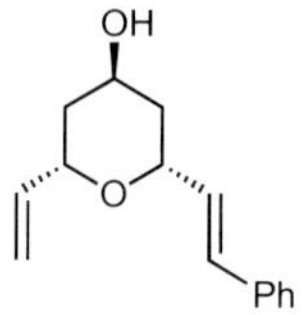	40a (5 %)	1.5, >98	80	94
9			40b (5 %)	2, >98	81	97
10			41a (5 %)	0.5, >98	83	74
11			41b (5 %)	1, >98	71	93

Scheme 8.33

The major advantage of Ru-catalysts **40** and **41** is their ability to promote AROM/CM in air, using undistilled solvents and with substrates that readily polymerize with chiral Mo-catalysts. Furthermore, around 90% of the catalyst can be recovered after catalysis, by chromatography on silica gel, and reused without significant loss in enantioselectivity.

8.8
Allylic Substitution Reaction

During the past decade, the enantiocontrol of metal-catalyzed allylic substitution reactions has been improved greatly. In contrast to most metal-catalyzed enantioselective processes, asymmetric allylic alkylations involve reaction at an sp^3 instead of an sp^2 center. The ability to transform achiral, prochiral or chiral racemic material to enantio-enriched product is unique to the asymmetric allylic alkylation reaction. However, two drawbacks exist in this catalytic process: the high substrate specificity and the slow reaction rates frequently observed. In addition to the generally used palladium, this reaction can be catalyzed by other transition metals, e.g., molybdenum, tungsten, iridium, nickel and copper [85].

8.8.1
Palladium Catalysis

Douthwaite et al. reported the synthesis of chiral bidentate silver(I)-NHC complexes **42** containing an imine function and an imidazolylidene linked by *trans*-1,2-diaminocyclohexane [33]. Some of these complexes can successfully catalyze the asymmetric allylic alkylation between (*E*)-1,3-diphenylprop-3-en-1-yl acetate and dimethyl malonate with high yields and excellent selectivities (up to 92%) (Scheme 8.34). A good level of enantioselectivity was observed when the imine group was derived from acetophenone while more bulky substituents led to a dramatic decrease of selectivity. Nevertheless, the NHC ligands do not seem to improve the activity of this catalytic process, which requires a high loading of catalyst (5 mol%) and elevated temperature.

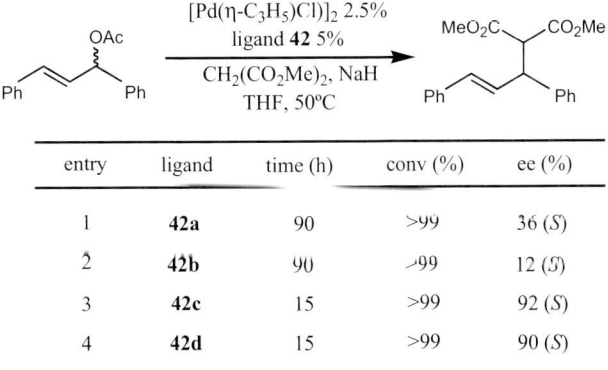

entry	ligand	time (h)	conv (%)	ee (%)
1	42a	90	>99	36 (*S*)
2	42b	90	>99	12 (*S*)
3	42c	15	>99	92 (*S*)
4	42d	15	>99	90 (*S*)

42a $R_1 = iPr, R_2 = Me, R_3 = H$
42b $R_1 = iPr, R_2 = tBu, R_3 = H$
42c $R_1 = iPr, R_2, R_3 = Me$
42d $R_1 = CHPh_2, R_2, R_3 = Me$

Scheme 8.34

8.8.2
Copper Catalysis

In 2004, Okamoto's group reported that copper-complexes **43a-b** bearing the Herrmann C_2-symmetrical NHC ligands were highly selective towards S_N2'-allylic substitution with Grignard reagents [86]. These chiral NHCs, which had previously been studied in the enantioselective conjugate addition and asymmetric hydrosilylation (Sections 8.4 and 8.6), gave moderate to good enantioselectivity, up to 70% e.e., in the allylic substitution of various substrates (Scheme 8.35). As mentioned above for the conjugate addition (Section 8.4), the steric hindrance positioned on the chiral center is crucial for stereocontrol of the substitution.

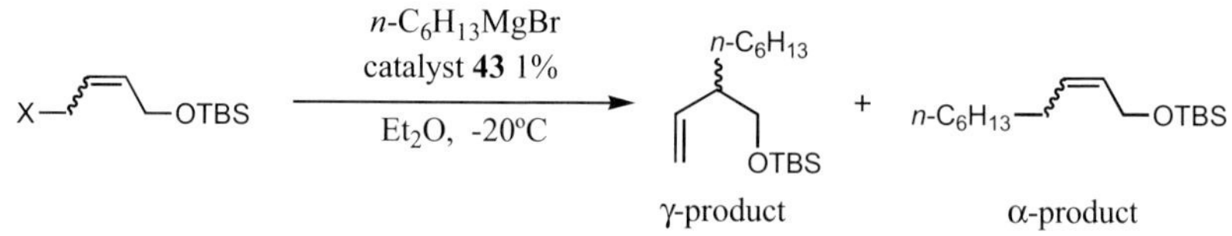

entry	catalyst	substrate (X =)	γ:α	ee (%)
1	43a	OAc (Z)	87:13	40 (R)
2	43b	OAc (Z)	95:5	60 (R)
3	43a	o-OPy (Z)	91:9	36 (R)
4	43b	o-OPy (Z)	98:2	70 (R)
5	43b	o-OPy (E)	86:14	60 (S)

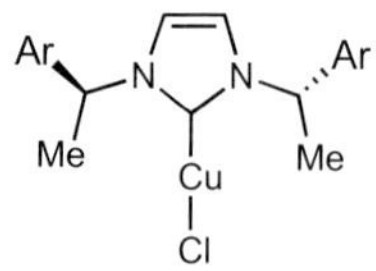

43a Ar = Ph
43b Ar = 1-Np

Scheme 8.35

Hoveyda and coworkers have achieved the best improvement for copper-catalyzed asymmetric allylic alkylation using chiral N-heterocyclic carbene complexes. Firstly, they used the chiral bidentate silver-NHC **44**, incorporating the binaphthyl axial chirality previously developed for asymmetric metathesis [87]. Enantiomeric excesses obtained for allylic alkylation are excellent for a large range of phosphated substrates and various dialkylzincs (Scheme 8.36). Secondly, they examined the potential of their second generation chiral bidentate silver-NHC (**45**) based on a chiral diamine backbone in combination with an achiral biphenol N-substituent [46]. Despite a slight reduction in activity, silver-NHC **45** in the presence of copper chloride complex catalyzed very efficiently the allylic substitution of trisubstituted olefins, using sterically hindered dialkylzincs, to generate quaternary stereogenic centers in high yields and excellent optical purities (up to 98% e.e.). Furthermore, the chiral copper-NHC complex **46** could also be isolated, showing similar high activity and selectivity. These chiral alkylation catalysts are exceptionally useful as they can catalyze the reaction using a small amount of metal (0.5–1 mol%) at low temperature (−15 °C).

8.9 Asymmetric α-Arylation

Scheme 8.36

entry	R_1, R_2	(alkyl)$_2$Zn	catalyst	time (h) conv (%)	ee (%)
1	Ph, H	Et$_2$Zn	CuCl$_2$.2H$_2$O/**44** (2/1)	4, >98	98
2	o-NO$_2$Ph, H	Et$_2$Zn	CuCl$_2$.2H$_2$O/**44** (2/1)	4, >98	97
3	Ph, Me	i-Pr$_2$Zn	CuCl$_2$.2H$_2$O/**45** (2/1)	12, >98	98
4	Ph, Me	i-Pr$_2$Zn	**46**	12, >98	97
5	o-OMePh, Me	Et$_2$Zn	CuCl$_2$.2H$_2$O/**45** (2/1)	24, 75	95
6	o-OMePh, Me	Et$_2$Zn	**46**	40, >98	98

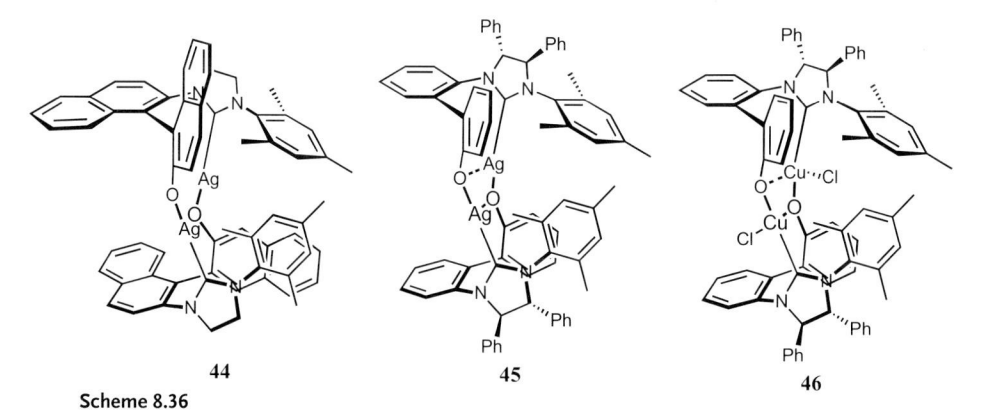

8.9
Asymmetric α-Arylation

Among the many palladium-catalyzed coupling reactions allowing the formation of C–C bonds, the α-arylation of carbonyl compounds has been the least investigated transformation, although it represents a simple and useful method to obtain interesting structures. This simple cross coupling reaction, involving a ketone enolate and an aryl halide in the presence of a transition metal catalyst, was independently reported in 1997 by the groups of Hartwig [88] and Buchwald [89]. At the same time, Muratake and Natsume reported the intramolecular version of the α-arylation, which is a convenient, short route to form cyclic complex molecules [90]. Several groups have studied asymmetric intramolecular α-arylation using chiral N-heterocyclic carbenes.

In 2001, Hartwig and coworkers were the first to attempt the enantioselective palladium-catalyzed intramolecular reaction of enolates and aromatic bromide [25], using three different monodentate chiral imidazolium salts (**47–49**). Enantioenriched oxindoles were thus formed in moderate to good yields (Scheme 8.37).

The highest selectivity (entry 3, 76% e.e.) was obtained using imidazolium salt **49**, which contains the (+)-bornyl pattern, in which the stereogenic center was closer than the (−)-isopinocamphenyl structure. Importantly, albeit the stereoselectivity observed with these chiral NHC-ligands remains modest, it was the best result in this field in comparison with well-established chiral phosphines such as BINAP, Duphos and Josiphos.>

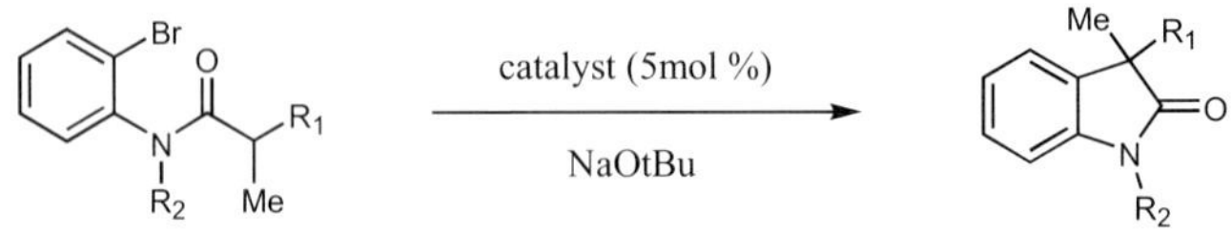

entry	substrate (R_1, R_2)	catalyst (Pd/L)	yield (%)	ee (%)
1	1-Np, Bn	Pd(OAc)$_2$/**47**	35	4
2	1-Np, Bn	Pd(dba)$_2$/**48**	88	67
3	1-Np, Bn	Pd(dba)$_2$/**49**	75	76
4	1-Np, Me	Pd(dba)$_2$/**49**	91	69
5	Ph, Me	Pd(dba)$_2$/**49**	74	57
6	Ph, Me	Pd(dba)$_3$/**50**	95	43
7	1-Np, Me	PdCl$_2$/**51**	90	11

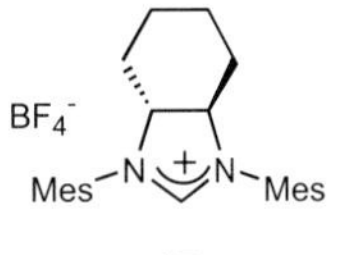

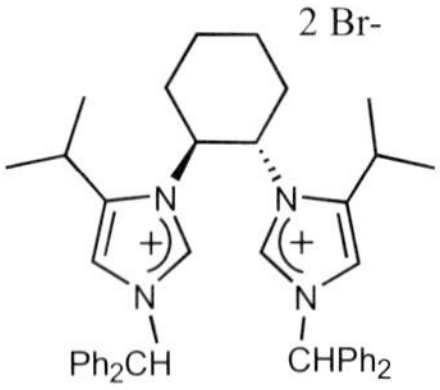

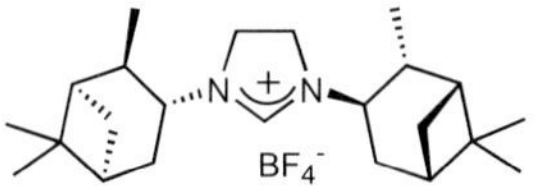

Scheme 8.37

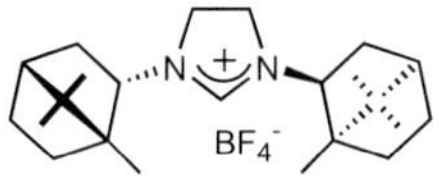

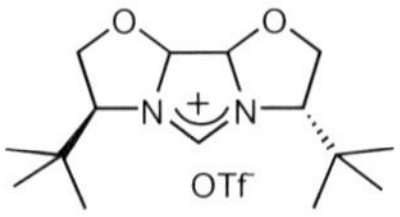

In 2002, Glorius and coworkers synthesized an original C_2-symmetric monodentate NHC (**50**) that contains a two-chiral oxazoline pattern [27]. Nevertheless, its use in enantioselective palladium-catalyzed intramolecular α-arylation gave only moderate results (e.e.s < 43%, Scheme 8.37, entry 6).

Finally, Douthwaite and coworkers have also carried out this reaction, using a C_2-symmetric diimidazolium salt **51** that contains a *trans*-1,2-diaminocyclohexane unit as linker/spacer. However, its palladium complex PdCl$_2$/**51** showed low selectivity (11% e.e., Scheme 8.37, entry 7) [34].

8.10
Palladium-catalyzed Kinetic Resolution

Sigman and coworkers have developed an unusual application of chiral N-heterocyclic carbenes. The aim is to use the NHC-Pd(II) complexes to catalyze aerobic oxidation for the kinetic resolution of secondary alcohols [42].

A wide variety of process have been developed for the oxidation of alcohols and a few of them are catalytic, using transition metals such as Ru, Cr, Co, Mo, Fe, Ir, Ce, Pd. A catalytic asymmetric version for secondary alcohols using Pd(II) and a chiral base was reported independently by Stoltz [91] and Sigman [92]. The best results were obtained with (–)-sparteine, which acts as both a ligand and a base (Scheme 8.38, entry 1, 98% e.e.), as showed by previous mechanistic studies [93]. However, only one enantiomer of sparteine is available and structural modifications to optimize the enantioselectivity are difficult. Because the N-heterocyclic carbene is strongly bound to the metallic center it cannot be displaced by (–)-sparteine, allowing the study of the exact role of (–)-sparteine in the selectivity of the aerobic oxidation [42].

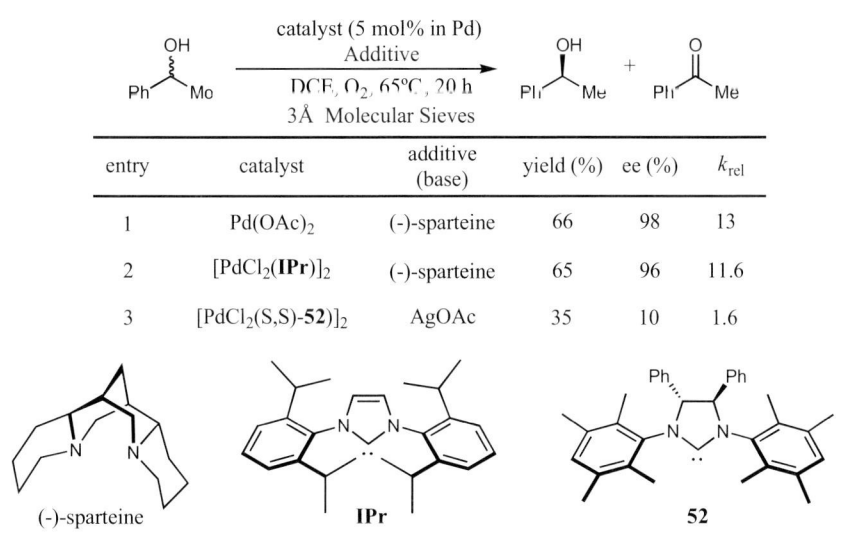

entry	catalyst	additive (base)	yield (%)	ee (%)	k_{rel}
1	Pd(OAc)$_2$	(-)-sparteine	66	98	13
2	[PdCl$_2$(**IPr**)]$_2$	(-)-sparteine	65	96	11.6
3	[PdCl$_2$(S,S)-**52**)]$_2$	AgOAc	35	10	1.6

Scheme 8.38

Firstly, use of the achiral [PdCl$_2$(**IPr**)]$_2$ catalyst in combination with (–)-sparteine gave similar selectivity (entry 2, 96% e.e.). Secondly, only 10% e.e. is obtained with oxidation in the presence of the optically pure [PdCl$_2$(S,S)-**52**)]$_2$ and silver acetate as base. These results clearly demonstrated that enantioselective induction during the catalytic transformation is principally related to the chiral base.

8.11
Conclusion and Outlook

Among their wide and valuable properties as transition metal ligands, the most important feature of N-heterocyclic carbenes (NHCs) is their well-established resistance toward dissociation, making them ideal candidates for asymmetric catalysis. Owing to their easy access and the versatility of their architecture, many optically pure NHCs have emerged within a short period. Albeit in most cases the phosphine based-metal catalysts have been surpassed by their NHC homologues in both activity and scope of applications, this is not yet the case with asymmetric catalysis. Furthermore, the strategy based on the phosphine mimics has been often used in the pioneering developments of chiral NHCs. Nevertheless, although a few examples vindicated this approach through high selectivities, most of reactions examined so far with early architectures give disappointing results. To develop more efficient chiral NHC ligands, several well-established and common structural motifs (or elements of symmetry) in ligand design for asymmetric catalysis have been screened (such as the oxazoline ring, the binaphthyl or the paracyclophane pattern). At this stage, of all the chiral NHC ligands indexed and summarized in this chapter, 1/3 of them are efficient in asymmetric catalysis, giving excellent selectivities up to 90% e.e., but only five types of diaminocarbene induce more than 98% selectivity.

Despite the low success rate in this area, research has revealed useful information on the structure–selectivity relationship, and the influence of several structural factors has been clearly identified. For example, the steric hindrance displayed by aromatic N-substituent units of the diaminocarbene (mesityl and 2,6-$iPr_2C_4H_6$) shows a beneficial effect on the selectivity while the orthogonal structural feature between the imidazole (or imidazoline) ring and the aromatic N-substituents can lead to the formation of useful chiral atropisomers. Actually, novel chiral NHC designs tend to use new and original stable NHC frameworks, as recently reported in the literature (Lassaletta [29, 94] and Stahl [95]), showing that NHC ligands are continually under investigation and have tremendous potential in the development of "next generation" asymmetric catalysts.

References

1 K. Öfele, *J. Organomet. Chem.* **1968**, *12*, 42.
2 H.W. Wanzlick, H.J. Schonherr, *Angew. Chem. Int. Ed.* **1968**, *7*, 141.
3 A.J. Arduengo III, R.L. Harlow, M. Kline, *J. Am. Chem. Soc.* **1991**, *113*, 361.
4 (a) W.A. Herrmann, *Angew. Chem. Int. Ed.* **2002**, *41*, 1291. (b) A.C. Hillier, S.M. Nolan, *Platinium Met. Rev.* **2002**, *46*, 50. (c) D. Bourissou, O. Guerret, F.P. Gabbai, G. Bertrand, *Chem. Rev.* **2000**, *100*, 39.
5 (a) T.M. Trnka, R.H. Grubbs, *Acc. Chem. Res.* **2001**, *34*, 18. (b) D. Astruc, *New J. Chem.* **2005**, *29*, 42.
6 A.W. Coleman, P.B. Hitchcock, M.F. Lappert, R.K. Maskell, J.H. Müller, *J. Organomet. Chem.* **1983**, *250*, C9.
7 (a) A.W. Coleman, P.B. Hitchcock, M.F. Lappert, R.K. Maskell, J.H. Müller, *J. Organomet. Chem.* **1985**, *296*, 173. (b) M.F. Lappert, *J. Organomet. Chem.* **1988**, *358*, 185.
8 W.A. Herrmann, L.J. Goossen, C. Köcher, G.R.J. Artus, *Angew. Chem. Int. Ed. Engl.* **1996**, *35*, 2805.
9 D. Enders, H. Gielen, K. Breuer, *Tetrahedron: Asymmetry* **1997**, *8*, 3571.
10 M.C. Perry, K. Burgess, *Tetrahedron: Asymmetry* **2003**, *14*, 951.
11 V. César, S. Bellemin-Laponnaz, L.H. Gade, *Chem. Soc. Rev.* **2004**, *33*, 619.
12 H.M.J. Wang, I.J.B. Lin, *Organometallics* **1998**, *17*, 972.
13 J.K. Whitesell, *Chem. Rev.* **1989**, *89*, 1581.

14 H.B. Kagan, J.P. Dang, *Chem. Commun.* **1971**, 481.
15 R. Noyori, H. Takaya, *Acc. Chem. Res.* **1990**, *23*, 345.
16 (a) A.K. Ghosh, P. Mathivanan, J. Cappiello, *Tetrahedron: Asymmetry* **1998**, *9*, 1. (b) K.A. Jørgensen, M. Johannsen, S. Yao, H. Audrain, J. Thorhauge, *Acc. Chem. Res.* **1999**, *32*, 605. (c) J.S. Johnson, D.A. Evans, *Acc. Chem. Res.* **2000**, *33*, 325.
17 E.M. McGarrigle, D.G. Gilheany, *Chem. Rev.* **2005**, *105*, 1563.
18 (a) G. Helmchen, A. Pfaltz, *Acc. Chem. Res.* **2000**, *33*, 336. (b) A. Pfaltz, J. Blankenstein, R. Hilgraf, E. Hörmann, S. McIntyre, F. Menges, M. Schönleber, S.P. Smidt, B. Wüstenberg, N. Zimmermann, *Adv. Synth. Catal.* **2003**, *345*, 33.
19 A. Togni, C. Breutel, A. Schnyder, F. Spindler, H. Landert, A. Tijiani, *J. Am. Chem. Soc.* **1994**, *116*, 4062.
20 A.M. Porte, J. Reibenspies, K. Burgess, *J. Am. Chem. Soc.* **1998**, *120*, 9180.
21 H. Steinhagen, M. Reggelin, G. Helmchen, *Angew. Chem. Int. Ed.* **1997**, *36*, 2108.
22 H. Seo, H.-J. Park, B.Y. Kim, J.H. Lee, S.U. Son, Y.K. Chung, *Organometallics* **2003**, *22*, 618.
23 M.T. Powell, D.-R. Hou, M.C. Perry, X. Cui, K. Burgess, *J. Am. Chem. Soc.* **2001**, *123*, 8878.
24 (a) D.-R. Hou, K. Burgess, *Org. Lett.* **1999**, *1*, 1745. (b) D.-R. Hou, J. Reibenspies, K. Burgess, *J. Org. Chem.* **2001**, *66*, 206.
25 S.L. Lee, J.F. Hartwig, *J. Org. Chem.* **2001**, *66*, 3402.
26 J. Huang, L. Jafarpour, A.C. Hillier, E.D. Stevens, S.P. Nolan, *Organometallics* **2001**, *20*, 2878.
27 F. Glorius, G. Altenhoff, R. Goddard, C. Lehmann, *Chem. Commun.* **2002**, 2704.
28 H. Seo, B.Y. Kim, J.H. Lee, H. J. Park, S.U. Son, Y.K. Chung, *Organometallics* **2003**, *22*, 4783.
29 M. Alcarazo, S.J. Roseblade, E. Alonso, R. Fernandez, E. Alvarez, F.J. Lahoz, J.M. Lassaletta, *J. Am. Chem. Soc.* **2004**, *126*, 13 242.
30 W.A. Herrmann, L.J. Goossen, M. Spiegler, *Organometallics* **1998**, *17*, 2162.
31 (a) V. César, S. Bellemin-Laponnaz, L.H. Gade, *Organometallics* **2002**, *21*, 5204. (b) V. César, S. Bellemin-Laponnaz, L.H. Gade, *Eur. J. Inorg. Chem.* **2004**, 3436. (c) V. César, S. Bellemin-Laponnaz, L.H. Gade, *Angew. Chem. Int. Ed.* **2004**, *43*, 1014. (d) V. César, S. Bellemin-Laponnaz, H. Wadepohl, L.H. Gade, *Chem. Eur. J.* **2005**, 11, 2862.
32 M.C. Perry, X. Cui, K. Burgess, *Tetrahedron: Asymmetry* **2002**, *13*, 1969.
33 L.G. Bonnet, R.E. Douthwaite, B.M. Kariuki, *Organometallics* **2003**, *22*, 4187.
34 L.G. Bonnet, R.E. Douthwaite, R. Hodgson, *Organometallics* **2003**, *22*, 4384.
35 L. Ren, A.C. Chen, A. Decken, C.M. Crudden, *Can. J. Chem.* **2004**, *82*, 1781.
36 P.L. Arnold, M. Rodden, K.M. Davis, A.C. Scarisbrick, A.J. Blake, C. Wilson, *Chem. Commun.* **2004**, 1612.
37 H. Clavier, L. Coutable, J.-C. Guillemin, M. Mauduit, *Tetrahedron: Asymmetry* **2005**, *16*, 921.
38 W. Zhang, Y. Qin, S. Zhang, M. Luo, *Arkivoc* **2005**, *xiv*, 39.
39 F. Guillen, C.L. Winn, A. Alexakis, *Tetrahedron: Asymmetry* **2001**, *12*, 2083.
40 J. Pytkowicz, S. Roland, P. Mangeney, *Tetrahedron: Asymmetry* **2001**, *12*, 2087.
41 T.J. Seiders, D.W. Ward, R.H. Grubbs, *Org. Lett.* **2001**, *3*, 3225.
42 D.R. Jensen, M.S. Sigman, *Org. Lett.* **2003**, *5*, 63.
43 A. Alexakis, C.L. Winn, F. Guillen, J. Pytkowicz, S. Roland, P. Mangeney, *Adv. Synth. Catal.* **2003**, *345*, 345.
44 A. Fürstner, G. Seidel, D. Kremzow, C.W. Lehmann, *Organometallics* **2003**, *22*, 907.
45 E. Bappert, G. Helmchen, *Synlett* **2004**, 1789.
46 J.J. Van Veldhuizen, J.E. Campbell, R.E. Guidici, A.H. Hoveyda, *J. Am. Chem. Soc.* **2005**, *127*, 6877.
47 C. Bolm, M. Kesselgruber, G. Raabe, *Organometallics* **2002**, *21*, 707.
48 S. Gischig, A. Togni, *Organometallics* **2004**, *23*, 2479.
49 (a) Y. Ma, C. Song, C. Ma, Z. Sun, Q. Chai, M.B. Andrus, *Angew. Chem. Int. Ed.* **2003**, *42*, 5871. (b) C. Song, C. Ma, Y. Ma, W. Feng, S. Ma, Q. Chai, M.B. Andrus, *Tetrahedron Lett.* **2005**, *46*, 3241.
50 C. Bolm, T. Focken, G. Raabe, *Tetrahedron: Asymmetry* **2003**, 14, 1733.

51 S. Gischig, A. Togni, *Organometallics* **2004**, *23*, 2479.
52 T. Focken, G. Raabe, C. Bolm, *Tetrahedron: Asymmetry* **2004**, *15*, 1693.
53 T. Focken, J. Rudolph, C. Bolm, *Synthesis* **2005**, *3*, 429.
54 D.S. Clyne, J. Jin, E. Genest, J.C. Gallucci, T.V. Rajanbabu, *Org. Lett.* **2000**, *2*, 1125.
55 J.J. Van Veldhuizen, S.B. Garber, J.S. Kingsbury, A.H. Hoveyda, *J. Am. Chem. Soc.* **2002**, *124*, 4954.
56 W.-L. Duan, M. Shi, G.-B. Rong, *Chem. Commun.* **2003**, 2916.
57 M.C. Perry, X. Cui, M.T. Powell, D.-R. Hou, H.J. Reibenspies, K. Burgess, *J. Am. Chem. Soc.* **2003**, *125*, 113.
58 R. Crabtree, *Acc. Chem. Res.* **1979**, *12*, 331.
59 (a) P. Perlmutter in *Conjugate Addition Reaction in Organic Synthesis*, Tetrahedron Organic Chemistry Series, no. 9, Pergamon Press, Oxford, **1992**. (b) Y. Yamamoto in *Methods Org. Chem. (Houben-Weyl)*, vol. 4 ("Stereoselective Synthesis"), **1995**, pp. 2041–2057.
60 A. Alexakis, S. Mutti, J.F. Normant, *J. Am. Chem. Soc.* **1991**, *113*, 6332.
61 A. Alexakis, J. Frutos, P. Mangeney, *Tetrahedron: Asymmetry* **1993**, *4*, 2427.
62 A. Alexakis, C. Benhaim, *Eur. J. Org. Chem.* **2002**, 3221.
63 A.J. Arduengo III, H.W.R. Diaz, J.C. Calabrese, F. Davidson, *Organometallics* **1993**, *12*, 3405.
64 P.K. Fraser, S. Woodward, *Tetrahedron Lett.* **2001**, *42*, 2747.
65 H. Clavier, L. Coutable, L. Toupet, J.-C. Guillemin, M. Mauduit, *J. Organomet. Chem.* **2005**, *690*, 5237.
66 Y. Takaya, M. Ogasawara, T. Hayashi, *J. Am. Chem. Soc.* **1998**, *120*, 5579.
67 A. Fürstner, H. Krause, *Adv. Synth. Catal.* **2001**, *343*, 343.
68 P. Barbaro, C. Blanchi, A. Togni, *Organometallics* **1997**, *16*, 3004.
69 L.-W. Xu, C.-G. Xia, *Eur. J. Org. Chem.* **2005**, 633.
70 M. Sakai, M. Ueda, N. Miyaura, *Angew. Chem. Int. Ed.* **1998**, *37*, 3279.
71 S. Diez-Gonzàlez, H. Kaur, F. Kauer Zinn, E.D. Stevens, S.P. Nolan, *J. Org. Chem.* **2005**, *70*, 4784.
72 J.E. Hill, T.A. Nile, *J. Organomet. Chem.* **1977**, *137*, 293.
73 W.A. Herrmann, L.J. Goossen, G.R.J. Artus, C. Köcher, *Organometallics* **1997**, *16*, 2472.
74 D. Enders, H. Gielen, J. Runsink, K. Breuer, S. Brode, K. Boehn, *Eur. J. Inorg. Chem.* **1998**, 913.
75 D. Enders, H. Gielen, *J. Organomet. Chem.* **2001**, *617–618*, 70.
76 (a) M. Gomez, G. Muller, M. Rocamora, *Coord. Chem. Rev.* **1999**, *193–195*, 769. (b) A.K. Ghosh, P. Mathivanan, J. Cappiello, *Tetrahedron: Asymmetry*, **1998**, *9*, 1.
77 I. Nakamura, Y. Yamamoto, *Chem. Rev.* **2004**, *104*, 2127.
78 S. Blechert, S.J. Connon, *Angew. Chem., Int. Ed.* **2003**, *42*, 1900.
79 R.R. Schrock, A.H. Hoveyda, *Angew. Chem. Int. Ed.* **2003**, *42*, 4592.
80 J.B. Alexander, D.S. La, D.R. Cefalo, A.H. Hoveyda, R.R. Schrock, *J. Am. Chem. Soc.* **1998**, *120*, 4041.
81 J.J. Van Veldhuizen, D.G. Gillingham, S.B. Garber, O. Kataoka, A.H. Hoveyda, *J. Am. Chem. Soc.* **2003**, *125*, 12502.
82 A.H. Hoveyda, D.G. Gillingham, J.J. Van Veldhuizen, O. Kataoka, S.B. Garber, J.S. Kingsbury, J.P.A. Harrity, *Org. Biomol. Chem.* **2004**, *2*, 8.
83 A.H. Hoveyda, R.R. Schrock, *Chem. Eur. J.* **2001**, *7*, 945.
84 D.G. Gillingham, O. Kataoka, S.B. Garber, A.H. Hoveyda, *J. Am. Chem. Soc.* **2004**, *126*, 12288.
85 B.M. Trost, M.L. Crowley, *Chem. Rev.* **2003**, *103*, 2921.
86 S. Tominaga, Y. Oi, T. Kato, D.K. An, S. Okamoto, *Tetrahedron Lett.* **2004**, *45*, 5585.
87 A.L. Larsen, W. Leu, C. Nieto Oberhuber, J.E. Campbell, A.H. Hoveyda, *J. Am. Chem. Soc.* **2004**, *126*, 11130.
88 B.C. Hamann, J.F. Hartwig, *J. Am. Chem. Soc.* **1997**, *119*, 12382.
89 M. Palucki, S.L. Buchwald, *J. Am. Chem. Soc.* **1997**, *119*, 11108.
90 H. Muratake, M. Natsume, *Tetrahedron Lett.* **1997**, *38*, 7581.
91 E.M. Ferreira, B.M. Stoltz, *J. Am. Chem. Soc.* **2001**, *123*, 7725.
92 D.R. Jensen, J.S. Pugsley, M.S. Sigman, *J. Am. Chem. Soc.* **2001**, *123*, 7475.
93 J.A. Mueller, D.R. Jensen, M.S. Sigman, *J. Am. Chem. Soc.* **2002**, *124*, 8202.
94 M. Alcarazo, S.J. Roseblade, A.R. Cowley, R. Fernandez, J.M. Brown, J.M. Lassaletta, *J. Am. Chem. Soc.* **2005**, *127*, 3290.
95 C.C. Scarborough, M.J.W. Grady, I.A. Guzei, B.A. Gandhi, E.E. Bunel, S.S. Stahl, *Angew. Chem. Int. Ed.* **2005**, *44*, 5269.

9
Chelate and Pincer Carbene Complexes

Guillermina Rivera and Robert H. Crabtree

9.1
Introduction

The rise of phosphines, PR_3, as a ligand set in organometallic chemistry was based on their easy, predictable steric and electronic tunability [1] by variation of the substituents, R, combined with their success in mediating a wide variety of homogeneous catalytic reactions [2]. The much more recent rise of N-heterocyclic carbenes (NHCs) as a ligand set of comparable importance [3] to phosphines was initiated by the isolation [4] of the free carbenes and encouraged by their success [5] as ligands in homogeneous catalysis versus the phosphine analogues. Perhaps the most notable example was the great improvement [6] brought about in the Grubbs metathesis catalyst by replacement of a phosphine by an NHC.

Like PR_3, NHCs (**1**) are sterically tunable by alteration of the wingtip groups at N1 and N3 and have an electron-donor power [7, 8] considerably greater than that of phosphines – a useful property in many contexts, although it would be desirable to have NHCs of lower donor power, comparable to that of phosphines. Tuning the electron-donor power is possible by altering the nature of the azole, as in the less powerful donor benzimidazole (**2**) and more powerful donor imidazoline (**3**) based NHCs, but so far essentially only imidazole-2-ylidene (**1**) has been used in chelate or pincer ligands. Considering the many nitrogen-containing heterocycles available, the opportunities for future development are vast.

The phosphine ligand family was greatly enriched by adoption of a wide variety of chelate and pincer ligand architectures [9]. The same developments are likely to arise in with NHCs [10], but the big differences between PR_3 and NHC structures

N-Heterocyclic Carbenes in Synthesis. Edited by Steven P. Nolan
Copyright © 2006 WILEY-VCH Verlag GmbH & Co. KGaA, Weinheim
ISBN: 3-527-31400-8

lead to differences in design strategy. The relatively few published examples of chelate bis-NHCs makes their promise very clear, because considerable modification of complex properties has been noted with relatively small changes in architecture. Chelation may also be useful in helping guard against decomposition pathways of NHC complexes, such as reductive elimination [11].

Many recent reviews overlap with the present chapter, including ones on chelate and pincer NHCs [10]; palladacycles in catalysis [12]; and decomposition pathways of metal-NHC complexes [13]. An exhaustive review of Ag(I) complexes also includes a discussion of many of the chelate NHC complexes that can be synthesized from them [14]. One review – perhaps the best available – covers the whole NHC field [5a]. Several recent monographs also include a treatment of closely related topics [15, 16]. We therefore concentrate here on topics less fully covered before, such as ligand design and synthesis, as well as discussing most recent developments. We also emphasize discussion of bis- and tris-NHC ligands over the many cases where chelation has been achieved with one NHC and one or more P- or N-donors.

9.2
Design Strategy

In a chelate phosphine, such as $Ph_2PCH_2CH_2PPh_2$ (**4**, $n = 2$, R = Ph), the linker is directly attached to the donor atoms, so for a linker of length n the resulting chelate ring size is $(n + 3)$. Typical linkers have $n = 1$–4, resulting in chelate ring sizes of 4–7. Furthermore, there is free rotation about all of the bonds in the chelate ring. In a chelating imidazole-2-ylidene NHC (**5**), in contrast, the linker is attached not to the donor atom at C2 but to the adjacent N1 or N3 atoms. The chelate ring size in the resulting complex is now $(n + 5)$, leading to typical ring sizes of 6–9 for linkers of length $n = 1$–4. In addition, the rigid azole ring imposes a near-coplanar conformation on both sets of M–C2–N1–C(linker) bonds, resulting in much less conformational flexibility for the chelate ring as a whole than is the case for diphosphines having the same value of n.

4 **5**

Recent work has shown strong effects of linker length on the orientation of the azole rings in a complex of type **5** [17]. For example, when a chelating NHC forms M–C bonds in the MC_2 (xy) plane of an octahedral or square-planar complex, a short linker will constrain the azole rings to adopt a conformation that aligns them close to the xy plane. A long linker of length 3–4, in contrast, produces a chelate ring size of 8–9 that aligns the azoles along the z direction and tends to lock them in that conformation.

Synthetic work with different length linkers (Scheme 9.1) [17] has shown that Rh(I) complexes formed by transmetallation to [Rh(cod)Cl]$_2$ of the silver NHC complex tend to prefer a conformation that aligns the azole rings in the z direction normal to the square plane for all linkers of length 1–4. This conformation places the bulky wingtip R groups in the otherwise vacant z direction normal to the square plane (xy) of the complex. For $n = 3$ and 4, this is possible in a chelate because the long linker permits the adoption of the favored azole orientation. For $n = 1$ and 2, however, a bis-Rh(I) nonchelate complex is formed because only in this way can both NHCs adopt the favored azole ring orientation – the chelate form would have too short a linker for both azole rings to align with the z direction.

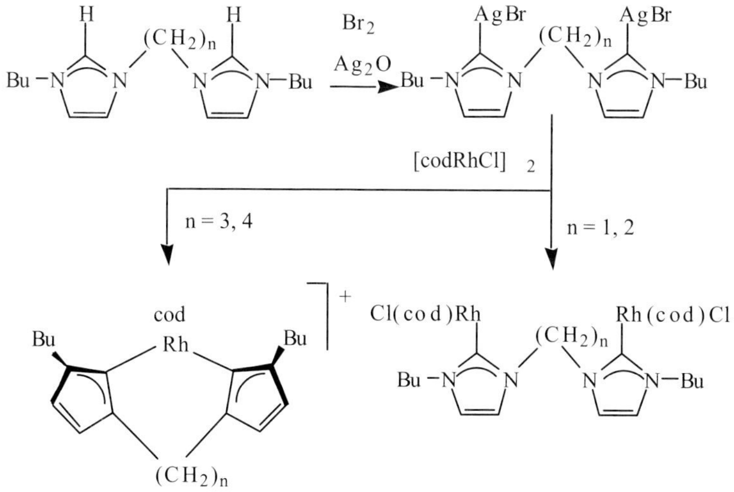

Scheme 9.1

In contrast, direct metallation of the imidazolium salt without Ag$^+$ (Scheme 9.2) gives a Rh(III) species where the z direction, now occupied by iodide ions, is more hindered than the xy plane, occupied by a small acetate group. Here, the shortest linker gives the highest yield and the longer linkers are less favored: $n = 1$, 74% yield; $n = 2$, 42%; $n = 3$, 10%; $n = 4$, 10%. Large linker effects are again seen, but in the yield, not in the nature, of the product [17].

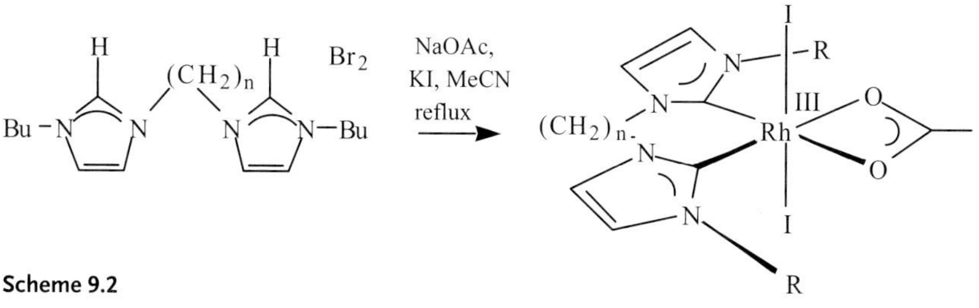

Scheme 9.2

Slaughter and coworkers have reported a counter-example [18] in which a dimeric silver complex of the bulky bis(N-heterocyclic carbene) ligand 1,1′-dimesityl-3,3′-methylenediimidazol-2,2′-diylidene gave a chelate Rh(I) derivative after transmetallation from the Ag complex even though an $n = 1$ linker was present. It is not yet clear, however, if the difference in outcome is due to the bulky mesityl linker or, more likely, to the use of a noncoordinating counter-ion, BF$_4$. The coordinating counter-ion, Br$^-$, always used by Mata et al. [17], may allow access to a reactive NHC-Ag-Br as transient intermediate, which could have different kinetic preferences to Slaughter's [Ag(NHC)$_2$]BF$_4$ intermediate. Alternatively, the presence of excess silver in soluble form (e.g., AgBF$_4$) allows reversible transmetallation that tends to yield thermodynamic products. Once this is better understood, it may be possible to choose synthetic conditions that permit high selectivity in such cases. In future, both coordinating and noncoordinating anions should be tried as counter-ions in the azolium precursor and in the intermediate Ag carbenes for the transmetallation synthetic route to metal-NHC complexes, in case different results are obtained.

A disadvantage of linker placement at N is the possibility of wingtip or linker cleavage via Hoffmann elimination when a β-H is present, or by attack of a nucleophile with N–C bond cleavage where the α-C is an sp^3 carbon (Scheme 9.3). The resulting intermediate (**6**) would probably easily decompose. The relative lability of different groups is not fully documented but, among alkyls, neopentyl [19, 20] seems resistant to both cleavage processes, and both -CH$_2$-CMe$_2$-CH$_2$- and o-C$_6$H$_4$ linkers seem very stable.

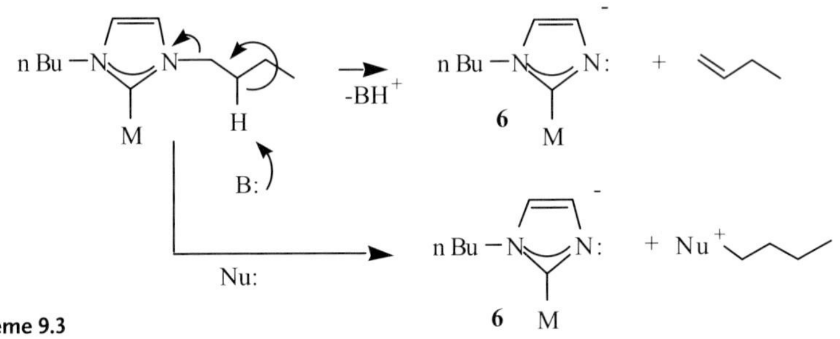

Scheme 9.3

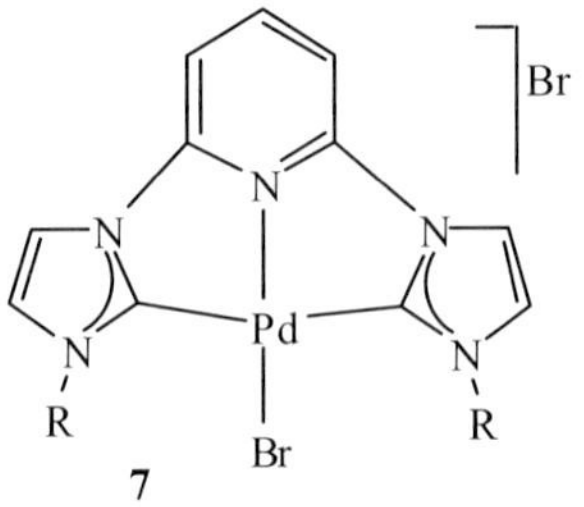

Pincer ligands [10] are subject to similar effects, except that now a direct bond is possible between the central arene ring and the lateral azole rings, resulting in a minimum chelate ring size of 5 and favoring planarity of the resulting pincer (e.g., **7**). Methylene-linked pincers are also common but give ruffled geometries [21]; pincers with longer linkers would be of interest. A PCP pincer with phosphine arms and an NHC core has been reported for Pd(II) [22].

9.2.1
Bite Angle

A leading variable affecting the properties of chelate phosphine complexes is the P–M–P angle or ligand bite angle [23]. When the linker enforces a bis-phosphine bite angle markedly different from the usual 90°, big reactivity differences have been noted in both stoichiometric and catalytic reactions. This aspect has not been much developed in the NHC case, where bite angles have always been close to 90°, but this is clearly an aspect worthy of future study.

In a rare experimental case where bite angle changes were associated with changes in reactivity, an N,C chelating 2-pyridyl substituted NHC was compared with that of a 2-pyridylmethyl substituted NHC. On metallation with Ni(II), the former gave a six-coordinate complex while the latter, where the CH_2 linker allows the ligand to adapt to the much more open tetrahedral case (ideal angle 109°), gave a tetrahedral complex [24]. Many complexes are known where the conformation of a 2-pyridylmethyl substituted NHC adapts to fit a near-90° angle so this ligand can be considered to have a flexible bite angle. The same argument should hold for bis-NHCs with $(CH_2)_n$ linkers but this has not yet been tested.

9.2.2
Tripod Ligands

Few facial tridentate (tripod) NHC ligands have yet been reported. Fehlhammer's chelating C,C,C tripod triscarbene ligand, hydrotris(3-alkyl-imidazoline-2-yliden-1-yl)borate, is noteworthy [25, 26]. It forms a hexacarbene iron complex, dating from 1996, and a cobalt complex, dating from 2001. A related tripod ligand [27] binds just once to Fe(II) to give tetrahedral [HB(NHC)$_2$FeBr]. Here, the bulky Bu^t wingtip substituents no doubt help prevent formation of the often undesired bis-tripod complex.

Peris, Meyer and coworkers have reported the reaction of the potentially tripodal ligand MeC(CH$_2$CH$_2$L)$_3$, with [M(cod)Cl]$_2$, where L is an N,N-dialkylated imidazolium group [28]. Instead of the tripodal binding that would have been expected for L = PMe$_2$, each ligand chelates to one metal center and binds in monodentate fashion to a third. Either irreversibility of M–NHC binding prevents "annealing" of the structure to give the tripod product or the bulky isopropyl wingtip groups prevent the facial tris-NHC structure from forming in this six-coordinate situation.

Once the synthetic difficulties are resolved, tripodal NHCs are likely to be of great interest because the rigid structure is likely to strongly disfavor decomposition pathways that cleave the NHC from the metal.

9.3
Synthetic Strategies

N,N′-Dialkyl imidazolium salts can be very readily synthesized by any of a wide variety of procedures, but most simply by alkylation of the parent imidazole or, to achieve placement of mixed alkyl or alkyl/aryl groups, of the appropriate N-alkyl or N-aryl imidazole. N,N′-Diaryl imidazolium salts are typically synthesized by forming the heterocyclic ring *de novo* from a glyoxal, formaldehyde and an arylammonium salt. Almost any desired functionalized NHC precursor can be prepared by one or more of the standard organic procedures that have been extensively discussed in a monograph [29].

Synthesis of phosphines can be challenging but, once achieved, binding to the metal is usually undemanding; in contrast, the synthesis of the imidazolium salts that are typical NHC precursors is usually much more easy and reliable than the subsequent metallation step. Typical metallation routes involve deprotonation of the imidazolium salt to give the free NHC with a strong base like BuLi, followed by interaction with the metal salt. This, the *free carbene route*, can be difficult if the salt has reactive functionality, such as a $-CH_2-$ linker, that is sensitive to BuLi. For chelate carbenes, *cyclometallation* of the imidazolium salt has often been used as a synthetic route [30–32]. The wingtip functionality may contain a group like a pyridine that first binds and brings the imidazolium salt C–H bond into the coordination sphere of the metal, where it can be cleaved by CH activation; alternatively, the NHC is installed first and a wingtip CH bond is subsequently cleaved. Ag_2O has proved to be an efficient reagent for metallation of the imidazolium salt, after which *transmetallation* to the desired transition metal can often be carried out. Originally developed by Lin [14], this procedure has proved very broadly useful. Other Ag(I) salts have also proved useful, such as the triflate, acetate and carbonate. *Direct metallation* of the imidazolium salt is sometimes possible using a mild base such as sodium acetate [33]. In such a case the imidazolium salt C–H bond may oxidatively add to the metal or form an agostic intermediate, followed in either case by loss of AcOH [34]. Acetate is considered a particularly good base for this purpose, perhaps because, in the η^1-form, the unbound acetate oxygen is correctly positioned to deprotonate the CH proton at C2 of an azolium, while at the same time the C2 begins to form a bond to the metal. In this way, free carbene need never be formed. *Oxidative addition* of the C–H bond can be useful [35], particularly where a low-valent metal precursor, such as $Pd_2(dba)_3$, is available. In such a case the Pd(II) NHC complex is normally obtained. In one case, formation of a bis-NHC Pd(II) compound was noted, probably indicating the loss of H_2 in the reaction.

9.3 Synthetic Strategies

In the cyclometallation route, imidazolium salts do not always give normal (C2 bound) carbenes on metallation [30], and the chemistry of these ligands can be much more complicated than previously thought. N,N'-Disubstituted imidazolium salts of type [(2-py)(CH$_2$)$_n$(C$_3$H$_3$N$_2$)R]BF$_4$ react with IrH$_5$(PPh$_3$)$_2$ to give N,C-chelate products (Scheme 9.4. n = 0, 1; 2-py = 2-pyridyl; C$_3$H$_3$N$_2$ = imidazolium; R = mesityl, n-Bu, i-Pr, Me). Depending on the circumstances, three types of kinetic products can be formed: in one, the imidazole metallation site is the normal C2, as expected; in the second, metallation occurs at the abnormal C5 site; and in the third, C5 metallation is accompanied by hydrogenation of the imidazolium ring. The bonding mode was confirmed by structural studies, but spectroscopic criteria were proposed for distinguishing the cases. Initial hydrogen transfer can take place from the metal polyhydride to the carbene to give the imidazolium ring hydrogenation product, as shown by isotope labeling; this hydrogen transfer proves reversible on reflux when the abnormal aromatic carbene is obtained as final product. Care may therefore be needed in future in verifying the structure(s) formed in cases where a catalyst is generated *in situ* from imidazolium salt and metal precursor.

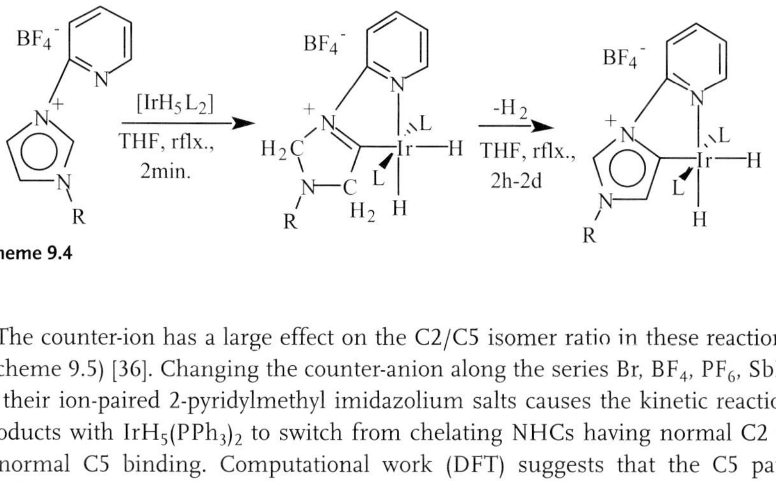

Scheme 9.4

The counter-ion has a large effect on the C2/C5 isomer ratio in these reactions (Scheme 9.5) [36]. Changing the counter-anion along the series Br, BF$_4$, PF$_6$, SbF$_6$ in their ion-paired 2-pyridylmethyl imidazolium salts causes the kinetic reaction products with IrH$_5$(PPh$_3$)$_2$ to switch from chelating NHCs having normal C2 to abnormal C5 binding. Computational work (DFT) suggests that the C5 path involves C–H oxidative addition to Ir(III) to give Ir(V) with little anion dependence. The C2 path, in contrast, goes by heterolytic C2–H activation with proton transfer to the adjacent hydride.

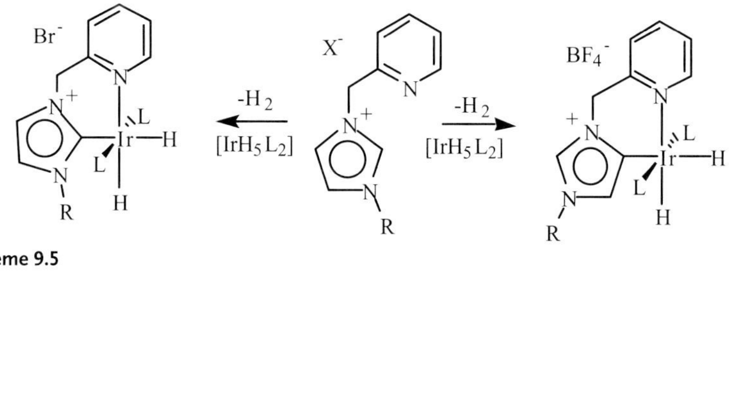

Scheme 9.5

The proton that is transferred is accompanied by the counter-anion in an anion-coupled proton transfer, leading to an anion dependence of the C2 path and, therefore, of the C2/5 selectivity. The C2 path goes via Ir(III), not Ir(V), because the normal C2 NHC is a much less strong donor ligand than the abnormal C5 NHC. PGSE NMR experiments support the formation of ion-pair in both the reactants and the products. ^{19}F,^{1}H-HOESY NMR experiments indicate an ion pair structure for the products that is consistent with the computational prediction [ONIOM(B3PW91/UFF)].

Nolan and coworkers have shown how IBu^t [1,3-bis(t-butyl)imidazole-2-ylidene] can cyclometallate in its reactions with Ir(I) and Rh(I) to give chelating NHCs (Scheme 9.6) [37]. When such a cyclometallation happens twice over, the very unusual 14-electron bis-chelate M(III) species **8** is formed. DFT calculations suggest that the NHC can act as a π-donor, thus stabilizing the 14e system. Meyer and coworkers [38] had previously suggested that NHCs can be significant π-acceptors; thus, depending on the metal fragment involved, this may mean that NHCs have an unusual degree of electronic flexibility. Since the ligand atom, carbon, is part of an aromatic ring, the imidazole-2-ylidene, there may be greater opportunity for charge distribution to and from the ring compared with conventional ligands such as PPh$_3$, where the arene ring is only a substituent at the ligand atom, and so is more electronically isolated.

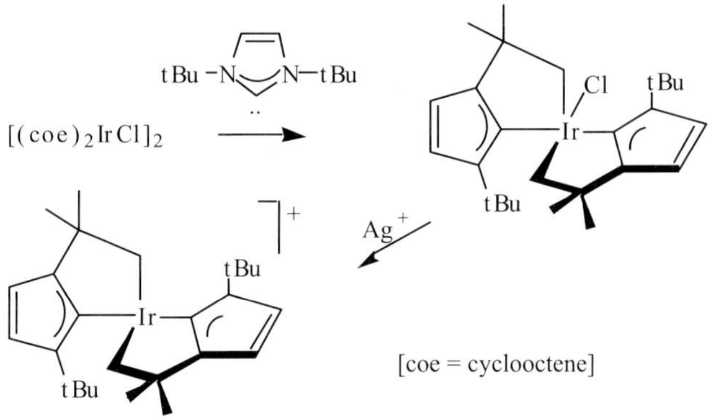

Scheme 9.6

As, occasionally, the Ag$_2$O reagent can give unexpected products in its reaction with an imidazolium salt, care needs to be taken to verify the desired reaction has indeed occurred. Oxidative C–C cleavage has been seen, for example (Scheme 9.7) [39]. In another case, deprotonation occurred on the wingtip substituent a to nitrogen to give an entirely unexpected chelate after transfer to Pd(II) (Scheme 9.8).

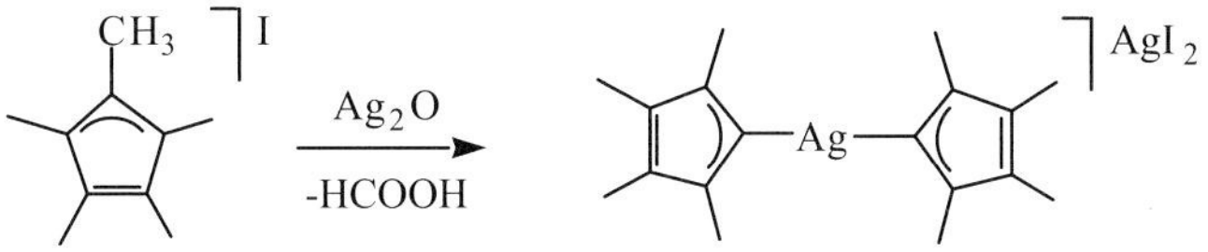

Scheme 9.7

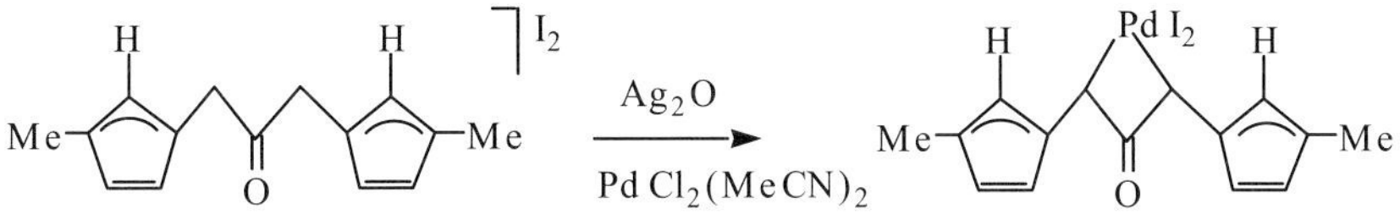

Scheme 9.8

9.4
Failure to Chelate

Potentially chelating di- or triphosphines almost always chelate if suitable labile sites are present at the metal. The lability of the M–P bond presumably allows the system to attain the thermodynamically preferred chelate product. A common problem in chelate and pincer NHC chemistry is that the NHC precursor often gives a kinetic product in which each NHC azole ring binds to a different metal center, as in {M-NHC-linker-NHC-M′}. This occurs even when the stoichiometry and order of addition is chosen to favor chelation. This can result in the formation of M_2L_2 complexes such as **9**, formed by transmetallation from the silver NHC; the normal chelate was also formed along with the 2:2 complex [40].

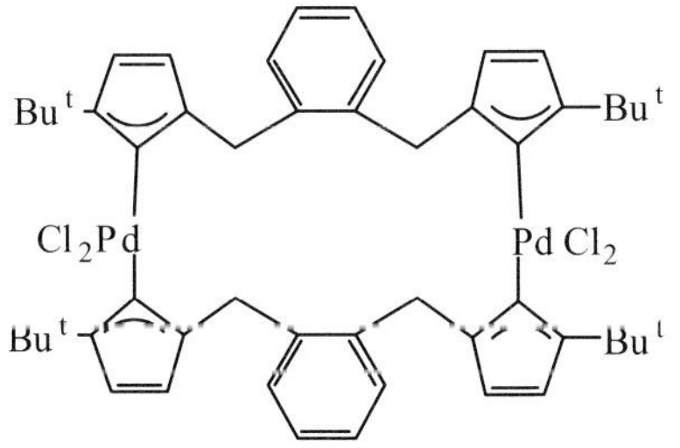

Since Ag(I) prefers a linear geometry, a potential chelator NHC forms, unsurprisingly, a nonchelating Ag-NHC-linker-NHC-Ag′ complex, and many examples exist [41]; however, this also happens with metals for which a square-planar or octahedral geometry is preferred, such as Rh(I), where a chelate would have been expected by analogy with phosphine coordination chemistry [17].

9 Chelate and Pincer Carbene Complexes

In one case, the presence of excess Ag(I) appears to have mediated the conversion of the 2:1 into the chelate complex. In the case shown, if the Ag-NHC, having an unknown structure, is isolated, the resulting transfer gives the 2:1 complex but if the excess Ag_2O is retained in the reaction mixture the chelate complex is formed (Scheme 9.9).

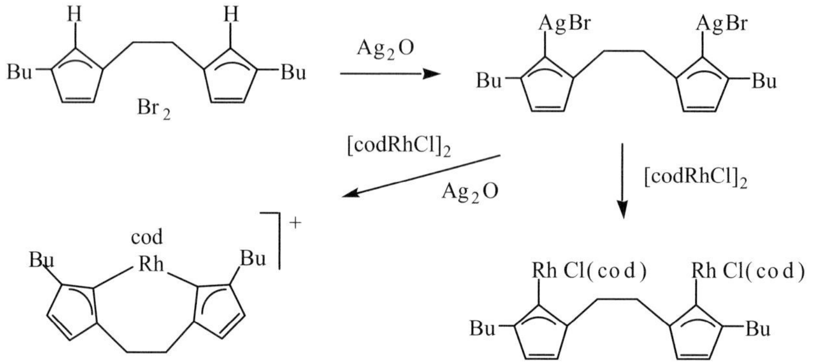

Scheme 9.9

The same problems often arise with potentially pincer or tripod ligands, as with **10** [42]. Here, the Rh(I) would lack π-acceptor ligands if the ligand did form a pincer, and so the system is strongly biased towards the 2:1 complex.

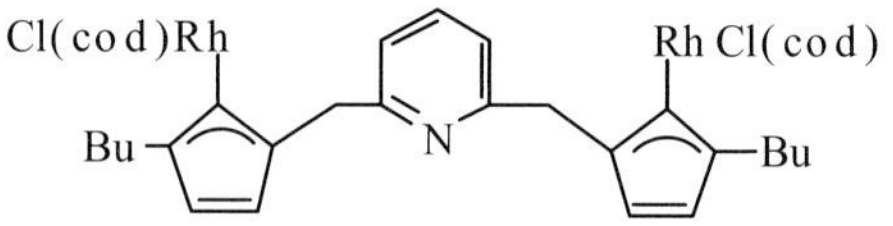

10

In a potential tripodal tris-NHC ligand with a $MeC(CH_2)_3$ linker, Peris and coworkers showed that *in situ* formation of the Ag(I) complex could be successfully followed up with transmetallation to Rh(I) and Ir(I); however, only two of the three NHC arms chelated to a single metal center, with the third arm being bound by a second metal in monodentate fashion (Scheme 9.10) [43].

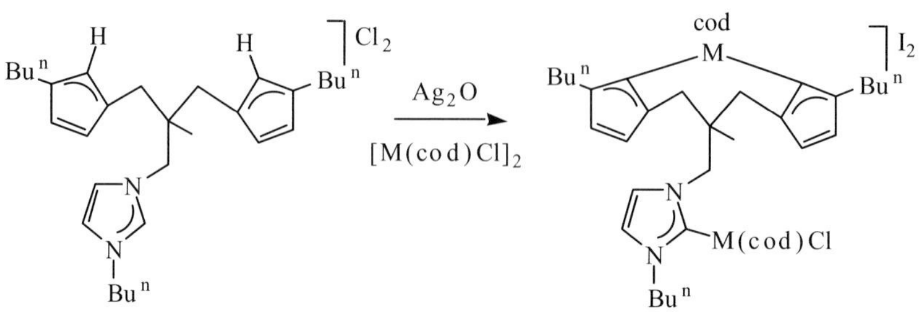

Scheme 9.10

So far no general solution has been found for this difficult problem, although there are cases where changing the synthetic method leads to successful chelate formation. For example, Scheme 9.11 shows a case where chelation fails with silver transmetallation, but is successful with direct metallation [17]. Here, a Rh(III) complex is formed where the linker-dependent bias towards the 2:1 complex, mentioned above, is now absent.

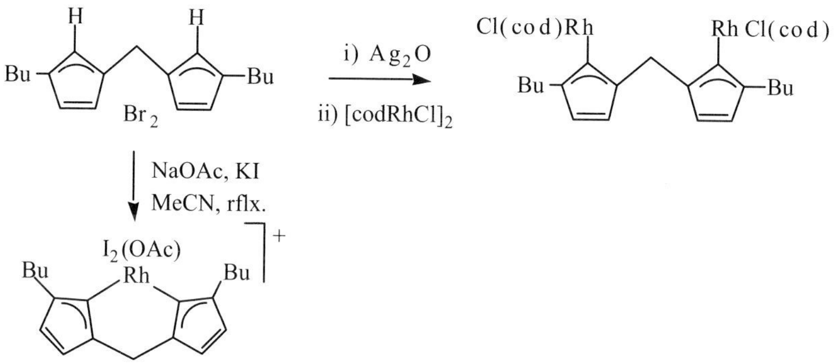

Scheme 9.11

9.5
Ligand Properties

NHCs have generally been considered much stronger donor ligands than phosphines. Chianese et al. have quantified the donor properties experimentally by looking at the ν(CO) IR frequencies for (L)Ir(CO)$_2$Cl in a Tolman [1]-like approach (L = NHC or PR$_3$) [44]. Abnormal imidazol-5-ylidenes are much stronger electron donors than their ubiquitous imidazol-2-ylidene normal NHC counterparts. After conversion of the Ir-based IR data into the Tolman scale, the Tolman electronic parameter for 1-isopropyl-2,4-diphenyl-3-methylimidazolin-5-ylidene was shown to be 2039 cm^{-1}, compared with ca. 2050 cm^{-1} for typical NHCs. Data on chelating (bis-NHC)Ir(CO)$_2$ complexes show that the electronic parameter is not significantly affected by chelation [17].

Quantification of the donor properties has also proved possible computationally. For example, Clot et al. have calculated ν(CO) for LNi(CO)$_3$ [45]. Nolan and coworkers made very similar LNi(CO)$_3$ derivatives experimentally – satisfactory accord was obtained between the prior predictions and measured values and the strong donor character of the NHC ligand was confirmed [46].

Their strong donor power makes NHCs useful for stabilizing complexes of ligands like N$_2$ that demand very strong back-donation from the metal. An excellent example is the Danopoulos–Motherwell [47] Fe(0) bis-N$_2$ complex stabilized by a CNC pincer. Fe(C–N–C)(N$_2$)$_2$ was obtained by the reduction of Fe(C–N–C)Br$_2$ with Na(Hg). The particular C–N–C involved, 2,6-bis(aryl-imidazol-2-ylidene)pyridine, also employs steric protection in the form of the bulky wingtip 2,6-(Pr^i)$_2$C$_6$H$_3$ groups.

Simple dissociation of NHCs must be much disfavored versus phosphines by the exceptionally high basicity of the free NHC. Nevertheless, direct loss of the NHC by substitution with a triarylphosphine, followed by reaction with the halocarbon solvent, was proposed [48] in the case of [(Ar$_3$P)$_2$RhCl(IMes)], although the process may be more complicated than simple displacement.

Loss of the NHC in the form of a 2-substituted derivative is also possible by reductive elimination with an adjacent alkyl or aryl group [49]. This is problematic in that catalytic cycles often involve alkyl or aryl intermediates that can lead to catalyst deactivation.

9.5.1
Types of Ligand

9.5.1.1 C,C Chelates

The linker-dependent conformational properties of bis-NHC chelates have been discussed earlier.

In a system reported by Tan, Bergman, and Ellman, a benzimidazole can cyclometallate, presumably after binding of the C=C group to Rh [50]. The resulting NHC has a wingtip proton, unusual in NHCs described to date (Scheme 9.12). The system can then go on to catalytically cyclize the benzimidazole via intramolecular addition of the C–H bond at the 2-position of the azole across the C=C double bond.

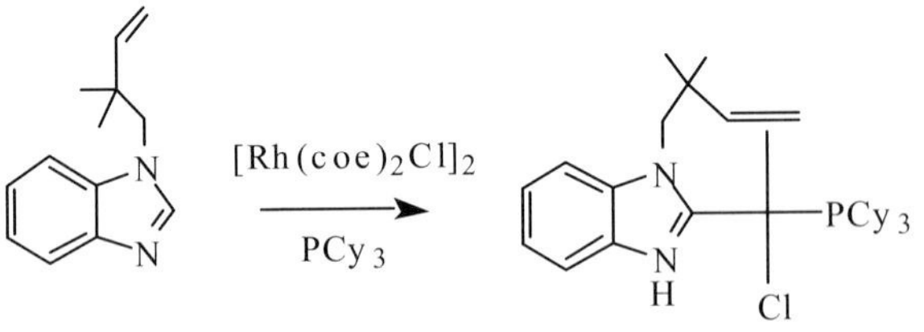

Scheme 9.12

9.5.1.2 N,C Chelates and Pincers

Numerous N,C chelates, most often bound to Pd, have been made and employed in catalysis. The synthesis is particularly easy because several routes are viable. With imidazolium salts as NHC precursors, Pd(OAc)$_2$ readily metallates, and Pd$_2$(dba)$_3$ readily undergoes CH activation. The transmetallation route from silver also works well. Pincers are also strongly represented in this class. Compounds 7 and 11–13 are typical illustrations [51–54]. The M–N bond is often labile, leading to fluxionality or binding of counter-ions.

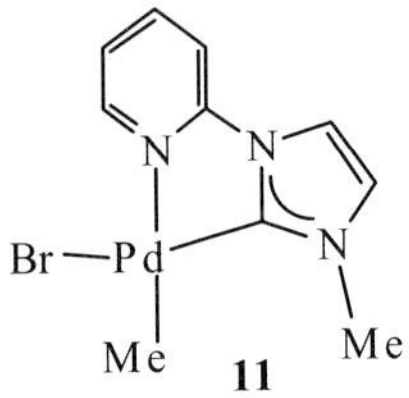

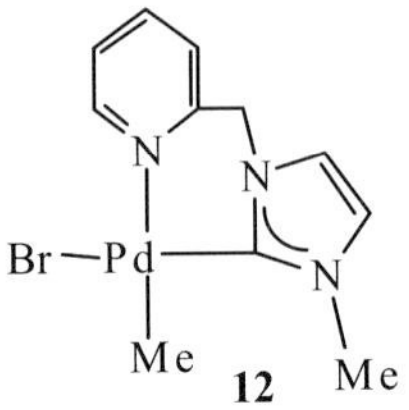

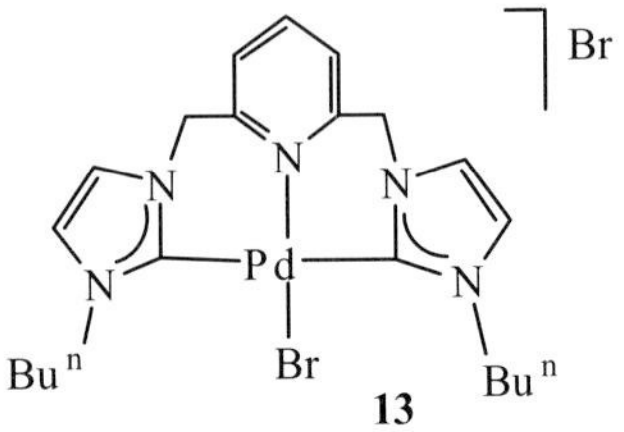

9.5.1.3 P,C Chelates and Pincers

N-heterocyclic carbenes are better donors and more strongly bound than phosphines, so a chelate or pincer having both groups can in principle give hemilabile behavior. P,C chelates and pincers, mainly of palladium, have been synthesized (e.g., **14–17**; X = halide; L = phosphine) by several different groups [55–58] and applied to C–C coupling catalysis.

The complexes were synthesized by typical methods – transfer from the silver carbene and direct metallation with $PdCl_2$ (both methods gave good results). Fluxional behavior similar to that of the CCC and CNC Pd(II) pincer analogues is observed [21].

Ruthenium complexes of this class are synthesized by a different route. First, the precursor silver P,C chelate carbene, studied in detail by Lee et al., gives a trinuclear silver complex [59], which upon transfer to ruthenium gives dimer complex **18**, which is active in transfer hydrogenation catalysis.

The cyclooctadiene ligand in rhodium and iridium complexes of the type [M(P-C)(cod)]BPh$_4$, can be displaced by CO to give [Rh(P-C)(CO)$_2$]BPh$_4$. Both were active in intramolecular hydroamination catalysis, with [Rh(P-C)(CO)$_2$]BPh$_4$ being more active than the bis-carbene analogue [60].

9.6 Catalysis

It is unclear how far the ligand loss by M–C bond cleavage mentioned in earlier sections can be inhibited by incorporation of the NHC into a chelate or pincer. In the phosphine series, a cyclopalladated tri-*o*-tolylphosphine Pd(II) complex introduced by Herrmann et al. for Pd-catalyzed Heck and cross-coupling reactions seemed an excellent example of catalyst stabilization by chelate formation [61]. However, doubts were expressed about the homogeneity of these catalysts, notably by Beletskaya and Cheprakov [62]. On this interpretation, not only the Pd–C bonds but also the Pd–P bonds of the precursor are cleaved on catalyst activation, and a nanoparticulate Pd(0) is the true catalyst. Beletskaya and Cheprakov have even gone so far as to classify pincer Pd catalysts in general as a subgroup of phosphine-free, nanoparticulate catalysts [12].

For many cases the balance of evidence supports the Beletskaya–Cheprakov arguments in favor of catalyst heterogeneity, but there do appear to be NHC-based pincer catalysts that resist decomposition even under harsh conditions. For example [63], the C,N,C pincer **7** maintains full Heck coupling activity even in the presence of Hg(0), an additive that normally poisons any Pd-based heterogeneous catalysis [64]. Immobilization of complex **7** on Montmorillonite K-10 afforded effective catalysts that show catalytic activity similar to their homogeneous counterparts. XPS data support the proposal that elemental Pd(0) is not the active catalyst in these systems. The authors were able to recycle the catalyst at least ten times, without significant loss of activity [65]. The same catalyst, supported on various clays, was also used for Sonogashira coupling; high yields and good recyclability were found [66].

9.6.1
Medicinal Applications

Youngs and coworkers have developed pincer Ag-NHC complexes as useful antimicrobials [67]. Silver(I)-2,6-bis(ethanolimidazolemethyl)pyridine hydroxide (**19**) and the propanolimidazolemethyl analogue have been synthesized in high yield by reaction of silver(I) oxide with appropriate N-substituted pincer ligand precursors. The bactericidal activity of the water-soluble silver(I)-carbene complexes proved to be better than that of silver nitrate.

19

9.7
Conclusions

NHCs have sometimes been considered phosphine analogues, but in fact they provide an entirely new set of properties and problems. They are more basic than phosphines and much less prone to simple dissociation, but this non-lability also makes it harder to construct chelates and pincers because initial "errors" are not corrected. NHCs can be irreversibly cleaved from the metal under circumstances that need to be better defined. Chelating bis-NHCs introduce various considerations that do not arise for phosphines: the strong linker-dependent chemistry seen to date has been ascribed to the conformational properties of the 6–9-membered rings that are formed for typical linker lengths ($n = 1$–4). Linker dependence is likely to be found for many other properties of these compounds in future. Synthetic methods need to be improved and a wider variety of heterocycles need to be examined. Much needs to be done in this rapidly developing area.

Acknowledgments

We thank the US DOE, Johnson Matthey, BP, Sanofi-Aventis, and CONACYT of Mexico (GR) for support of our work in this area.

References

1 Tolman CA, *Chem. Rev.*, **1977**, *77*, 313–348.
2 Cornils B, Herrmann WA, *Applied Homogeneous Catalysis with Organometallic Compounds*, Wiley-VCH, New York, **2002**; Crabtree RH, *The Organometallic Chemistry of the Transition Metals*, 4th edn., Wiley, New York, **2005**.
3 Weskamp T, Bohm VPW, Herrmann WA, *J. Organometal. Chem.*, **2000**, *600*, 12–22.
4 Arduengo AJ, Harlow RL, Kline M, *J. Am. Chem. Soc.*, **1991**, *113*, 361–363; Igau A, Grutzmacher H, Baceiredo A, Bertrand G, *J. Am. Chem. Soc.*, **1988**, *110*, 6463–6466.
5 (a) Bourissou D, Guerret O, Gabbai FP, Bertrand G, *Chem. Rev.*, **2000**, *100*, 39–91. (b) Hillier AC, Grasa GA, Viciu MS, Lee HM, Yang CL, Nolan SP, *J. Organometal. Chem.*, **2002**, *653*, 69–82.
6 Chatterjee AK, Toste FD, Choi TL, Grubbs RH, *Adv. Synth. Catal.*, **2002**, *344*, 634–637.
7 Huang JK, Stevens ED, Nolan SP, Petersen JL, *J. Am. Chem. Soc.*, **1999**, *121*, 2674–2678.
8 Chianese AR, Kovacevic A, Zeglis BM, Faller JW, Crabtree RH, *Organometallics*, **2004**, *23*, 2461–2468.
9 van der Boom ME, Milstein D, *Chem. Rev.*, **2003**, *103*, 1759–1792.
10 Peris E, Crabtree RH *Coord. Chem. Rev.*, **2004**, *248*, 2239–2246.
11 Cavell KJ, McGuinness DS, *Coord. Chem. Rev.*, **2004**, *248*, 671–681.
12 Beletskaya IP, Cheprakov AV, *J. Organometal. Chem.*, **2004**, *689*, 4055–4082.
13 Crudden CM, Allen DP, *Coord. Chem. Rev.*, **2004**, *248*, 2247–2273.
14 Lin IJB, Vasam CS, *Comments Inorg. Chem.*, **2004**, *25*, 75–129.

15 *Metal Carbenes in Organic Synthesis (Topics in Organometallic Chemistry)*, ed. Dötz, KH, Springer, New York, **2004**.

16 *Carbene Chemistry: From Fleeting Intermediates to Powerful Reagents*, ed. Bertrand, G, Dekker, New York, **2002**.

17 Mata JA, Chianese AR, Miecznikowski JR, Poyatos M, Peris E, Faller JW, Crabtree RH, *Organometallics*, **2004**, *23*, 1253–1263.

18 Wanniarachchi YA, Khan MA, Slaughter LM, *Organometallics*, **2004**, *23*, 5881–5884.

19 Miecznikowski JR, Crabtree RH, *Organometallics*, **2004**, *23*, 629–631.

20 Albrecht M, Miecznikowski JR, Samuel A, Faller JW, Crabtree RH, *Organometallics*, **2002**, *21*, 3596–3604.

21 Miecznikowski JR, Grundemann S, Albrecht M, Mégret C, Clot E, Faller JW, Eisenstein O, Crabtree RH, *J. Chem. Soc., Dalton Trans.*, **2003**, 831–838.

22 Lee HM, Zeng JY, Hu CH, Lee MT, *Inorg. Chem.*, **2004**, *43*, 6822–6829.

23 Dierkes P, van Leeuwen PWNM, *J. Chem. Soc., Dalton Trans.*, **1999**, 1519–1529.

24 Winston S, Stylianides N, Tulloch AAD, Wright JA, Danopoulos AA, *Polyhedron*, **2004**, *23*, 2813–2820.

25 Kernbach U, Ramm M, Luger P, Fehlhammer WP, *Angew. Chem. Int. Ed. Engl.*, **1996**, *35*, 310–312.

26 Frankel R, Kernbach U, Bakola-Christianopoulou M, Plaia U, Suter M, Ponikwar W, Noth H, Moinet C, Fehlhammer WP, *J. Organometal. Chem.*, **2001**, *617*, 530–545.

27 Nieto I, Cervantes-Lee F, Smith JM, *Chem. Commun.*, **2005**, 3811–3813.

28 Mas-Marza E, Peris E, Castro-Rodriguez I, Meyer K, *Organometallics*, **2005**, *24*, 3158–3162.

29 Grimmett MR, *Imidazole and Benzimidazole Synthesis*, Academic Press, San Diego, **1997**.

30 Gründemann S, Kovacevic A, Albrecht M, Faller JW, Crabtree RH, *J. Am. Chem. Soc.*, **2002**, *124*, 10 473–10 481.

31 Grundemann S, Kovacevic A, Albrecht M, Faller JW, Crabtree RH, *Chem. Commun.*, **2001**, 2274–2275.

32 Gründemann S, Albrecht M, Kovacevic A, Faller JW, Crabtree RH, *J. Chem. Soc., Dalton Trans.*, **2002**, 2163.

33 Albrecht M, Crabtree RH, Mata J, Peris E, *Chem. Commun.*, **2002**, 32–33.

34 Viciano M, Mas-Marza E, Poyatos M, Sanau M, Crabtree RH, Peris E, *Angew. Chem. Int. Ed.*, **2005**, *44*, 444–447.

35 Gründemann S, Albrecht M, Kovacevic A, Faller JW, Crabtree, RH, *J. Chem. Soc., Dalton Trans.*, **2002**, 2163.

36 Kovacevic A, Grundemann S, Miecznikowski JR, Clot E, Eisenstein O, Crabtree RH, *Chem. Commun.*, **2002**, 2580–258; Appelhans LN, Zuccaccia D, Kovacevic A, Chianese A, Miecznikowski M, Macchioni A, Clot E, Eisenstein O, Crabtree RH, *J. Am. Chem Soc.*, **2005** *127*, 16299–16311.

37 Scott NM, Dorta R, Stevens ED, Correa A, Cavallo L, Nolan SP, *J. Am. Chem. Soc.*, **2005**, *127*, 3516–3526.

38 Hu X, Castro-Rodriguez I, Olsen K, Meyer, K, *Organometallics*, **2004**, *23*, 755–764.

39 Chianese AR, Zeglis BM, Crabtree RH, *Chem. Commun.*, **2004**, 2176–2177.

40 Magill AM, McGuinness DS, Cavell KJ, Britovsek GJP, Gibson VC, White AJP, Williams DJ, White AH, Skelton BW, *J. Organometal. Chem.*, **2001**, *617*(1), 546–560.

41 Lee KM, Wang HMJ, Lin IJB, *J. Chem. Soc., Dalton Trans.*, **2002**, 2852–2856.

42 Simons RS, Custer P, Tessier CA, Youngs WJ, *Organometallics*, **2003**, *22*, 1979–1982.

43 Mas-Marza E, Poyatos M, Sanau M, Peris E, *Inorg. Chem.*, **2004**, *43*, 2213–2219.

44 Chianese AR, Kovacevic, A, Zeglis, BM, Faller JW, Crabtree RH,*Organometallics*, **2004**, *23*, 2461–2468.

45 Perrin L, Clot E, Eisenstein O, Loch J, Crabtree RH, *Inorg. Chem.*, **2001**, *40*, 5806–5811.

46 Dorta R, Stevens ED, Scott NM, Costabile C, Cavallo L, Hoff CD, Nolan SP, *J. Am. Chem. Soc.*, **2005**, *127*, 2485–2495.

47 Danopoulos AA, Wright JA, Motherwell WB, *Chem. Commun.*, **2005**, 784–786.

48 Allen DP, Crudden CM, Calhoun LA, Wang RY, *J. Organometal. Chem.*, **2004**, *689*, 3203–3209.
49 McGuinness DS, Saendig N, Yates BF, Cavell KJ, *J. Am. Chem. Soc.*, **2001**, *123*, 4029–4040.
50 Tan KL, Bergman RG, Ellman JA, *J. Am. Chem. Soc.*, **2002**, *124*, 3202–3203.
51 Ketz BE, Cole AP, Waymouth RM, *Organometallics*, **2004**, *23*, 2835–2837.
52 Peris E, Loch JA, Mata J, Crabtree RH, *Chem. Commun.*, **2001**, 201–202.
53 Tulloch AAD, Danopoulos AA, Tooze RP, Cafferkey SM, Kleinhenz S, Hursthouse MB, *Chem. Commun.*, **2000**, 1247–1248.
54 McGuinness DS, Cavell KJ, *Organometallics* **2000**, *19*, 741–748.
55 Yang C, Lee HM, Nolan SP, *Org. Lett.* **2001**, *3*, 1511.
56 Lee HM, Zeng JY, Hu CH, Lee M, *Inorg. Chem.* **2004**, *43*, 6822–6829.
57 Tsoureas N, Danopoulos AA, Tulloch AA, Light ME, *Organometallics*, **2003**, *22*, 4750–4758.
58 Lee HM, Chiu PL, Zeng JY, *Inorg. Chim. Acta*, **2004**, *357*, 4313–4321.
59 Chiu PL, Lee HM, *Organometallics*, **2005**, *24*, 1692–1702.
60 Field LD, Messerle BA, Vuong KQ, Turner P, *Organometallics*, **2005**, *24*, 4241–4250.
61 Herrmann WA, Brossmer C, Öfele K, Reisinger CP, Priermeier T, Beller M, *Angew. Chem. Int. Ed.*, **1995**, *34*, 1844–1848.
62 Beletskaya IP, Cheprakov AV, *Chem. Rev.*, **2000**, *100*, 3009–3059.
63 Peris E, Loch JA, Mata J, Crabtree RH, *Chem. Commun.*, **2001**, 201–202.
64 Anton DR, Crabtree RH, *Organometallics*, **1983**, *2*, 855–859.
65 Poyatos M, Marquez F, Peris E, Claver C, Fernandez E, *New J. Chem.*, **2003**, *27*, 425–431.
66 Mas-Marza E, Segarra AM, Claver C, Peris E, Fernandez E, *Tetrahedron Lett.* **2003**, *44*, 6595–6599.
67 Melaiye A, Simons RS, Milsted A, Pingitore F, Wesdemiotis C, Tessier CA, Youngs WJ, *J. Med. Chem.*, **2004**, *47*, 973–977.

10
The Quest for Longevity and Stability of Iridium-based Hydrogenation Catalysts: N-Heterocyclic Carbenes and Crabtree's Catalyst

Leslie D. Vazquez-Serrano and Jillian M. Buriak

10.1
Introduction: Rhodium and Iridium-based Hydrogenation Catalysts

Reductive hydrogenation of unsaturated organic compounds is an extremely common and highly utilized reaction in synthetic organic chemistry [1]. As eloquently described by Trost a decade ago, hydrogenation reactions are an ideal demonstration of atom economy, whereby both hydrogen atoms in the H_2 molecule appear in the reduction product (Fig. 10.1a), assuming that no side-reactions such as isomerization or coupling occur [2–4]. As direct proof of the importance of homogeneous hydrogenation in both academia and industry, homogeneous hydrogenation has directly contributed to two Nobel Prizes (1973 and 2001), and is the subject of at least 1000 papers since the mid-1960s, when the field literally exploded concurrently with the expansion of the broader area of organometallic chemistry.

Work around that time showed that late transition metal complexes based on Pt(II), Ru(II), Co(I) and (II), Fe(0), Ir(I) and Rh(I) were capable of hydrogenating olefins, some of them exclusively "activated" olefins with an adjacent polar group [5]. The earliest well-characterized homogeneous hydrogenation catalyst platform was a neutral rhodium-based Wilkinson's catalyst, $Rh(PPh_3)_3Cl$ (Fig. 10.1b) [6]. The complex is active at 1 atm of H_2 pressure, at room temperature, and can easily reduce monosubstituted and cis-disubstituted alkenes such as cyclohexene. Wilkinson's catalyst system was investigated through detailed kinetics and NMR studies, and a mechanism that cycled between Rh(I) and Rh(III) intermediates proposed; interestingly, this mechanism has essentially been verified, decades later, despite several proposed variations on the theme [7]. The next generation of this catalyst, the Osborn–Schrock catalyst series, [Rh(diene)(phosphine)$_2$]X (X = PF_6, BF_4 or another non-coordinating anion, excluding BPh_4), eliminated the halide ligand (chloride with Wilkinson's catalyst), yielding a cationic complex, and thus removing the necessity of dissociation of a phosphine ligand for catalytic activity [8]. The diene ligand is hydrogenated initially, leading to release of the active, solvated catalytic species. After a thorough examination of the high reactivity of the cationic rhodium(I) complexes, Osborn and coworkers looked to iridium

N-Heterocyclic Carbenes in Synthesis. Edited by Steven P. Nolan
Copyright © 2006 WILEY-VCH Verlag GmbH & Co. KGaA, Weinheim
ISBN: 3-527-31400-8

complexes of the general makeup [Ir(diene)(phosphine)$_2$]X but were discouraged due to the low apparent reactivity compared with Rh(I) [9]. In 1977, however, Crabtree and coworkers reported that by changing the solvent of the system from polar coordinating (like methanol or THF) to polar non-coordinating (like dichloromethane) an iridium-based catalyst system be could obtained that was more reactive towards the hydrogenation of olefins with different degrees of substitution, even more hindered than cis-disubstituted alkenes [10]. The best catalyst precursor in the series is known as Crabtree's catalyst, [Ir(COD)(PCy$_3$)(py)]PF$_6$ (Fig. 1b). This compound leads, under very gentle hydrogenation conditions of 1 atm pressure and room temperature, to the most active and general homogeneous hydrogenation catalyst – it can tackle even highly hindered tetrasubstituted olefins (Fig. 10.1c).

a) Hydrogenation scheme:

b) Parent hydrogenation catalysts:

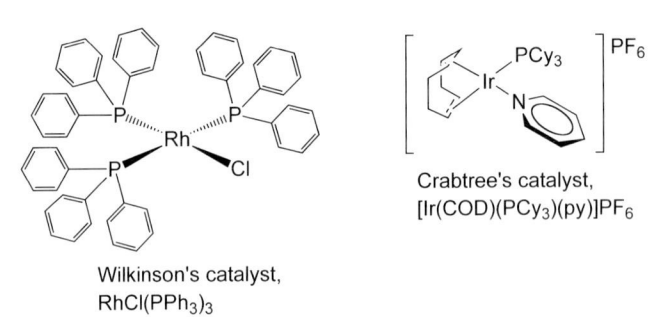

Wilkinson's catalyst, RhCl(PPh$_3$)$_3$

Crabtree's catalyst, [Ir(COD)(PCy$_3$)(py)]PF$_6$

c) Typical non-prochiral substrates for hydrogenation:

1-octene   cyclohexene   1-methyl cyclohexene   2,3-dimethyl-2-butene

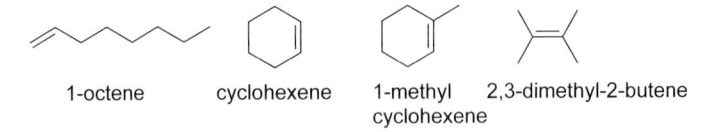

Fig. 10.1 (a) Scheme for alkene hydrogenation. Both H atoms from H$_2$ end up in the alkane product. (b) Structures of Wilkinson's and Crabtree's catalysts. (c) Commonly utilized alkene substrates for fundamental catalytic hydrogenation studies. The alkenes range from unhindered primary to very sterically crowded tetrasubstituted alkenes.

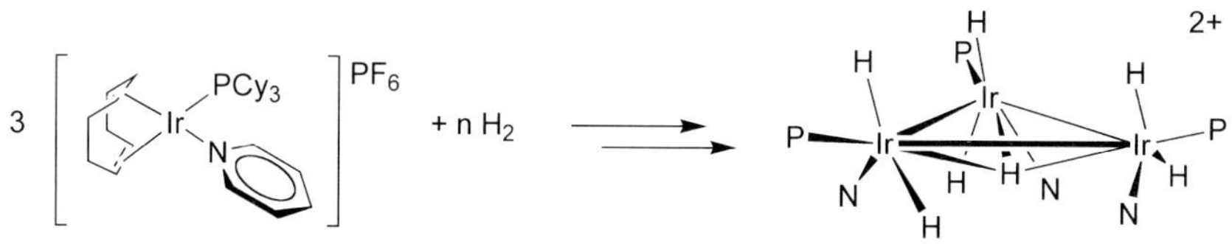

Fig. 10.2 Structure of the trinuclear hydride-bridged cluster that forms upon deactivation of Crabtree's catalyst, [Ir(COD)(PCy$_3$)(py)]PF$_6$. The P represents the phosphorous of PCy$_3$, and N the nitrogen of the pyridine ligand. This structure was assigned based on crystal structure and NMR data [11].

Crabtree's catalyst is, without question, extremely rapid for the hydrogenation of a wide range of alkenes, but its main downfall is that it rapidly deactivates (Fig. 10.2) [11].

In under an hour during the catalytic reaction, and particularly with sterically hindered olefins that are poor ligands, the catalyst oligomerizes to irreversibly form inactive hydride-bridged trimers [11]. These dimers/trimers are also formed with secondary and primary olefins after complete consumption of substrate. Multiple additions of catalyst are required to reach total conversion of sterically unhindered alkenes at higher substrate/catalyst ratios due to deactivation via this process. This step increases the total amount of catalyst and manipulations required but, nonetheless, the catalyst is still extremely useful since it is the only heavy hitter, in terms of homogeneous metal complexes, that can rapidly tackle sterically hindered alkene substrates such as tetrasubstituted alkenes [12].

10.2
Building upon Crabtree's Catalyst with N-Heterocyclic Carbenes

N-Heterocyclic carbenes (NHC) have ignited a renaissance in organometallic catalysis [13–16]. By providing viable and stable alternatives to the classic phosphine ligands, many catalytic systems that had been thoroughly studied in their phosphine-substituted forms were brought back to the forefront, and resynthesized with NHC ligands. The results have been quite varied, some with stellar results, and others showing lessened reactivity. The best known result is that of the Grubb's olefin ruthenium-based metathesis catalyst; the latest generation contains both a PCy$_3$ phosphine and a saturated, bulky carbene ligand [17]. The NHC-substituted Ru(II) catalyst is both more reactive and tolerant of polar functional groups than the parent bis-phosphine ruthenium complex. Other late transition metal-NHC complexes serve as pre-catalysts for numerous catalytic reactions such as Heck [18], Suzuki [19] and Sonogashira coupling [20], transfer hydrogenation [21], hydroformylation [22], hydrosilylation [23, 24], and olefin metathesis [25], among others [26, 27].

It is therefore natural that Crabtree's catalyst would be reinvestigated in the NHC renaissance; its unique activity for rapid hydrogenation of hindered alkene substrates but low stability provided much incentive and room for improvement. The parent Crabtree's catalyst (Fig. 10.1b) has two disparate ligands that could be replaced with an NHC ligand, i.e., the bulky PCy$_3$ phosphine and/or the small pyridine [28, 29]. The question is, which ligand(s) should be substituted to lead to the most active catalyst precursor? First, consider what is known about the characteristics of NHCs as ligands. NHCs can be considered to be strong σ-donor ligands with little π-accepting capabilities. The sterics of the NHC ligands can be tailored and controlled by changing the substituents on the nitrogen atoms, but differ greatly from the classic cone angle analysis of phosphines [30] as they have a "fan" or "fence"-type structure [31]. To simplify the description of NHC ligands, Fig. 10.3 shows all of the ligands mentioned in this chapter.

Achiral unsaturated carbenes:

IMes   IMe

Achiral saturated carbenes:

SIMes   SIMe   SIiPr

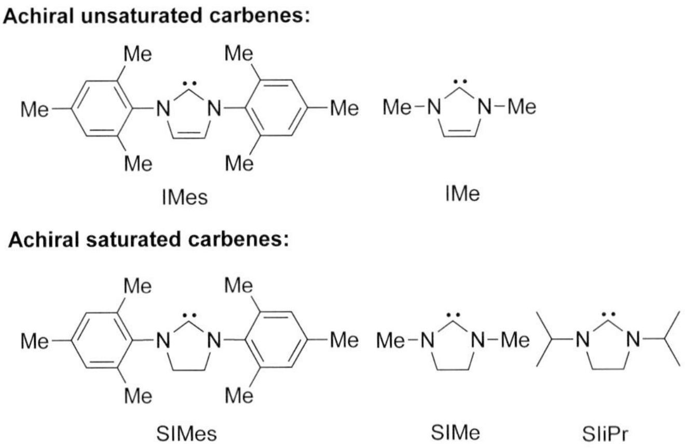

Fig. 10.3 N-Heterocyclic carbene (NHC) ligands mentioned in this chapter.

It was Nolan's group that first re-investigated Crabtree's catalyst with an NHC ligand, in 2001, and naturally they looked first to the phosphine ligand for substitution, therefore keeping the pyridine intact, producing [Ir(COD)(SIMes)(py)]PF$_6$ as the catalyst precursor (Fig. 10.4a), a complex that was characterized crystallographically [32]. The complex was synthesized in a manner analogous to the approach used for Crabtree's catalyst, starting from the bis-pyridine complex [Ir(COD)(py)$_2$]PF$_6$.

The new catalyst was examined for the hydrogenation of simple, aliphatic alkenes. The hydrogenation activity was lower than that of the parent Crabtree's catalyst but was thermally more stable. Table 10.1 summarizes selected results; for the most highly substituted alkene in this study, 1-methylcyclohex-1-ene, 42% conversion is achieved with [Ir(COD)(SIMes)(py)]PF$_6$ at 1 mol% catalyst precursor and 1 atm H$_2$, room temperature, which is a lower yield than that observed for Crabtree's catalyst (entry 9 versus 13). The incomplete hydrogenation is attributed to irreversible hydride-bridged cluster formation. Where the [Ir(COD)(SIMes)(py)]PF$_6$ complex shows improvement over Crabtree's catalyst is at higher temperatures (50 °C) and higher pressures (60 psi). Whereas Crabtree's catalyst deactivates upon reaching 34% yield at 1 mol% catalyst for the 1-methylcyclohex-1-ene substrate, [Ir(COD)(SIMes)(py)]PF$_6$ continues on, reaching 100% yield after 7 h.

The next derivative of Crabtree's catalyst, prepared by the Buriak group, substituted not the phosphine, but the small, labile pyridine group, and the overall catalyst motif in this case can be summarized as [Ir(COD)(phosphine)(NHC)]X (X = counter-anion) (Fig. 10.5) [33].

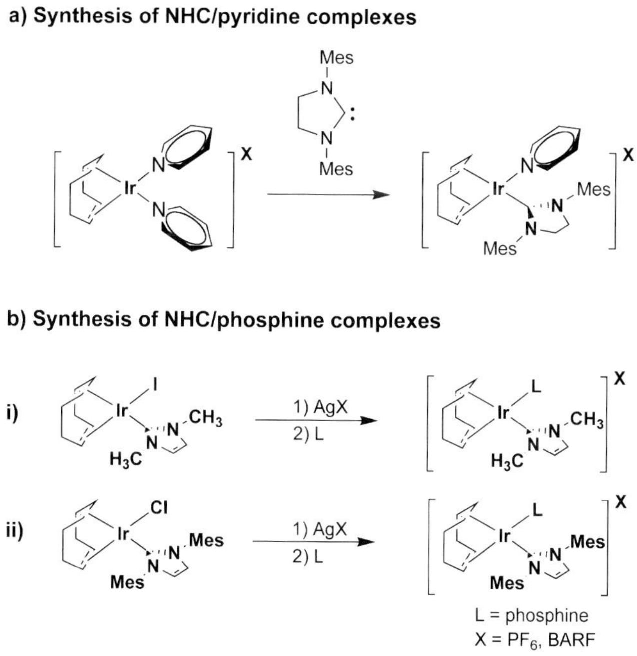

a) Synthesis of NHC/pyridine complexes

b) Synthesis of NHC/phosphine complexes

L = phosphine
X = PF$_6$, BARF

Fig. 10.4 Synthetic procedures to prepare Ir(I) NHC/pyridine and Ir(I) NHC/phosphine complexes from accessible intermediates.

Figure 10.4(b) shows representative synthetic strategies for these complexes. Both bulky and small NHC ligands were tested, including IMes and IMe, as well as different phosphines, such as PCy$_3$, P(n-Bu)$_3$, and PPh$_3$. The combination of the phosphine and NHC ligand leads to a highly active catalyst, with activity comparable to Crabtree's catalyst. These catalysts have a very obvious and sharp visual endpoint at the end of the reaction — the reaction solution changes abruptly from red-orange to colourless upon consumption of all available olefin (when the yield is >98%). The best overall catalyst in the series is [Ir(COD)(IMe){P(n-Bu)$_3$}]PF$_6$, which can hydrogenate all substrates tested at 1 mol% catalyst precursor, 1 atm H$_2$, and room temperature, from primary to tetrasubstituted alkenes (Table 10.1). For the most demanding alkene, 2,3-dimethylbut-2-ene, Crabtree's catalyst and [Ir(COD)(IMe){P(n-Bu)$_3$}]PF$_6$ are essentially equivalent, yielding 100% conversion in 40 min (compare entries 14 and 15, Table 10.1). At 0.1 mol% catalyst, however, Crabtree's catalyst is still superior, reaching 49% conversion before deactivating, while [Ir(COD)(IMe){P(n-Bu)$_3$}]PF$_6$ can only reach 5%. To summarize, [Ir(COD)-(phosphine)(NHC)]X (X = PF$_6$) is as reactive as Crabtree's catalyst at 1 mol% loadings in terms of rate, even with tetrasubstituted olefins, but still has the same stability problems that affect the parent catalyst.

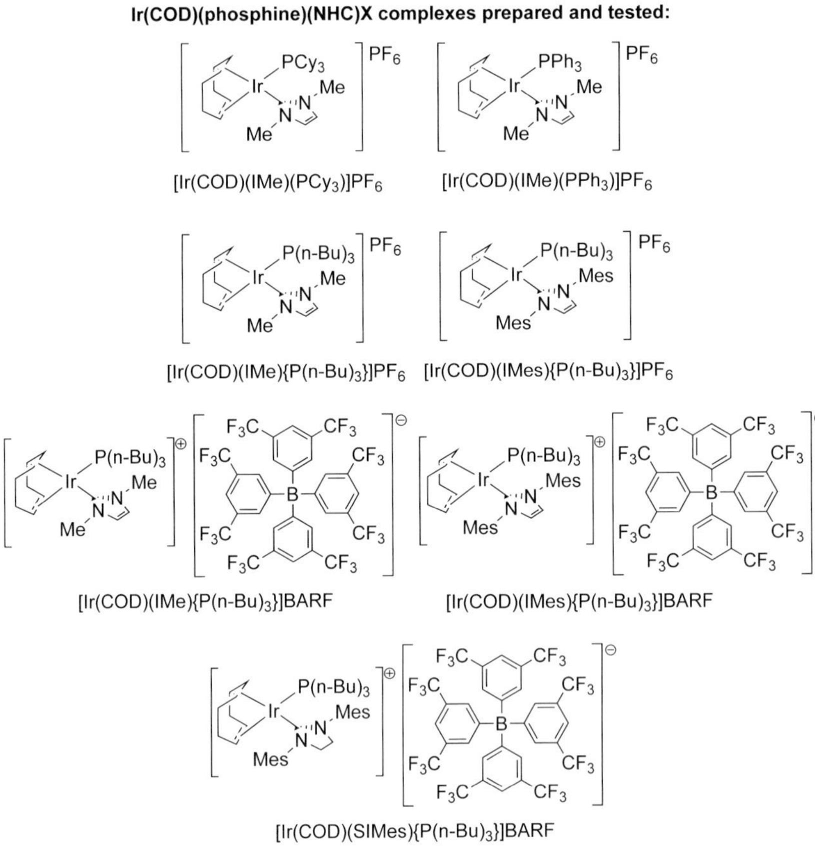

Ir(COD)(phosphine)(NHC)X complexes prepared and tested:

[Ir(COD)(IMe)(PCy$_3$)]PF$_6$ [Ir(COD)(IMe)(PPh$_3$)]PF$_6$

[Ir(COD)(IMe){P(n-Bu)$_3$}]PF$_6$ [Ir(COD)(IMes){P(n-Bu)$_3$}]PF$_6$

[Ir(COD)(IMe){P(n-Bu)$_3$}]BARF   [Ir(COD)(IMes){P(n-Bu)$_3$}]BARF

[Ir(COD)(SIMes){P(n-Bu)$_3$}]BARF

Fig. 10.5 [Ir(COD)(NHC)(phosphine)]X complexes prepared and tested for alkene hydrogenation.

While [Ir(COD)(phosphine)(NHC)]PF$_6$ was comparable to the Crabtree's catalyst benchmark in terms of rates and ability to reduce hindered alkenes, further tailoring of the ligand substituents was carried out, leading to faster and more stable catalysts [34]. Saturated NHC ligands such as SIMe, SIMes and SIiPr were examined, but the catalysts were not faster in the series [Ir(COD)(saturated NHC){P(n-Bu)$_3$}]PF$_6$. These results are in contrast with those of the Grubbs ruthenium metathesis system where replacement of the unsaturated NHC ligand with a saturated variety improved the reactivity profile of the catalyst [17]. In the iridium system here, saturated NHCs do not appear to improve either yield or stability.

Table 10.1 Selected results for the catalytic hydrogenation of olefins using Ir(I) NHC/pyridine (py) and Ir(I) NHC/phosphine complexes, with a PF_6 counter-anion. All catalytic results are from either Refs. [33] and [34] (NHC/phosphine combination) or Ref. [32] (NHC/pyridine combination).

Entry	Catalyst[a]	Substrate	Cat. (mol%)	Time (min)	Yield (%)	Rate [(mol substrate reduced) (mol of catalyst)$^{-1}$ h^{-1}]
1	Crabtree's catalyst	Oct-1-ene	1	9	100	1146
			0.1	41	>99	2437
3	[Ir(COD)(IMe){P(n-Bu)$_3$}]PF$_6$	Oct-1-ene	1	13	100	797
			0.1	62	100	1445
4	[Ir(COD)(IMes){P(n-Bu)$_3$}]PF$_6$	Oct-1-ene	1	11	100	1037
			0.1	45	100	3894
5	Crabtree's catalyst	Cyclohexene	1	9	100	925
			0.1	36	99	2290
6	[Ir(COD)(IMe){P(n-Bu)$_3$}]PF$_6$	Cyclohexene	1	15	100	505
			0.1	100	>99	1168
7	[Ir(COD)(IMes){P(n-Bu)$_3$}]PF$_6$	Cyclohexene	1	14	100	614
			0.1	75	100	1963
8	[Ir(COD)(SIMes)(py)]PF$_6$	Cyclohexene	1	2	100	n.r.
9	Crabtree's catalyst	1-Methylcyclohex-1-ene	1	16	99	516
			0.1	60	70	1496
10	[Ir(COD)(IMe)(PCy$_3$)]PF$_6$	1-Methylcyclohex-1-ene	1	60	100	63
11	[Ir(COD)(IMe)(PPh$_3$)]PF$_6$	1-Methylcyclohex-1-ene	1	30	100	259
12	[Ir(COD)(IMe){P(n-Bu)$_3$}]PF$_6$	1-Methylcyclohex-1-ene	1	21	100	278
			0.1	130	71	649
13	[Ir(COD)(SIMes)(py)]PF$_6$	1-Methylcyclohex-1-ene	1	210	42	n.r.
14	Crabtree's catalyst	2,3-Dimethylbut-2-ene	1	40	95	208
			0.1	50	49	1057
15	[Ir(COD)(IMe){P(n-Bu)$_3$}]PF$_6$	2,3-Dimethylbut-2-ene	1	39	>99	165
			0.1	30	5	70
16	[Ir(COD)(IMes){P(n-Bu)$_3$}]PF$_6$	2,3-Dimethylbut-2-ene	1	220	19	30

a) Conditions: Please see Refs. [32–34] for exact experimental conditions. Catalysis carried out at 1 atm H_2, room temperature, in CH_2Cl_2. Crabtree's catalyst is [Ir(COD)(PCy$_3$)(py)]PF$_6$.

The important breakthrough was reached when PF_6 was replaced with the much more non-coordinating anion tetrakis[3,5-bis(trifluoromethylphenyl)]borate (BARF); Fig. 10.5 shows the complexes prepared with this counter-anion. Previous work by the groups of Pfaltz and Burgess has shown that the BARF counter-anion is superior to traditional "non-coordinating" ions such as PF_6 and BF_4 in their iridium-phosphine/oxazoline alkene hydrogenation catalytic systems [35, 36]. When the [Ir(COD)(saturated NHC){P(n-Bu)$_3$}]BARF catalyst was tested, the rates and yields improve uniformly compared to the PF_6 versions, and significant enhancement can even be observed for Crabtree's catalyst when PF_6 is replaced with BARF to produce [Ir(COD)(PCy$_3$)(py)]X (X = BARF instead of PF_6). Most notably, the improvement can be seen with the most demanding substrate, 2,3-dimethyl-but-2-ene. Some of the best results are summarized in Table 10.2; these results show, among others, that [Ir(COD)(IMe){P(n-Bu)$_3$}]BARF is the only catalyst capable of hydrogenating this tetrasubstituted alkene at a catalyst loading of 0.1 mol%, to 100% yield. Crabtree's catalyst deactivates after 49% conversion under these conditions (compare entry 14, Table 10.1, with entry 14, Table 10.2).

As shown in entries 4, 7, 11 and 14 of Table 10.2, the [Ir(COD)(IMe){P(n-Bu)$_3$}]-BARF complex tolerates sloppy catalytic conditions, in which the glassware is not pre-dried (taken directly from the drawer), and using reagent grade dichloromethane taken straight from an opened bottle, in air. The rates of the reactions were slightly slower but complete conversion can be reached under all circumstances. The [Ir(COD)(IMe){P(n-Bu)$_3$}]BARF complex can also give high yields at only 0.01 mol% catalyst loading with respect to cyclohexene, achieving >90% conversion for this substrate, a value not attainable with Crabtree's catalyst [34].

In terms of the catalytic mechanism of the [Ir(COD)(NHC)(phosphine)]X complexes, para-hydrogen induced polarization (PHIP) ^{1}H NMR experiments were carried out with d^8-styrene as the substrate [33]. An enhancement was observed that can only be due to both H atoms from one H_2 molecule adding across the styrene olefin. This indicates that at least a fraction, if not all, of the reaction proceeds via a dihydride mechanism, although a competing monohydride mechanism cannot be discounted. Kinetic studies on three different complexes of this series indicate that the rate is related to $[Ir]^{1/2}$ [34]. An order of one-half with respect to catalyst implies a dimeric resting state that dissociates into two monomeric species – the proposed active catalyst [37]. The proposed catalytic cycle therefore involves an iridium dimer catalyst resting state, presumably with hydride ligands, that dissociates into catalytically active monomeric species. The stability of the catalyst may therefore be attributed to the reversibility of this dimer formation under catalytic conditions, since with Crabtree's catalyst these higher ordered bridged hydride species are inactive complexes, and represent a dead-end for catalysis.

In conclusion, the solution to the iridium bridging hydride species may not be to prevent their formation, but to ensure that their formation is reversible. If the equilibrium constant is such that some small amount of active species can be released via splitting of the dimeric resting state, then these oligomers may actually be favorable by providing a stable resting state environment. This conclusion

Table 10.2 Selected results for the hydrogenation of alkenes using the Ir NHC/phosphine complexes with a BARF counter-anion. Crabtree's catalyst with a BARF counter-anion is shown for comparison.

Entry	Catalyst[a]	Substrate	Cat. (mol%)	Time (min)	Yield (%)	Rate [(mol substrate reduced) (mol of catalyst)$^{-1}$ h^{-1}]
1	Crabtree's catalyst with PF$_6$	Oct-1-ene	1	9	100	1146
			0.1	41	>99	2437
2	Crabtree's catalyst with BARF	Oct-1-ene	1	7	100	1566
			0.1	28	100	3426
3	[Ir(COD)(IMe){P(n-Bu)$_3$}]BARF	Oct-1-ene	1	10	100	966
			0.1	28	100	3426
4	[Ir(COD)(IMes){P(n-Bu)$_3$}]BARF	Oct-1-ene	1	10	100	939
			1 (air)	11	100	715
			0.1	42	100	1615
5	[Ir(COD)(SIMes){P(n-Bu)$_3$}]BARF	Oct-1-ene	1	10	100	1194
			0.1	38	100	3989
6	Crabtree's catalyst with BARF	Cyclohexene	1	7	100	1522
			0.1	26	100	3398
7	[Ir(COD)(IMe){P(n-Bu)$_3$}]BARF	Cyclohexene	1	10	100	765
			1 (air)	13	100	543
			0.1	47	100	1602
8	[Ir(COD)(IMes){P(n-Bu)$_3$}]BARF	Cyclohexene	1	11	100	787
			0.1	75	100	2011
10	Crabtree's catalyst with BARF	1-Methylcyclohex-1-ene	1	13	100	753
			0.1	60	78	1396
11	[Ir(COD)(IMe){P(n-Bu)$_3$}]BARF	1-Methylcyclohex-1-ene	1	16	100	375
			1 (air)	23	100	279
			0.1	86	100	850
12	[Ir(COD)(IMes){P(n-Bu)$_3$}]BARF	1-Methylcyclohex-1-ene	1	60	85	82
13	Crabtree's catalyst with BARF	2,3-Dimethylbut-2-ene	1	21	>99	381
			0.1	65	51	788
14	[Ir(COD)(IMe){P(n-Bu)$_3$}]BARF	2,3-Dimethylbut-2-ene	1	25	100	217
			1 (air)	32	100	157
			0.1	120	100	493
15	[Ir(COD)(IMes){P(n-Bu)$_3$}]BARF	2,3-Dimethylbut-2-ene	1	25	27	87
16	[Ir(COD)(SIMes){P(n-Bu)$_3$}]BARF	2,3-Dimethylbut-2-ene	1	262	15	3

a) Conditions: Please see Ref. [34] for exact experimental conditions. Catalysis carried out at 1 atm H$_2$, room temperature, in CH$_2$Cl$_2$.
Crabtree's catalyst is [Ir(COD)(PCy$_3$)(py)] (with PF$_6$ or BARF as counter-anion as indicated).
Reaction carried out in air, in undistilled, undegassed reagent grade CH$_2$Cl$_2$ where specified.

demands further study for verification since, if it is valid, it may be the ideal solution to hydride-bridged oligomers to use them to our advantage instead of trying to fight their formation.

10.3
Chiral Iridium N-Heterocyclic Catalysts

As can be seen from Figure 10.1(a), alkenes can be prochiral, depending upon the substituents, meaning that, upon reduction, chiral products are rendered. Whereas alkenes such as the highly substituted prochiral cinnamic acid derivatives that are amino acid precursors can be successfully hydrogenated with high enantioselectivity [38], simpler (and more general) aliphatic alkenes are much more challenging at present. Chiral titanium, zirconium and lanthanide catalysts have achieved some success but are saddled with high catalyst loadings, extreme air- and water-sensitivity, and moderate enantiomeric excesses (e.e.s) [39]. The high reactivity and relative air insensitivity of Crabtree's catalyst was not forgotten, however, and in 1998 Pfaltz and coworkers directed their attention back to iridium [35]. A series of chiral bidentate ligands based on a phosphinooxazoline framework (Fig. 10.6) were used to synthesize a series of successful enantioselective iridium-based hydrogenation catalysts, capable of reducing simple, unfunctionalized alkenes with moderate to high enantioselectivities.

The phosphine-oxazoline combination mimics the critical P-N binding motif in Crabtree's catalyst. Necessary for obtaining lower catalyst loadings (<1%) was the use of the BARF counter-anion (vide supra). Burgess and coworkers followed up this work with a related library of ligands based on the phosphine-oxalozoline combination and extensively mapped out the enantioselectivity/reactivity space for this catalyst class [36].

The Pfaltz–Burgess phosphinooxazoline iridium catalysts showed good to excellent enantioselectivities, but required high pressures ($p \geq 50\,atm$) of H_2 to function on the hour timescale, and had some deactivation problems. To attempt to remedy such low reactivity, Burgess reported the substitution of the phosphine portion of a bidentate chiral phosphinooxazoline ligand by an NHC group, and its subsequent binding to an Ir(I) centre (Fig. 10.6b) [40–42]. This new catalyst system is much more reminiscent of Crabtree's catalyst as it can efficiently hydrogenate primary, secondary and tertiary olefins with good to high enantioselectivities, often at just 1 atm H_2 (Table 10.3).

The Burgess group then proceeded to synthesize a library of 17 ligands for screening for their catalytic activities [43]. One complex in particular, highlighted in Fig. 10.6, emerged as the best overall catalyst in terms of enantioselectivity for the hydrogenation of alkenes such as (E)-α-methylstilbene, a fact that may be related to steric hindrance effects, observable when the steric environment around the metal center is divided into quadrants. Synthesis of small ligand libraries such as these is a very practical approach when little is known about the mechanism, or about the steric and chiral requirements necessary for high enantioselectivity.

10.3 Chiral Iridium N-Heterocyclic Catalysts | 251

a) Pfaltz's and Burgess' chiral non-NHC ligands

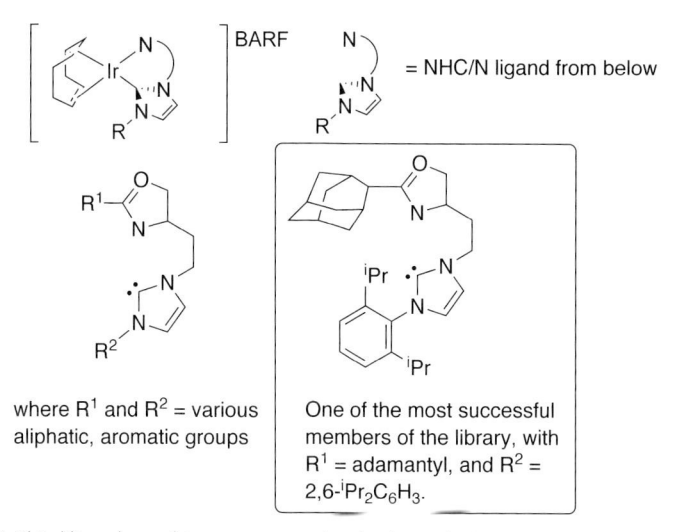

Fig. 10.6 Chiral ligands used to prepare enantioselective variants of Crabtree's catalyst. (a) Non-NHC ligands and (b) chiral bidentate NHC ligands for iridium-based enantioselective hydrogenation.

Determination of the mechanism is complicated by the fact that some substrates show significant dependence of enantioselectivity on temperature and hydrogen pressure, whereas others do not. For instance, 2-phenylbut-1-ene is hydrogenated to yield a product with a 64% e.e. (R) under high pressure/low temperature conditions, and 89% e.e. (S) under low pressure/high temperature conditions (Table 10.3, entries 2 and 3) [41]. Careful experiments examining the location of deuter-

Table 10.3 Selected results for the hydrogenation of various prochiral unsubstituted alkenes with the Burgess iridium catalyst highlighted in Fig. 10.6(b) (R^1 = adamantyl, and $R^2 = 2,6\text{-}^iPr_2C_6H_3$).[a]

Entry	Substrate	Pressure (bar)	Temperature (°C)	e.e. (%)[b]	Yield (%)
1	A	90	−30	−61	76
2	A	85	−15	−64	100
3	A	1	25	+89	100
4	A	1	30	+89	100
5	A	1	40	+88	100
6	B	50	25	+96	99
7	C	50	25	+79	99
8	D	50	25	+84	95
9	E	50	25	+49	58
10	F	50	25	+89	100
11	G	50	25	+54	54

a) Conditions: see Ref. [41] for exact experimental conditions.
b) + or − refers to opposite enantiomers.

ium in the final products using D_2 reveal that these results are not consistent with prior isomerization. The results do, however, indicate a switch of mechanism as the concentration of hydrogen in solution changes from high to low. In addition, deuterium labeling experiments suggest competing Ir-allyl intermediates in the presence of D_2/H_2, leading to hydrogenated alkenes product with deuterium in unexpected positions [41].

To further investigate the reactivity of these chiral Ir-NHC complexes, Burgess and coworkers looked to prochiral dienes [42, 44]. Asymmetric diene hydrogenation has seen little investigation, but can provide new information into the mechanism since it can yield various products, depending upon the reaction mechanisms and accessible pathways. Hydrogenation of a diene can lead to matched or mismatched products, and an internal alkene that is an isomerization product. Kinetic profiling of the reaction with GC reveals that after the initial hydrogenation and release of the COD ligand the next dominant step is a poorly enantioselective hydrogenation of one of the alkene substituents. When all the diene has been transformed into monoene, the second alkene is then hydrogenated, with the enantioselectivity still mainly under catalyst control. Andersson and coworkers suggest that a shift in mechanism may occur at this critical juncture, from Ir(I)/Ir(III) intermediates to Ir(III)/Ir(V) [45]; the latter appears to be faster and more selective, but further fundamental mechanistic studies are required at this point to truly elucidate the nature of the catalytically active organometallic species.

10.4 Conclusions

The area of NHC ligands with respect to iridium catalysis is just opening up. Promising results, including high stability and reactivity, excellent enantioselectivities, and hints of unique and surprising reaction mechanisms point to a rich and exciting area of research. Detailed mechanistic studies are required and should be carried out to buttress the development of very practical and useful catalysts that can enantioselectively reduce prochiral tri- and tetrasubstituted olefins lacking a specific arrangement of polar functional groups. After close to 30 years since the development of Crabtree's catalyst, the ideal combination of ligands, thanks to straightforward access to NHC ligands from both commercial and synthetic sources, may be within reach to allow harnessing of the high reactivity of this catalyst system.

Acknowledgments

LDVS thanks Purdue University and the National Institutes of Health (NIH) for support. JMB thanks the University of Alberta, the National Institute for Nanotechnology, and the Canada Research Chairs program for support.

References

1 Rylander, P. N. *Hydrogenation Methods*, Academic Press, New York, **1985**.
2 (a) Trost, B. M. *Science* **1991**, *254*, 1471. (b) Trost, B. M. *Angew. Chem. Int. Ed. Engl.* **1995**, *34*, 259.
3 Bläser, H.-U., Studer, M. *Appl. Catal. A: General* **1999**, *189*, 191.
4 Trost, B. M. *Pure Appl. Chem.* **1994**, *66*, 2007.
5 Ru(II): Halpern, J., Harrod, J. F., James, B. R. *J. Am. Chem. Soc.* **1966**, *88*, 5150. Pt(II): Cramer, R. D., Jenner, F. L., Lindsey, Jr., R. V., Stolberg, U. G. *J. Am. Chem. Soc.* **1963**, *85*, 1691. Co(II): Kwiatek, J., Mador, I. L., Seyler, J. K. *Reactions of Coordinated Ligands and Homogeneous Catalysis*, Advances in Chemistry Series, No. 37, American Chemical Society, Washington, D.C., **1963**. Ir(I): Vaska, L., Rhodes, R. E. *J. Am. Chem. Soc.* **1965**, *87*, 4970. Rh(I): Young, J. F., Osborn, J. A.,

Jardine, F. H., Wilkinson, G. *J. Chem. Commun.* **1965**, 131. Co(I): Marko, L. *Chem. Ind.* **1962**, 260. Fe(0): Frankel, E. N., Emken, E. A., Peters, H. M., Davidson, V. L., Butterfield, R. O. *J. Org. Chem.* **1964**, *29*, 3292.

6 Osborn, J. A., Jardine, F. H., Young, J. F., Wilkinson, G. J. *J. Chem. Soc. A* **1966**, 1711.

7 Duckett, S. B., Newell, C. L., Eisenberg, R. *J. Am. Chem. Soc.* **1994**, *116*, 10556.

8 (a) Schrock, R. R., Osborn, J. A. *J. Am. Chem. Soc.* **1976**, *98*, 2134. (b) Schrock, R. R., Osborn, J. A. *J. Am. Chem. Soc.* **1976**, *98*, 2143. (c) Schrock, R. R., Osborn, J. A. *J. Am. Chem. Soc.* **1976**, *98*, 4450.

9 (a) Shapley, J. R., Schrock, R. R., Osborn, J. A. *J. Am. Chem. Soc.* **1969**, *91*, 2816. (b) Shapley, J. R., Osborn, J. A. *J. Am. Chem. Soc.* **1970**, *92*, 6976.

10 Crabtree, R. H., Felkin, H., Morris, G. E. *J. Organomet. Chem.* **1977**, *141*, 205.

11 Crabtree, R. H. *Acc. Chem. Res.* **1979**, *12*, 331.

12 Burgess, K., Cui, X. *Chem. Rev.* **2005**, *105*, 3272.

13 Arduengo, A. J., Rasika Dias, H. V., Harlow, R. L., Kline, M. *J. Am. Chem. Soc.*, **1992**, *114*, 5530.

14 Tafipolsky, M., Scherer, W., Öfele, K., Artus, G., Pedersen, B., Herrmann, W. A., McGrady, G. S. *J. Am. Chem. Soc.* **2002**, *124*, 5865.

15 Arduengo, A. J., Goerlich, J. R., Marshall, W. J. *J. Am. Chem. Soc.* **1995**, *117*, 11027.

16 Bourissou, D., Guerret, O., Gabbaï, F. P., Bertrand, G. *Chem. Rev.* **2000**, *100*, 39.

17 Grubbs, R. H. *Tetrahedron* **2004**, *60*, 7117.

18 Crudden, C. M., Sateesh, M., Lewis, R. *J. Am. Chem. Soc.* **2005**, *127*, 10045.

19 Kim, J. H., Kim, J. W., Shokouhimehr, M., Lee, Y. S. *J. Org. Chem.* **2005**, *70*, 6714.

20 Peris, E., Crabtree, R. H. *Coord. Chem. Rev.* **2004**, *248*, 2239.

21 Albrecht, M., Miecznikowski, J. R., Samuel, A., Faller, J. W., Crabtree, R. H. *Organometallics* **2002**, *21*, 3596.

22 Zarka, M. T., Bortenschlager, M., Wurst, K., Nuyken, O., Weberskirch, R. *Organometallics* **2004**, *23*, 4817.

23 Mas-Marzá, E., Poyatos, M., Sanaú, M., Peris, E. *Inorg. Chem.* **2004**, *43*, 2213.

24 Maifeld, S. V., Tran, M. N., Lee, D. *Tetrahedron Lett.* **2005**, *46*, 105.

25 Jafarpour, L., Nolan, S. P. *Organometallics* **2000**, *19*, 2055.

26 Köcher, C., Herrmann, W. A. *J. Organomet. Chem.* **1997**, *532*, 261.

27 For review of some of the application of NHC's see (a) Herrmann, W. A. *Angew. Chem. Int. Ed.* **2002**, *41*, 1290. (b) Herrmann, W. A., Köcher, C. *Angew. Chem. Int. Ed. Engl.* **1997**, *36*, 2162.

28 Lee, H. M., Jiang, T., Stevens, E. D., Nolan, S. P. *Organometallics* **2001**, *20*, 12.

29 Perry, M. C., Burgess, K. *Tetrahedron: Asymmetry* **2003**, *14*, 951.

30 Tolman, C. A. *Chem. Rev.* **1977**, *77*, 313.

31 Scott, N. M., Nolan, S. P. *Eur. J. Inorg. Chem.* **2005**, 1815.

32 Lee, H. M., Jiang, T., Stevens, E. D., Nolan, S. P. *Organometallics* **2001**, *20*, 1255.

33 Vazquez-Serrano, L. D., Owens, B. T., Buriak, J. M. *Chem. Commun.* **2002**, 2518.

34 Vazquez-Serrano, L. T., Owens, B. T., Buriak, J. M. *Inorg. Chim. Acta* **2006**, *359*, 2786.

35 Lightfoot, A., Schnider, P., Pfaltz, A. *Angew. Chem. Int. Ed. Engl.* **1998**, *37*, 2897.

36 Hou, D.-R., Reibenspies, J., Colacot, T. J., Burgess, K. *Chem. Eur. J.* **2001**, *7*, 5391.

37 Chan, Y. N. C., Osborn, J. A. *J. Am. Chem. Soc.* **1990**, *112*, 9400.

38 Noyori, R. *Angew. Chem., Int. Ed.* **2002**, *41*, 2008.

39 Halterman, R. K., Vollhardt, K. P. C., Welker, M. E. *J. Am. Chem. Soc.* **1987**, *109*, 8105. Halterman, R. L., Vollhardt, K. P. C. *Organometallics* **1988**, *7*, 883. Wild, R. W. P., Zsolnai, J., Huttner, G., Brintzinger, H. H. *J. Organomet. Chem.* **1982**, *232*, 233. Waymouth, R., Pino, P. *J. Am. Chem. Soc.* **1990**, *112*, 4911. Grossman, R. B., Doyle, R. A., Buchwald, S. L. *Organometallics* **1991**, *10*, 1501. Broene, R. D., Buchwald, S. L. *J. Am. Chem. Soc.* **1993**, *115*, 12569.

Troutman, V., Appella, D. H., Buchwald, S. L. *J. Am. Chem. Soc.* **1999**, *121*, 4916; Haar, C. M., Stern, C. L., Marks, T. J. *Organometallics* **1996**, *15*, 1765.

40 Powell, M. T., Hou, D.-R., Perry, M. C., Cui, X., Burgess, K. *J. Am. Chem. Soc.* **2001**, *123*, 8878.

41 Perry, M. C., Cui, X., Powell, M. T., Hou, D.-R., Reibenspies, J. H., Burgess, K. *J. Am. Chem. Soc.* **2003**, *125*, 113.

42 Cui, X., Burgess, K. *J. Am. Chem. Soc.* **2003**, *125*, 14 212.

43 Burgess, K., Perry, M. C., Cui, X. *Symposium Series No. 880 Methodologies in Asymmetric Catalysis*, ed. Malhotra, S., ACS Publications, Washington D.C., **2004**, p. 61.

44 Cui, X., Ogle, J. W., Burgess, K. *Chem. Commun.* **2005**, 672.

45 Brandt, P., Hedberg, C., Andersson, P. G. *Chem. Eur. J.* **2003**, *9*, 339.

11
Cu-, Ag-, and Au-NHC Complexes in Catalysis

Pedro J. Pérez and M. Mar Díaz-Requejo

11.1
Introduction

Perhaps the best known and general feature of N-heterocyclic ligands that an organometallic chemist can use to define them comes from their comparison with phosphine ligands, in the sense that the former provide the metal center with electron density through σ-donation with practically non-existent back-donation. Because of this, during the last decade of expansion of the chemistry of this family of ligands, most efforts have been directed toward the study of the effect of substituting phosphine with NHC ligands in the plethora of complexes already reported with such phosphorus donors. This role of phosphine surrogates attributed to N-heterocyclic carbenes has been applied not only from a purely synthetic organometallic point of view but also to the use of NHC-transition metal complexes as catalyst, or catalyst precursor, for several transformations.

In 2002 Herrmann reviewed the homogeneous catalysis area, dealing with the use of NHC complexes, under the title "N-Heterocyclic carbenes as ligands for metal complexes – challenging phosphane ligands in homogeneous catalysis" [1]. Not surprisingly, most of the work reviewed had been performed with group 8–10 metals, cross coupling (Suzuki, Heck, etc.) or olefin metathesis processes being, largely, the most studied examples. Very few systems of other transition metal-based catalysts appeared in that account, which could be understood in terms of the numerous already reported phosphine-containing catalysts of the iron, cobalt and nickel groups. This could also explain the relatively few reports on group 11 metal catalysts, given that relatively few examples of phosphine catalysts of copper, silver and gold have been described in the literature.

Inescapably, therefore, the first examples of copper and gold in homogeneous catalysis appeared as the result of the use of NHC-analogs of well-known phosphine catalysts. However, these metals have recently been freed from the plain phosphine surrogate role in the sense that, in some cases, other catalytic systems have been developed for reactions for which no phosphine catalysts had been previously reported. The case of silver is quite singular after the work by Lin and

N-Heterocyclic Carbenes in Synthesis. Edited by Steven P. Nolan
Copyright © 2006 WILEY-VCH Verlag GmbH & Co. KGaA, Weinheim
ISBN: 3-527-31400-8

Wang [2], which describes the facile synthesis of this class of compounds as well as their excellent capabilities in transferring the NHC ligand to another metal center, in the so-called transmetallation reaction. This strategy has been widely employed in the last few years to develop other (NHC)M catalysts for many reactions, where M can also stand for copper and gold.

After the rapid development of transition metal catalysts with NHC ligands, their application into the asymmetric catalysis area was also considered a priority for several researchers. A recent review by Gade and coworkers covers the relevant advances [3]. After a not very successful beginning (late 1990s), several highly enantioselective catalytic systems based on chiral NHC ligands have been described in the last few years; most of them, again, with the metals of groups 8–10. Among the examples of catalytic systems based on group 11 metals with NHC ligands, relatively few employ chiral N-heterocyclic carbenes, and therefore induce any enantiomeric excess, and they are still limited to copper.

This chapter brings together the reported catalytic systems based on the use of NHC-M complexes as the unambiguous catalyst, where the metal is one of the group 11 elements. There are four sections, one for each of the three metals, with a detailed explanation of the different catalytic systems, depending on reaction type; the fourth section describes the use of the three metals for the same transformation: the carbene transfer reaction, from diazo compounds, to organic substrates.

11.2
Copper

The first example of a copper-catalyzed reaction with a NHC-containing complex appeared in 2001, and corresponded to its use in the conjugate addition of organozinc reagents to enones [4]. This work opened the way to a series of related articles, including some with chiral catalysts and asymmetric induction. The reduction of carbonyl compounds constituted the second, chronologically speaking, use in catalysis of NHC-Cu complexes. Later, the carbene transfer reaction from ethyl diazoacetate to several saturated and unsaturated substrates was also reported (Section 11.5) before a group of papers related to the enantioselective copper-catalyzed allylic alkylation or substitution reactions were published.

11.2.1
Conjugate Additions

In the early 1990s, Alexakis and coworkers [5] reported the strong rate acceleration caused by the phosphoramidite ligand **1** in the copper-catalyzed addition of diethylzinc to cyclohexenone (Scheme 11.1). The proposal that this ligand accelerating effect (LAC) was related to strong σ-donation from the phosphine ligand inspired Woodward et al. [4] to employ an N-heterocyclic carbene instead of any phosphorus-containing donor. This seminal work employed the ligand SIMes (**2**), generated *in situ*, for the conjugate addition to cyclohexenone and other enones, with the observation of a strong LAC effect. The use of a 4.5% of Cu(OTf)$_2$ and 5% of the *in situ* generated NHC ligand led to quantitative conversion into the desired product within 30 min, whereas in the absence of added ligand the conversion at that time remained below 10%. Other cyclic and acyclic enones were also alkylated by this procedure with quantitative yields.

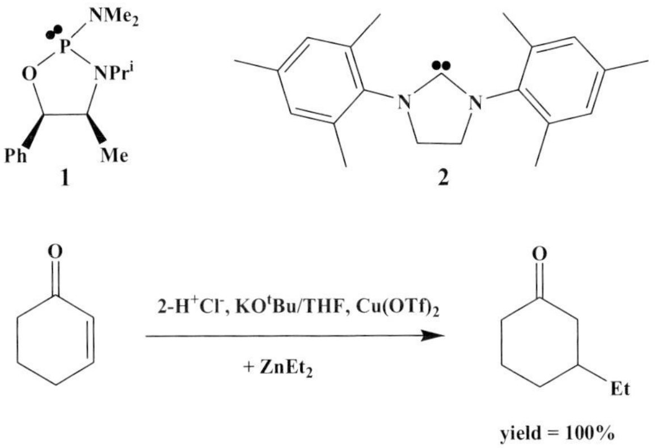

Scheme 11.1

A few months after Woodward's report, the first asymmetric versions of this catalytic system were independently, but simultaneously, reported by Alexakis [6] and Roland [7] in back-to-back articles published in the same issue of *Tetrahedron: Asymmetry*. Five different imidazolium salts containing chiral fragments along with a copper source were employed by Alexakis and coworkers to induce a certain level of asymmetry in this addition reaction. The best results, 50% e.e., were achieved with a saturated ligand (Table 11.1, entry 1) that contains both a chiral backbone and chiral substituents at nitrogen. Roland et al. employed the strategy of *in situ* transmetallation, using the corresponding silver complex (NHC)AgI (Table 11.1, entry 2). Control experiments revealed that this compound did not act as the catalyst, but merely as the carbene transfer agent to the copper ion. A comparison between these two methodologies seems to favor that using the silver reagent, not only because of the stabilities of these silver compounds but also because in this case there is no need to employ strong bases to deprotonate the imidazolium salts and, therefore, this system is not restricted to polar solvents to dissolve such precursors. However, only a modest 23% e.e. was obtained with this methodology.

Both research groups later published a joint article [8a] in which the two strategies were presented, together with various ligands and silver reagents. The best results with cyclohex-2-enone as the substrate are displayed in Table 11.1, entries 3–6. Previous data suggested that saturated carbenes could be the ligands of choice since the deviation of planarity of the backbone could be responsible for the enantioselection. However, an unsaturated ligand (entry 3) gave the best results (54%) among a series of eight imidazolium salt precursors, with copper(II)

Table 11.1 Enantioselective copper-catalyzed 1,4-addition of diethylzinc to cyclohexen-2-one.

Entry	Catalyst Precursor[a]	Temp (°C)	Yield	ee	Ref.
1	imidazolinium BF$_4$ with Ph groups + Cu(OTf)$_2$	−78 °C	91	50	[6]
2	imidazolidine with Bu^t groups, AgI + Cu(OTf)$_2$	0 °C	98	23	[7]
3	imidazolium BF$_4$ with Ph groups + Cu(OAc)$_2$	−78 °C	75	54	[8]
4	imidazol-2-ylidene with Ph groups, AgCl + CuTC[b]	−78 °C	99	62	[8]
5	imidazolidine with Bu^t groups, AgCl + CuTC[b]	−78 °C	100	58	[8]
6	imidazolidine with OMe/Bu^t groups, AgCl + CuTC[b]	−78 °C	99	69	[8]
7	Bu^t-oxazoline–imidazole Cu-Cl-thf complex	−30 °C	100	51	[9]
8	mesityl-imidazolinium PF$_6$ with hydroxyalkyl + Cu(OTf)$_2$	rt	99	83	[10]

a) Catalyst loading: 4 mol % with respect to the enone in all cases but entries 2 (2 mol %) and 8 (0.1 mol %).
b) TC = thiophenecarboxylate.

triflate or copper(I) thiopheneacetate as the copper source. The silver complex of that ligand was also tested as the carbene transfer agent, showing an increase in the e.e. up to 62% (entry 4). These results were obtained with chirality located at the nitrogen atoms in the carbene ring. However, when saturated ligands with chirality situated only in the backbone were used, further improvement in the enantioselection, to a maximum of 69%, was obtained with the ligand depicted in entry 6. The limited data yet available do not allow a clear proposal about the matching and mismatching effects between the chiral groups placed in the backbone and at the nitrogen atoms. In addition to cyclohex-2-enone, other enones were studied. Notably, with cyclohept-2-enone the very high 93% e.e. was observed with a silver complex as the carbene transfer agent. The scope of this family of catalysts has been extensively described for an array of copper sources, NHC ligands and substrates for the general 1,4-conjugate addition reaction [8b].

The above NHC ligands are monodentate, with no other donor atom interacting with the metal center. Alkoxy-containing bidentate NHC ligands have, however, been developed independently by Arnold [9] and Mauduit [10a]. Arnold et al. prepared and structurally characterized a series of Cu(II) complexes bearing these ligands, and tested them in the 1,4-addition of diethylzinc to cyclohex-2-enone. Modest enantioselectivities were obtained, the best results corresponding to the complex shown in Table 11.1, entry 7 (51% e.e.). Mauduit and coworkers have employed the same strategy, with great success, with the imidazolium salt shown in entry 8 and $Cu(OTf)_2$. These results are noteworthy for two reasons: (a) this system has provided the highest e.e. (83%), to date, for cyclohex-2-enone as the substrate and (b) such a degree of enantioselection has been obtained at room temperature, in contrast with the low temperatures employed in other cases. A complete study of the conditions in the catalytic reaction has shown the influence of several variables (temperature, ligand structure, precatalyst source, base and solvent) in the induced enantioselection [10b].

The already reported systems based on NHC-copper catalysts have shown the potential of this class of compounds for this transformation. The design of new catalysts to improve the degree of enantioselection is of interest and will probably be examined in the next few years.

11.2.2
Reduction of Carbonyl Compounds

Since the discovery by Stryker and coworkers [11] of the catalytic capabilities of the hexameric $[(Ph_3P)CuH]_6$ for the regioselective reduction of carbonyl derivatives, phosphine-containing copper complexes have been widely employed to induce the catalytic reduction of such substrates with an external source of hydride. Not surprisingly, this has inspired a series of articles in this area on the use of NHC ligands instead of phosphines.

Sadighi, Buchwald et al. [12] have applied this strategy to the conjugate reduction of a,β-unsaturated carbonyl compounds with the copper complex IPrCuCl (3) [IPr = 1,3-bis(2,6-diisopropylphenyl)imidazolium] as catalyst (Scheme 11.2). Cyclic enones and α,β-unsaturated esters were treated with poly(methylhydrosiloxane), PMHS, as the stoichiometric reductant in the presence of catalytic amounts of **3** (0.05–1 mol%) to afford the corresponding products in high yields. Notably, a four-fold excess of *t*-BuOH as an additive was needed with the addition to esters to ensure quantitative yields. This methodology has been used successfully with a series of substrates for which the reduction of the conjugate double bond was performed without affecting other functional groups in the molecule (cyano, other unsaturated bonds, etc.).

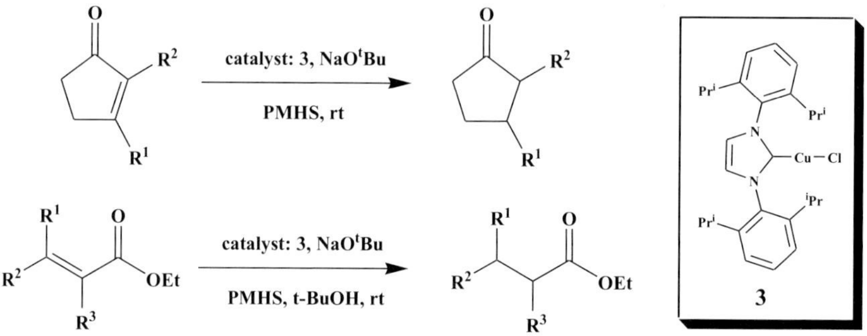

Scheme 11.2

The authors proposed that the active catalytic species is the copper-hydride IPr-CuH, formed upon consecutive reactions of **3** with NaOBu^t and of the subsequent alkoxide with PMHS, by means of a σ-bond metathesis reaction. Such a hydrido complex would react with the substrate to give the corresponding copper-enolate. At this point, two different pathways have been proposed to differentiate between enones and esters. In the former (Scheme 11.3, left), the copper-enolate would react with PMHS to yield the silyl enol ether (later hydrolyzed) and reintegrate the copper-hydrido complex. However, the observed enhancement of reaction rates when adding *t*-BuOH in the reduction of α,β-unsaturated esters suggested that the related copper enolate would undergo protonation by Bu^{t}OH, with concomitant formation of the final product and the copper *tert*-butoxide complex.

Complex **3** and a series of related complexes of general formula (NHC)CuCl have been described by Nolan and coworkers [13, 14] to catalyze the hydrosilylation of carbonyl compounds, mainly ketones, to yield the protected alcohols as the immediate products. As shown in Scheme 11.4, the reduction of cyclohexanone with Et$_3$SiH in the presence of *in situ* generated or previously isolated (NHC)CuCl complexes gives the corresponding hydrosilylated product in moderate to high yield, depending on the catalyst employed. The best results were obtained with **3**, which was then used to study the scope of this methodology with a series of ketones, which were converted in nearly quantitative yields into the desired products (Scheme 11.4). In all cases, NaOBu^t was added in 6–20 mol%, relative to the ketone: the use of lower amounts led to lower conversions and longer reaction

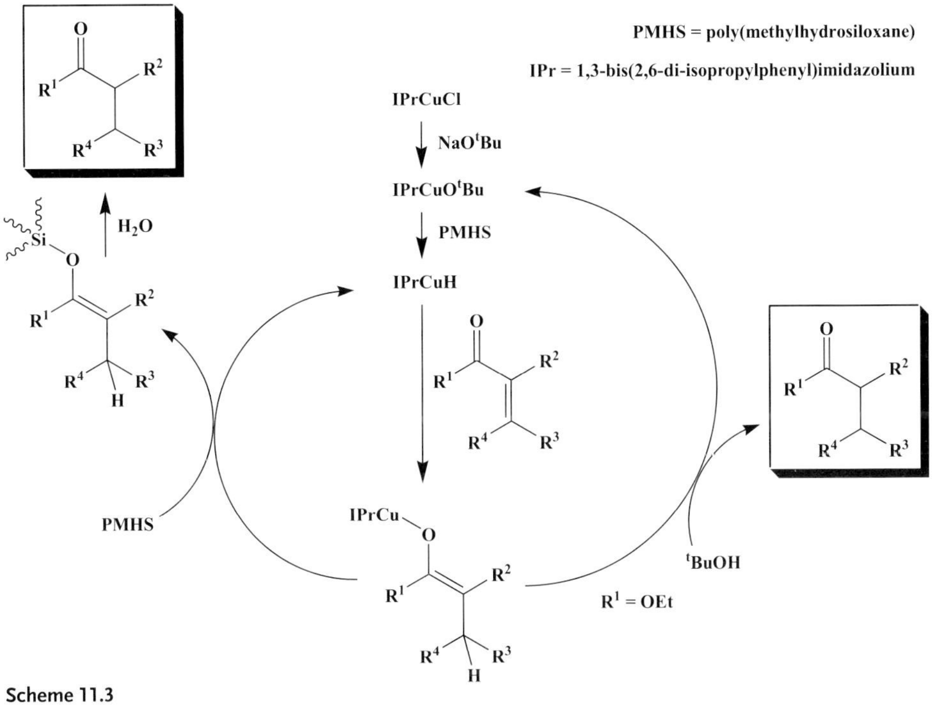

Scheme 11.3

times. These reactions were carried out at room temperature, but under these conditions hindered substrates such as dicyclohexyl ketone failed to react. However, upon raising the temperature to 80 °C this ketone underwent hydrosilylation in quantitative yield in 4 h with **3** as catalyst. Other bulky ketones were also screened with similar results, with reaction times within the range 1–21 h. Importantly, Stryker's reagent was not capable of exerting such a hydrosilylation reaction with the hindered substrates.

This work by Nolan et al. also described the role of the NHC ligand in the catalysis, which seemed to be affected mainly by electronic effects rather than by sterics. In addition, the use of well-defined catalysts, isolated and fully characterized, including X-ray studies, provided identical results to those obtained when the catalyst was generated *in situ*. In fact, the complex ICyCuCl (**4**) was employed to enlarge the scope of this system, with a series of substrates containing other functional groups such as amines, ethers, halogen- and CF_3-substituted phenyl groups, or heteroaromatic ketones, thiophene and pyridine derivatives, among others. In all cases yields were very high in the protected alcohol. Very interestingly, the authors compared these results with those obtained with a series of phosphino-copper catalysts, showing that, for dicyclohexyl ketone as the model substrate, the NHC-based catalyst were more efficient than the phosphine-containing catalysts studied. This was explained as a result of the effect of the shape of the NHC ligands, which may favor the coordination of the substrate to the

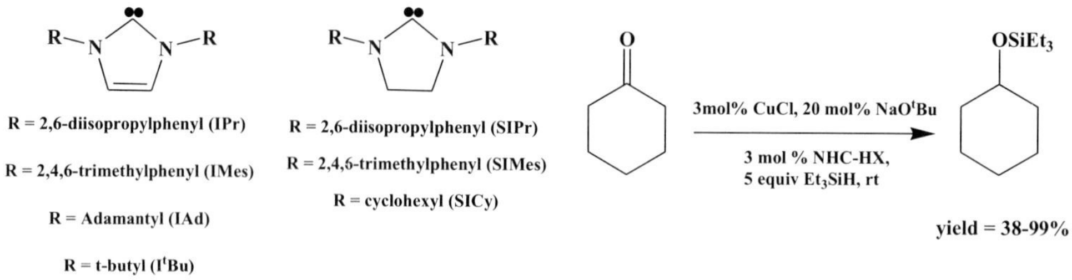
R = 2,6-diisopropylphenyl (IPr)
R = 2,4,6-trimethylphenyl (IMes)
R = Adamantyl (IAd)
R = t-butyl (I^{t}Bu)
R = 2,6-diisopropylphenyl (SIPr)
R = 2,4,6-trimethylphenyl (SIMes)
R = cyclohexyl (SICy)
3mol% CuCl, 20 mol% NaO^{t}Bu
3 mol % NHC-HX,
5 equiv Et_3SiH, rt
$OSiEt_3$
yield = 38-99%

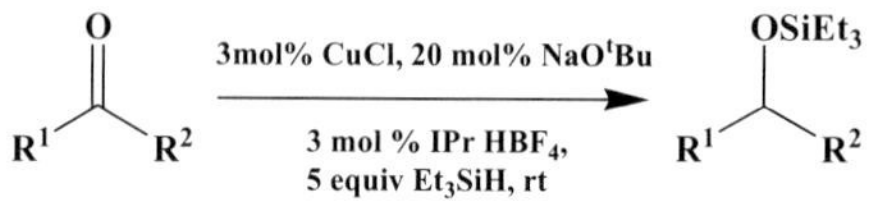
3mol% CuCl, 20 mol% NaO^{t}Bu
3 mol % IPr HBF_4,
5 equiv Et_3SiH, rt
$OSiEt_3$

R^1	R^2	Yield
Me	Et	96
Et	Et	99
n-hexyl	Me	98
Cyclopropyl	Cyclopropyl	90
$C_6H_5CH_2CH_2$	Me	96
Ph	Me	99
Ph	Et	98
α-naphthyl	Me	99

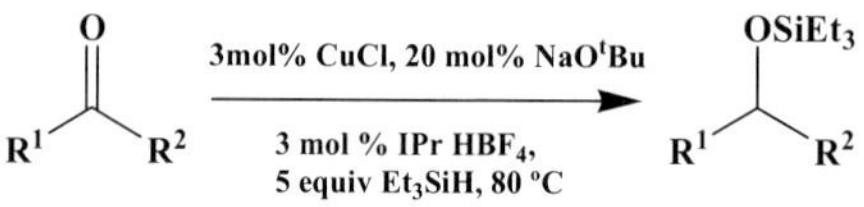
3mol% CuCl, 20 mol% NaO^{t}Bu
3 mol % IPr HBF_4,
5 equiv Et_3SiH, 80 °C
$OSiEt_3$

R^1	R^2	Yield
i-Pr	i-Pr	87
t-Bu	t-Bu	84
Cyclohexyl	Cyclohexyl	100
Cyclopropyl	Cyclopropyl	100
Ph	Ph	100
Ph	t-Bu	100

Scheme 11.4

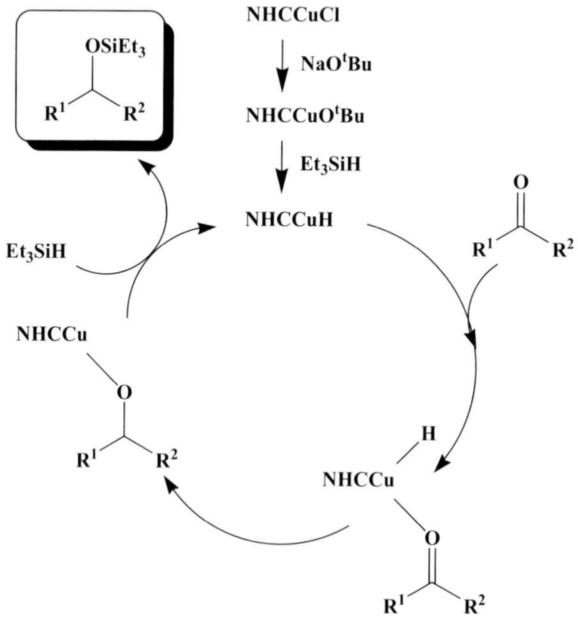
NHCCuCl
NaO^{t}Bu
NHCCuO^{t}Bu
Et_3SiH
NHCCuH
Et_3SiH
$OSiEt_3$
NHCCu
NHCCu

Scheme 11.5

metal center. The mechanistic proposal for these transformations is similar to that of Sadighi and Buchwald, and it is also based on the copper-hydrido complex (NHC)CuH as the real catalyst (Scheme 11.5).

In a separate report, Nolan et al. have described the use of a new family of complexes, the cationic bis(carbene) [(IPr)$_2$Cu]X (X = BF$_4$ or PF$_6$), as precatalysts for the hydrosilylation of carbonyl compounds [15]. These compounds, particularly the BF$_4$ salts, can induce such a transformation not only for simple ketones but also with hindered ones as well as aldehydes, in high yields, under similar conditions to those shown in Scheme 11.4, but using THF as solvent. To demonstrate the potential of these catalysts, a much less reactive carbonyl compound, an ester, was also reacted to give the corresponding protected alcohol in 69% yield. Overall, these complexes, which are very air- and moisture-stable, displayed remarkable activity for the hydrosilylation reaction.

11.2.3
Enantioselective Allylic Alkylations

Copper-catalyzed addition of organometallic reagents (Grignard, Et$_2$Zn) to allylic substrates has been known for years. In this transformation, two different products may form, coming from the so-called α- or γ-allylic substitutions, that correspond to the S$_N$2 or S$_N$2′ reactions, respectively (Scheme 11.6). Okamoto and co-workers were the first to describe the use of NHC-Cu catalysts for this transformation, including the chiral version [16a]. Thus, when complexes IPrCuCl (3) or IMesCuCl (4) were employed as catalysts in the reaction of (E)-nonenol derivatives with Pr^{i}MgCl, a preferential γ-substitution was observed with quantitative yields, with 4 displaying a higher reaction rate. The reaction was found to be general for a series of allylic substrates as well as for several Grignard reagents.

After these results, the authors looked for a chiral system to exert a certain asymmetry, using ligands with chirality located in the backbone and/or at the nitrogen atoms. With a similar procedure, and n-HexMgBr as the alkylating agent, they found that the reaction proceeded with very high regioselectivity through the S$_N$2′ pathway to give optically active γ-substituted products, although with moderate

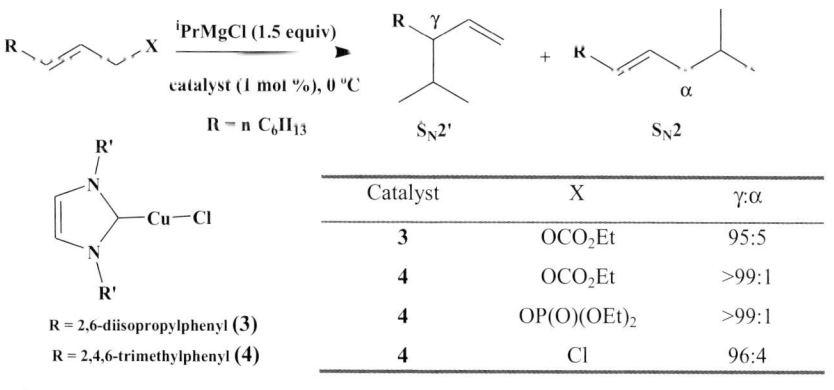

Catalyst	X	γ:α
3	OCO$_2$Et	95:5
4	OCO$_2$Et	>99:1
4	OP(O)(OEt)$_2$	>99:1
4	Cl	96:4

Scheme 11.6

e.e. s (Scheme 11.7). The very bulky-at-nitrogen ligand provided the best enantioselectivity, whereas ligands with chirality localized at the backbone induced low asymmetry. Recently, Okamoto et al. have postulated that the active species in this system is the ate-type complex of composition [(NHC)CuR$_2$]$^-$(MgX)$^+$ [16b].

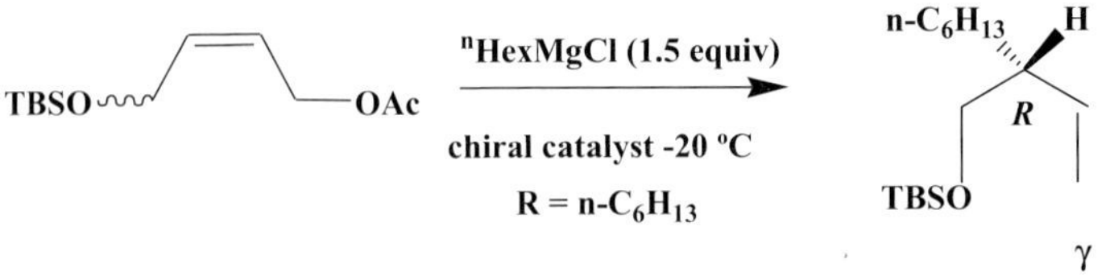

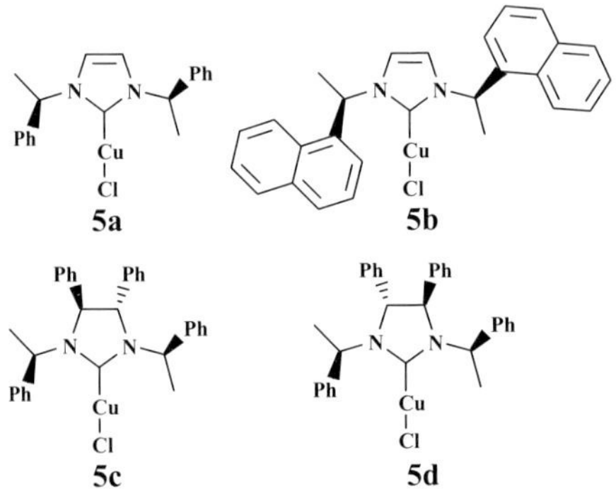

Catalyst	γ:α	ee %
5a	87:13	40
5b	95:5	60
5c	88:12	38
5d	84:16	6

Scheme 11.7

Shortly after this, Hoveyda and coworkers described a copper-based system with a chiral NHC ligand containing an optically pure binaphthyl fragment [17]. Initial attempts were made on the reaction of the allylic phosphate Ph-CH=CHCH$_2$O-PO(OEt)$_2$ with diethylzinc in the presence of *in situ* generated catalysts [Cu(OTf)$_2$ and the chiral NHC ligand, Scheme 11.8]. A very regioselective transformation towards the S$_N$2′ product (>98:2) was observed with a series of five ligands employed, although the asymmetric induction was quite distinct along the series (21–82%). The low conversions obtained with **6c**, bearing the methoxy group, could be a consequence of the need for a covalent oxygen–copper bond for the catalytic process to occur, attesting to the bidentate nature of these ligands during catalysis. The authors prepared the silver(I) complex with the more active ligand **6a** to give complex **7**, X-ray studies of which revealed its dimeric nature. This complex was inactive for catalysis *per se*, but in addition to CuCl$_2$ promoted the enantioselective allylic alkylation of a group of phosphates in very high yield, very regioselectively (γ substitution) and with high enantiomeric excess (up to 97%). Remarkably, a very low catalyst loading (2%) was employed, and this methodology could not only be applied to tertiary but also to quaternary carbon stereogenic centers. A Cu(II) dimeric complex containing the ligand **6a** was prepared by transmetallation from **7** and structurally characterized. The catalytic results obtained with these well-defined catalysts were similar to those obtained by *in situ* generation.

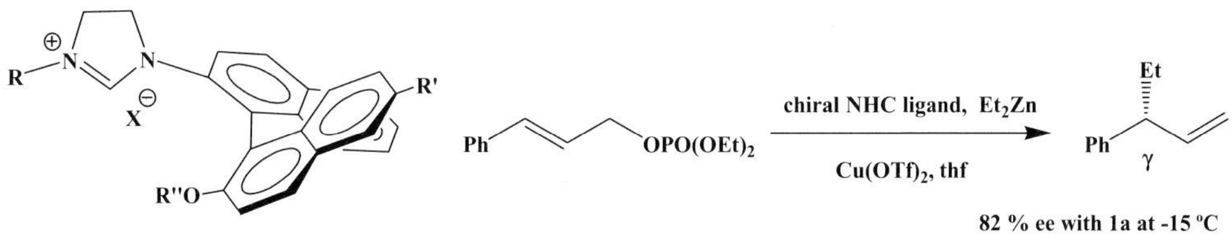

Scheme 11.8

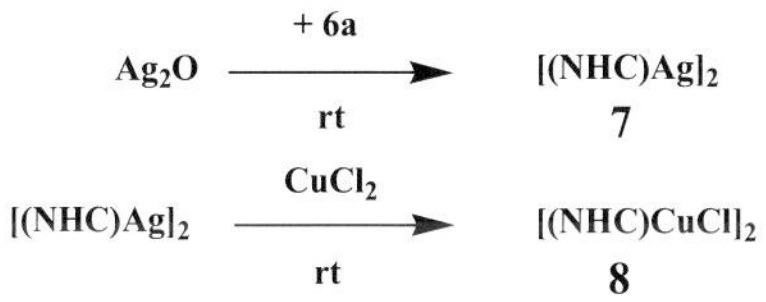

Catalyst	ee %
6a/Cu(OTf)$_2$	82
7/Cu(OTf)$_2$	89
7/CuCl$_2$	86
7/Cu(OTf)$_2$	91[a]
8	92[a]

[a]obtained with 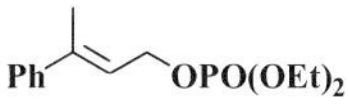 Ph-CH=CH-CH$_2$-OPO(OEt)$_2$

Scheme 11.9

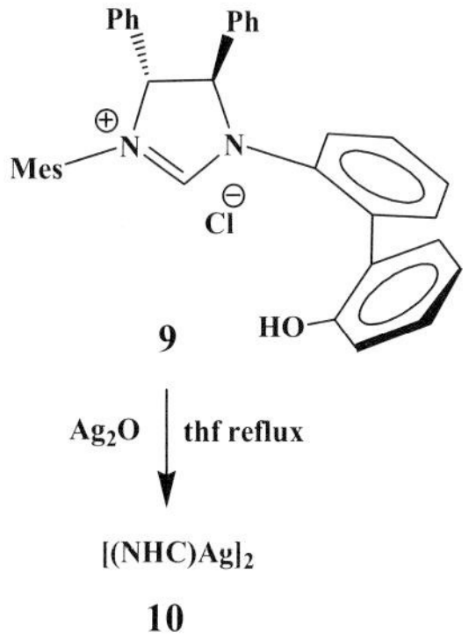

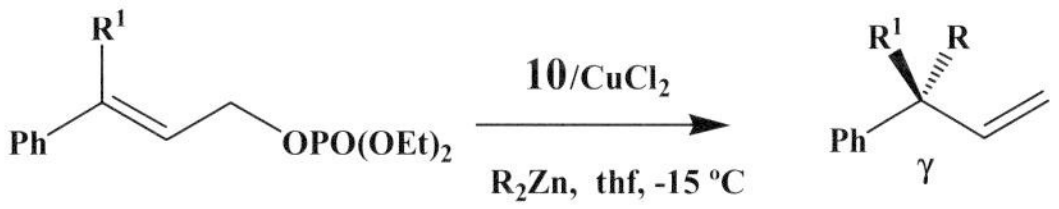

R^1	R	ee %
H	Et	90
H	iPr	86
H	nBu	89
Me	Et	97
Me	iPr	98

The previous work required the use of an optically pure binaphthyl-based amino alcohol. To avoid this, Hoveyda and coworkers approached the ligand design from the point of view of chirality located at the backbone, with the aid of chiral diamines [18]. With this procedure, they prepared the new NHC chiral ligand 9 (Scheme 11.9) as well as its silver complex and tested them for the copper-catalyzed allylic alkylation of disubstituted olefins. The use of the silver complex 10 and copper(II) chloride provided higher e.e.s than those previously obtained with 7/Cu salts, and no products derived from the S_N2 reaction were observed. Additionally, alkyl sources other than Et$_2$Zn were tested, such as (Pr^i)$_2$Zn or (Bu^n)$_2$Zn, and also gave very high e.e.s. When trisubstituted olefins were studied, leading to the synthesis of quaternary carbons, the results were also quite impressive, with enantiomeric excesses being found within the range 94–97%, including the bulky

(Pr^i)$_2$Zn as substrate. Complex **10** has also been employed as a carbene transfer agent to give a new NHC-Cu(II) complex that, similarly to **8**, also catalyzed the allylic alkylation, with activities and selectivities nearly identical to those of the **10**/CuCl$_2$ system.

11.3
Silver

Compared with copper, there are very few examples of NHC-Ag complexes with catalytic properties. Strictly speaking, we could say that there are only two cases: the preparation of 1,2-bisboronate esters [19] and the functionalization of C–H bonds with diazo compounds (see Section 11.5). However, the well-known transmetallation reaction from silver to other metals has been employed in some catalytic systems and although in such cases the silver center is not responsible for the catalytic reaction, owing to its use *in situ*, these systems can be considered here.

11.3.1
Synthesis of 1,2-bis(Boronate) Esters

Catalytic diboration of alkenes, a reaction of interest in the synthesis of diols, with a (NHC)Ag complex has been reported by Peris, Fernández and coworkers [19]. They have prepared a new silver complex of composition [(mentimid)$_2$Ag]AgCl$_2$ (**11**) containing the mentimid ligand (1-methyl-3-(+)-methylmenthoxide imidazolidene). This complex catalyzes the reaction of terminal, as well as internal, alkenes with bis(catecholato)diboron (Scheme 11.10) to give the corresponding diols after treatment with NaOH/H$_2$O$_2$, in the first example of a silver-based catalyst for this transformation. Although no asymmetric induction was observed, despite the chirality of the NHC ligand, the role of **11** as catalyst has been assessed by the fact that: (a) no reaction was observed in the presence of silver salts and (b) the use of related silver-phosphine complexes was also unsuccessful.

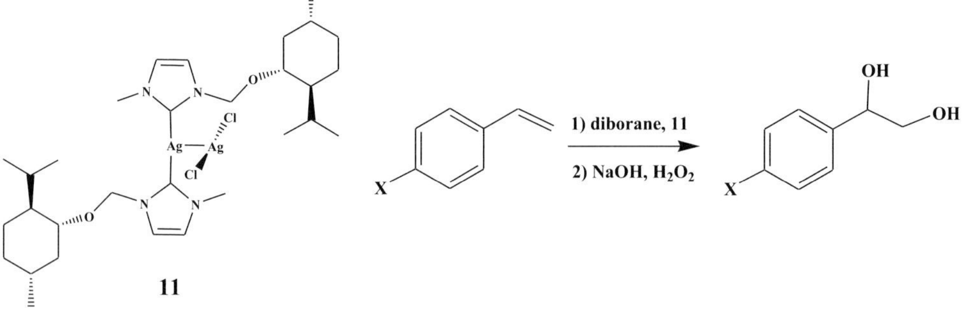

Scheme 11.10

11.3.2
NHC-Ag as Carbene Delivery Agents

An N-heterocyclic carbene ligand is readily transferred from the coordination sphere of silver in a given complex, a feature that has been applied in some catalytic systems. Based on it, Waymouth, Hedrick and coworkers [20] have induced the polymerization of lactide by a NHC-containing silver complex similar to **11**, but with methyl and ethyl as the substituents of the imidazolic nitrogens (Scheme 11.11). The silver complex actually serves as a reservoir for the free carbene ligand: the authors have proposed that an equilibrium of the silver complex **12** (an ionic liquid at room temperature) with the free carbene and silver chloride takes place, and that the carbene ligand is the real catalytic species in the cycle.

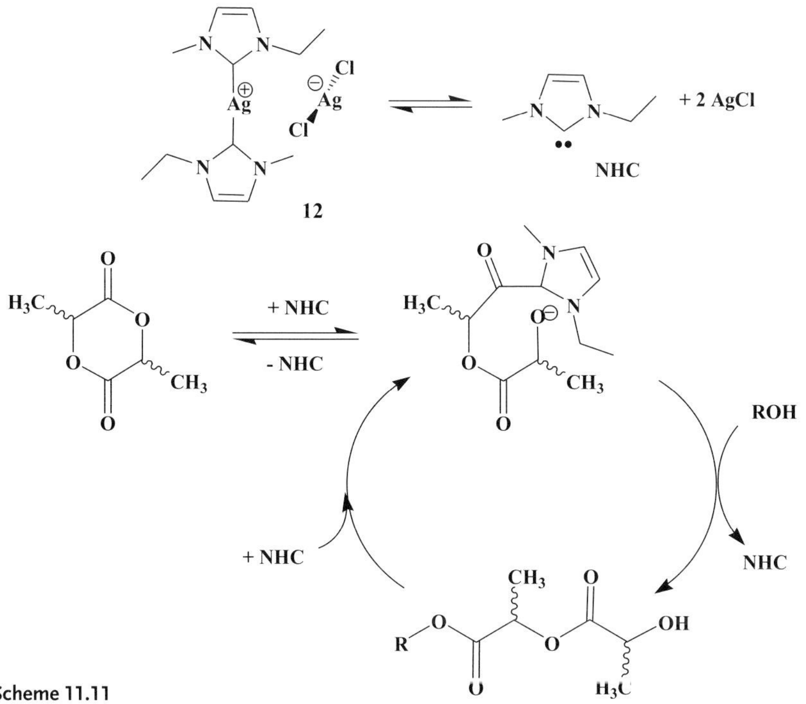

Scheme 11.11

In other cases, the NHC-Ag complex has been employed for the *in situ* transfer of the NHC ligand to another metal, which is the real catalyst. Some examples have been discussed in Section 11.2 on copper [7, 8, 10, 17, 18]. Dowthwaite et al. have applied this methodology to asymmetric allylic alkylation [21], with palladium as the metal carrying out the catalytic transformation (Scheme 11.12). The NHC-ligand was transferred *in situ* from the corresponding silver complex **13**. Charette and coworkers have described another example [22], with an anionic NHC ligand being transferred from silver (complex **14**) to rhodium to generate a catalyst for the reaction of imines and PhB(OH)$_2$ (Scheme 11.12) [22].

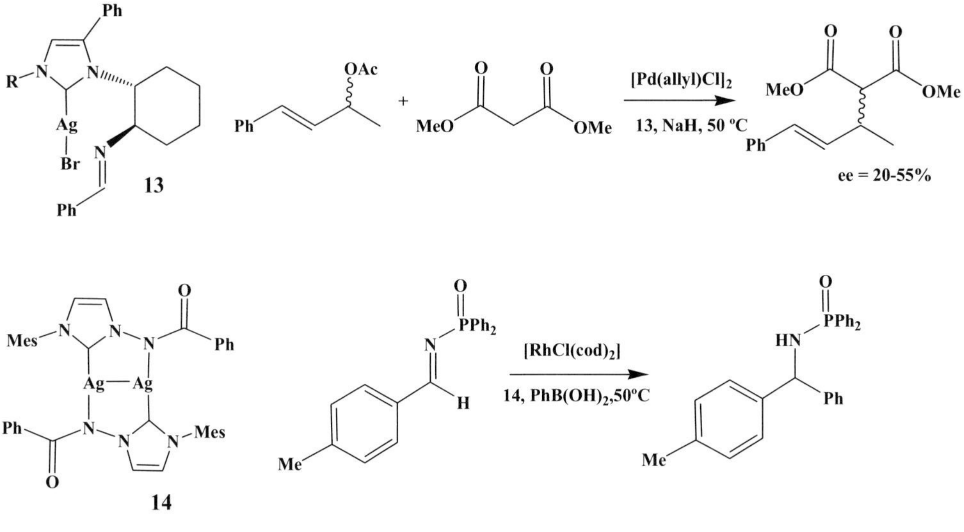

Scheme 11.12

11.4
Gold

It was not until 2003 that Herrmann et al. described the first catalytic use of a NHC-containing gold(I) complex [23]. Not surprisingly, this system was inspired by the well-known related phosphine gold(I) systems for the addition of nucleophiles to non-activated alkynes. In this work, the authors reported the use of the complex (NHC)Au(OAc) for the addition of water to hex-3-yne, in the presence of a Lewis acid as co-catalyst ($B(C_6F_5)_3$, Scheme 11.13). Notably, the related chloride complex (NHC)AuCl did not catalyze this reaction, emphasizing the need for a weakly coordinating anion, i.e., evidencing that a cationic species (NHC)Au$^+$ might be the real catalyst. However, the catalytic activity of this system was not as good as that with the related phosphine-based catalysts.

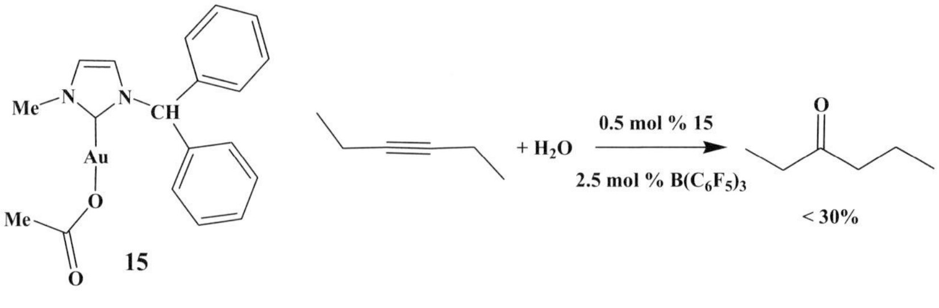

Scheme 11.13

11.5 Cu-, Ag-, and Au-NHC Complexes as Catalysts for Carbene Transfer Reactions...

Echavarren and coworkers have explored the use of NHC ligands as phosphine surrogates during their studies with gold(I)-based catalysts for intramolecular cycloaddition reactions [24, 25]. In addition to the use of biphenyl-derived phosphines, they have also employed N-heterocyclic ligands in the cyclization of 1,6-enynes (Scheme 11.14). Although the catalysts (NHC)AuCl are effective for this transformation, their activity is far from that observed with phosphine-containing catalysts. The same authors have also tested the catalytic capabilities of such complexes in the intramolecular cyclopropanation of dienynes (Scheme 11.14). Here, the activity was quite similar to that of a series of phosphine-gold catalysts [26]. This also applies to the selectivity of the reaction, since in some cases a skeletal rearrangement has been observed; no such by-product has been observed with the (NHC)AuCl catalysts, yielding >98% of the product derived by intramolecular cyclopropanation.

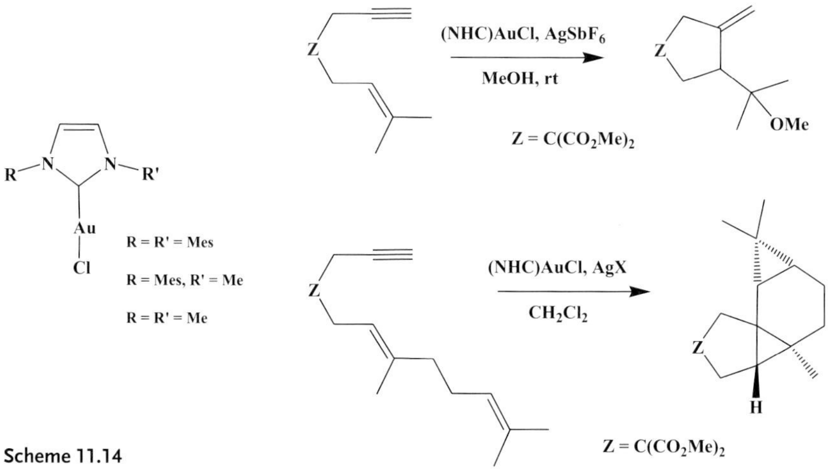

Scheme 11.14

11.5
Cu-, Ag-, and Au-NHC Complexes as Catalysts for Carbene Transfer Reactions from Ethyl Diazoacetate

In 2004, Pérez, Díaz-Requejo, Nolan and coworkers reported the catalytic properties of the complex IPrCuCl (**3**) toward the decomposition of ethyl diazoacetate (N_2CHCO_2Et, EDA) and the subsequent transfer of the carbene :$CHCO_2Et$ unit to saturated (amines, alcohols) and unsaturated (olefins) substrates [27]. These authors have extended that study to silver and gold in a series of IPrMCl complexes (M = Cu, **3**; Ag, **16**; Au, **17**) as pre-catalysts for such carbene transfer reactions, and with an array of substrates, with particular emphasis on insertion into C–H bonds [28–31]. In contrast to most catalytic systems commented on in this chapter, there were no related, phosphine-based catalytic systems of these metals for this type of transformation. Interestingly, the IPrMCl complexes did not react with EDA after a few hours in methylene chloride unless a donor substrate (amine, alcohol, olefin) was added or NaBAr′$_4$ was used as halide scavenger (Scheme 11.15). The latter additive was mandatory with hydrocarbons as substrates: cyclohexane was converted into ethyl cyclohexylacetate in 70–90% yield with IPrMCl + NaBAr′$_4$ as catalyst. These data imply that a cationic species IPrM$^+$ is the real catalytic species, which is accessible by displacement or replacement of the chloride by the donor substrates or the NaBAr′$_4$, respectively (Scheme 11.15). Such cationic intermediates would react with EDA to give the transient metallocarbene species, which would readily transfer the :CHCO$_2$Et group to the corresponding substrate. Notably, the use of a gold-based catalyst for diazo decomposition was previously unknown [28].

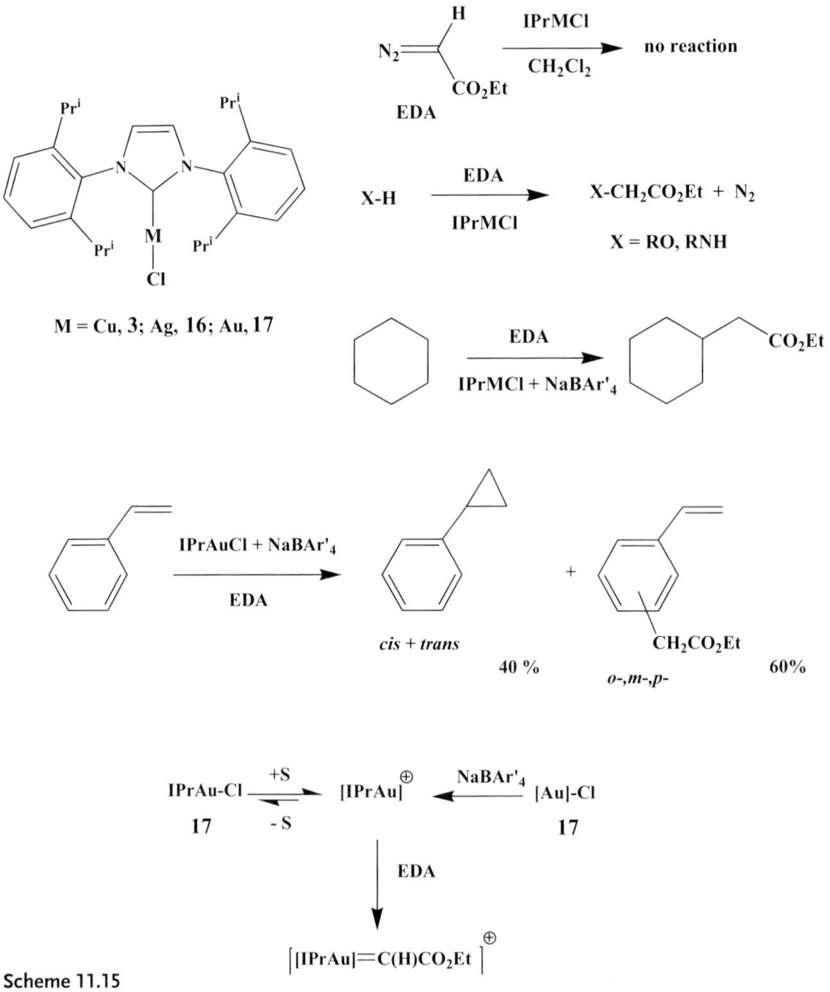

Scheme 11.15

11.5 Cu-, Ag-, and Au-NHC Complexes as Catalysts for Carbene Transfer Reactions...

A quite interesting reaction was observed when using the gold complex **17** as catalyst in the cyclopropanation of styrene with EDA. Along with the expected cyclopropanes, the major product obtained was styryl acetate (Scheme 11.15), as the result of formal insertion of the :CHCO$_2$Et fragment into the aromatic sp^2 C–H bond, with the alkenyl C=C bond remaining intact. Actually, this insertion reaction has also been observed with benzene and toluene, where the related products were obtained in addition to the expected cycloheptatriene derivatives (formed in the metal-catalyzed Büchner reaction) [28]. The use of the silver analog **16** for the same transformations gave minor, but detectable, amounts of the insertion products (<10%) for the three substrates [30].

This methodology has also been employed with success to the functionalization of sp^3 C–H bonds of alkanes. Thus, when **3** or **17** were employed as the catalyst in the reaction of EDA and 2,3-dimethylbutane, two products were formed with different ratios, depending of the catalyst. Insertion into the primary C–H bonds was favored with gold whereas a nearly opposite regioselectivity was observed with copper (Scheme 11.16). There is an alternative to the use of different metals in tuning the catalyst towards regioselective control: the use of different NHC ligands. The authors have studied a series of NHCAuCl complexes for this reaction, with 2,3-dimethylbutane as the substrate, and have found a strong effect of the ligand on both the chemo- and regioselectivity [29]. The IPr ligand provided the highest conversions as well as the best regioselectivity towards primary C–H activation products, whereas IMes gave the lowest in both cases, with IAd and IBu^t in the middle. The existence of a correlation between the conversion (chemoselectivity) and the ratio of products (regioselectivity) is remarkable in the sense that the higher the former the higher is the amount of product derived from insertion into the primary carbon–hydrogen bonds.

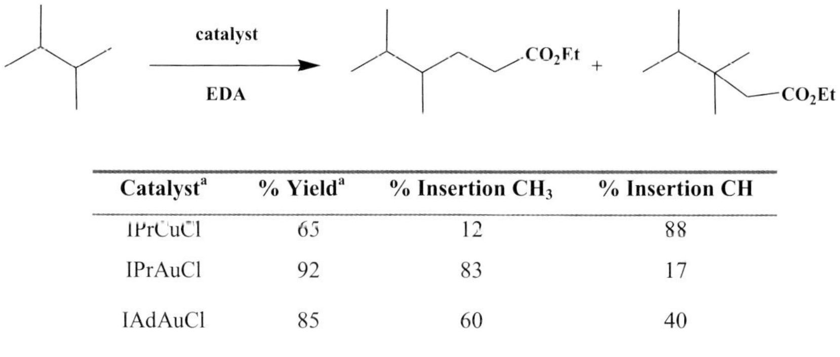

Catalyst[a]	% Yield[a]	% Insertion CH$_3$	% Insertion CH
IPrCuCl	65	12	88
IPrAuCl	92	83	17
IAdAuCl	85	60	40
ItBuAuCl	76	53	47
IMesAuCl	40	32	68

[a] NaBAr'$_4$ added in all cases. [b] Diethyl fumarate and maleate accounted for 100%.

Scheme 11.16

References

1. W. A. Herrmann, *Angew. Chem. Int. Ed.* **2002**, *41*, 1290.
2. H. M. J. Wang, I. J. B. Lin, *Organometallics*, **1998**, *17*, 972.
3. V. César, S. Bellemin-Laponnaz, L. H. Gade, *Chem. Soc. Rev.* **2004**, *33*, 619.
4. P. K. Fraser, S. Woodward, *Tetrahedron. Lett.* **2001**, *42*, 2747.
5. A. Alexakis, J. C. Frutos, P. Mangeney, *Tetrahedron: Asymmetry* **1993**, *4*, 2427.
6. A. Alexakis, C. L. Winn, F. Guillen, *Tetrahedron: Asymmetry* **2001**, *12*, 2083.
7. J. Pytkowicz, S. Roland, P. Mangeney, *Tetrahedron: Asymmetry* **2001**, *12*, 2087.
8. (a) A. Alexakis, C. L. Winn, F. Guillen, J. Pytkowicz, S. Roland, P. Mangeney, *Adv. Synth. Catal.* **2003**, *3*, 345. (b) C. L. Winn, F. Guillen, J. Pytkowicz, S. Roland, P. Mangeney, A. Alexakis, *J. Organomet. Chem.* **2005**, *690*, 5672.
9. P. L. Arnold, M. Rodden, K. M. Davis, A. C. Scarisbrick, A. J. Blake, C. Wilson, *Chem. Commun.* **2004**, 1612.
10. (a) H. Clavier, L. Contable, J.-C. Guillemin, M. Mauduit, *Tetrahedron: Asymmetry* **2005**, *16*, 921. (b) H. Clavier, L. Contable, L. Toupec, J.-C. Guillemin, M. Mauduit, *J. Organomet. Chem.* **2005**, *690*, 5237.
11. W. S. Mahoney, D. M. Brestensky, J. M. Stryker, *J. Am. Chem. Soc.* **1988**, *110*, 291.
12. V. Jurkauskas, J. P. Sadighi, S. L. Buchwald, *Org. Lett.* **2003**, *5*, 2417.
13. H. Kaur, F. K Zinn, E. D. Stevens, S. P. Nolan, *Organometallics* **2004**, *23*, 1157.
14. S. Díez-González, H. Kaur, F. K. Zinn, E. D. Stevens, S. P. Nolan, *J. Org. Chem.* **2005**, *70*, 4787.
15. S. Díez-González, N. M. Scott, S. P. Nolan, *Organometallics* **2000**, *25*, 2355.
16. (a) S. Tominaga, Y. Oi, T. Kato, D. K. An, S. Okamoto, *Tetrahedron Lett.* **2004**, *45*, 5585. (b) S. Okamoto, S. Tominaga, N. Saino, K. Kase, K. Shimoda, *J. Organomet. Chem.* **2005**, *690*, 6001.
17. A. O. Larsen, W. Leu, C. Nieto-Oberhuber, J. E. Campbell, A. H. Hoveyda, *J. Am. Chem. Soc.* **2004**, *126*, 11 130.
18. J. J. Van Veldhuizen, J. E. Campbell, R. E. Giudici, A. H. Hoveyda, *J. Am. Chem. Soc.* **2005**, *127*, 6877.
19. J. Ramirez, R. Corberán, M. Sanaú, E. Peris, E. Fernández, *Chem. Commun.* **2005**, 3056.
20. A. C. Sentman, S. Csihony, R. M. Waymouth, J. L. Hedrick, *J. Org. Chem.* **2005**, *70*, 2391.
21. L. G. Bonnet, R. E. Douthwaite, *Organometallics* **2003**, *22*, 4187.
22. C. Y. Legault, C. Kendall, A. B. Charette, *Chem. Commun.* **2005**, 3826.
23. S. K. Schneider, W. A. Herrmann, E. Herdtweck, *Z. Anorg. Allg. Chem.* **2003**, *629*, 2363.
24. C. Nieto-Oberhuber, S. López, A. M. Echavarren, *J. Am. Chem. Soc.* **2005**, *127*, 6178.
25. C. Nieto-Oberhuber, M. P. Muñoz, S. López, E. Jiménez, C. Nevado, E. Herrero-Gómez, M. Raducan, A. M. Echavarren, *Chem. Eur. J.* **2006**, *12*, 1677.
26. C. Nieto-Oberhuber, S. López, M. P. Muñoz, E. Buñuel, D. J. Cárdenas, A. M. Echavarren, *Chem. Eur. J.* **2006**, *12*, 1694.
27. M. R. Fructos, T. R. Belderráin, M. C. Nicasio, S. P. Nolan, H. Kaur, M. M. Díaz-Requejo, P. J. Pérez, *J. Am. Chem. Soc.* **2004**, *126*, 10 846.
28. M. R. Fructos, T. R. Belderráin, P. de Fremont, N. M. Scott, S. P. Nolan, M. M. Díaz-Requejo, P. J. Pérez, *Angew. Chem. Int. Ed.* **2005**, *44*, 5284.
29. M. R. Fructos, T. R. Belderráin, P. de Fremont, N. M. Scott, S. P. Nolan, H. Kaur, M. M. Díaz-Requejo, P. J. Pérez, *Organometallics* **2006**, *25*, 223.
30. M. R. Fructos, T. R. Belderráin, P. de Fremont, N. M. Scott, H. Kaur, P. Nolan, M. M. Díaz-Requejo, P. J. Pérez, unpublished results.
31. M. M. Díaz-Requejo, P. J. Pérez, *J. Organomet. Chem.* **2005**, *690*, 5441.

12
N-Heterocyclic Carbenes as Organic Catalysts

*Andrew P. Dove, Russell C. Pratt, Bas G. G. Lohmeijer, Hongbo Li,
Erik C. Hagberg, Robert M. Waymouth, and James L. Hedrick*

12.1
Introduction

Organocatalysis has a long and venerable history, as exemplified by Wohler and Liebig's pioneering demonstration that cyanide ions catalyze the benzoin condensation [1]. Recent advances in enantioselective organocatalysis have stimulated renewed interest in the application of organic catalysts for stereoselective reactions [2–4]. Chiral pyrroles and imidazolidinone have proven effective catalysts for reactions such as the Mannich reaction, asymmetric hetero-Diels–Alder reaction, 1,3-dipolar cycloadditions, and 1,4-conjugate Friedel–Crafts reactions [5–8]. Several groups have reported effective non-enzymatic catalyst for the kinetic resolution of secondary alcohols using chiral phosphines [9–11] or amine catalysts [6, 12]. To some degree this strategy mimics that carried out by enzymes [13].

The demonstration by Breslow that stabilized singlet carbenes derived from thiamine cofactors are nucleophilic catalysts [14] and the pioneering work by Wanzlick [15] on the chemistry of nucleophilic N-heterocyclic carbenes provided the inspiration for the development of N-heterocyclic carbenes as nucleophilic organic catalysts [2, 16–24]. NHCs have been widely employed as ligands in organometallic catalysis [25, 26]. The strong σ-donating properties of these stabilized carbenes lead to enhanced catalytic activity over phosphine or amine ligated species. In addition to their use as ligands for transition metal catalysts, it is clear from several studies that these stable carbenes are effective catalysts in their own right. Owing to their high reactivity and ease of structural diversification, they are proving to be powerful organic catalysts for various transformations.

N-Heterocyclic Carbenes in Synthesis. Edited by Steven P. Nolan
Copyright © 2006 WILEY-VCH Verlag GmbH & Co. KGaA, Weinheim
ISBN: 3-527-31400-8

12.2
In situ Generation of Free Carbenes

Carbenes have long been studied as reactive intermediates, and various strategies have been developed for the generation of these reactive species *in situ* [27]. The higher stability of N-heterocyclic carbenes has enabled their isolation and study [28–30]; nevertheless, the generation of these reactive species *in situ* from stable precursors is often experimentally expedient.

Of the various ways to generate N-heterocyclic carbenes, the most common strategy is the deprotonation of imidazolium, triazolium or thiazolium salts. Typically, the free carbenes can be generated using a base such as triethylamine or potassium *t*-butoxide, depending on the pK_a of the salt (Scheme 12.1A) [31]. For many reasons, e.g., the deprotonation can be carried out *in situ*, this technique is a versatile method of screening carbene compounds for activity and selectivity.

Because of the interest in NHCs as ligands for transition metals, various strategies have been devised for ligating carbenes to transition metals. The use of silver(I) carbene complexes as transfer agents in the generation of transition metal complexes prompted Waymouth and Hedrick to investigate these silver complexes

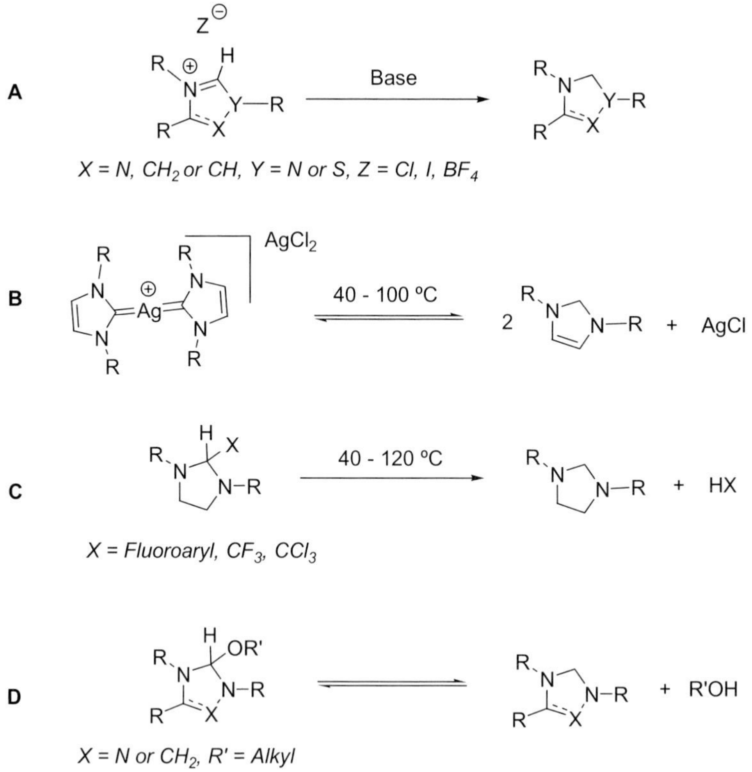

A: X = N, CH$_2$ or CH, Y = N or S, Z = Cl, I, BF$_4$

B: 40 - 100 °C

C: X = Fluoroaryl, CF$_3$, CCl$_3$; 40 - 120 °C

D: X = N or CH$_2$, R' = Alkyl

Scheme 12.1 Methods for *in situ* generation of free carbene.

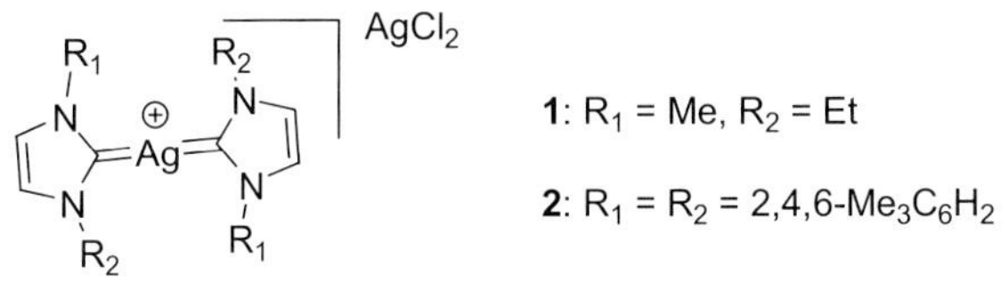

Fig. 12.1 Examples of silver(I) carbene complexes.

as potential delivery agents for free carbenes in ring-opening polymerization (ROP) and transesterification [32]. Complexes of this type are readily prepared by reaction of imidazolium chloride with silver(I) oxide, and this was shown to be a general method for various unsaturated imidazol-2-ylidene compounds (Scheme 12.1B). Delivery of free carbene was demonstrated by differential scanning calorimetry (DSC) and thermogravimetric analysis (TGA). Additionally, the imidazolium salt ionic liquids could be used as an effective pre-catalyst reservoir. ROP was successful in both neat ionic liquid and a biphasic mixture of ionic liquid and THF [21].

A facet of imidazolin-2-ylidene and triazol-2-ylidene carbenes is their ability to form adducts of alkyl and aryl species containing acidic C–H bonds such as chloroform, pentafluorobenzene, methyl phenyl sulfone, acetylene or alcohol [26, 33, 34]. Such adducts have been utilized in the synthesis of transition-metal complexes [21, 25, 35–37] and more recently have been demonstrated to be useful precursors for the *in situ* generation of free carbenes for ring-opening polymerization reactions [38, 39].

Carbene adducts containing electron-withdrawing substituents were easily realized in good yield by the reaction of a diamine with a benzaldehyde derivative or a hemiacetal [39]. Upon heating, the fluoroaryl adducts degrade to free carbene and fluorobenzene, and the transition temperature depends on the degree of fluorination of the aryl group (Scheme 12.1C). Adducts with weak electron-withdrawing groups such as trifluoromethane, benzene, and 2,3,4-trifluorobenzene are thermally very stable and do not readily degrade to the free carbene even at 100 °C. The pentafluorobenzene adduct is readily generated from the aldehyde and the diamine and thermally generates the carbene at 60 °C. Chloroform adducts can be readily accessed through reaction of chloroform with the free carbene [33, 39] or by treatment of the imidazolinium salt with sodium hydroxide and chloroform [21]. Generation of the free carbene has been demonstrated by thermolysis [33, 39]

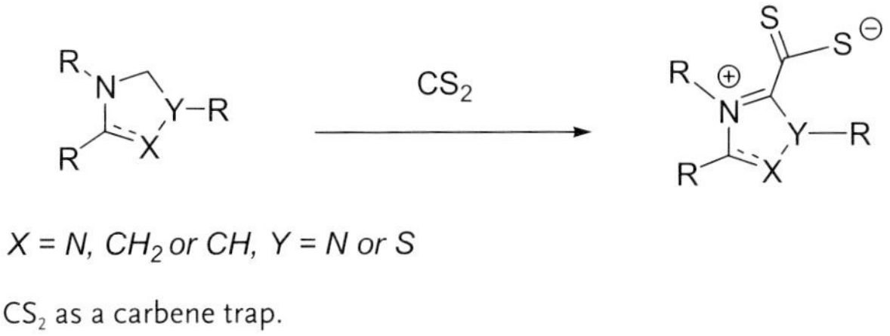

$X = N, CH_2$ or CH, $Y = N$ or S

Scheme 12.2 CS_2 as a carbene trap.

and trapping with CS_2 (Scheme 12.2). The use of alcohol adducts of carbenes has been more widely investigated as a delivery method for free carbenes (Scheme 12.1D). Both imidazolin- and triazolin-2-ylidenes form stable alcohol adducts. Enders demonstrated that heating the alkoxytriazolylidene under vacuum at 80 °C led to the generation of free carbene [34], and more recently Hedrick and Waymouth have shown that the alcohol adducts at elevated temperature in solution are useful precursors to the carbenes [19].

12.3
Small Molecule Transformations

12.3.1
Benzoin and Formoin Condensation

One of the interesting organocatalytic reactions involving N-heterocyclic carbenes is the self-condensation of benzaldehyde to benzoin. Although this reaction was already reported in 1832 to be catalyzed by cyanide anions [1] and its mechanism elucidated in 1903 [40]. The first report on using a carbene organocatalyst appeared in 1943 [41]. Thiazolium salts were the precatalysts and, in the presence of base, the catalytically active carbene compound was generated *in situ*. The mechanism of benzoin condensation, using thiazol-2-ylidenes, described in 1958 by analysis of deuterium exchange experiments [14] is still the generally accepted model (Scheme 12.3) despite much discussion in the 1980s and 1990s [42–47].

The free carbene reacts with first benzaldehyde molecule via nucleophilic addition to the carbonyl group, then subsequently rearranges to the hydroxyenamine (Breslow intermediate), which acts as an acyl anion equivalent for addition to the carbonyl of

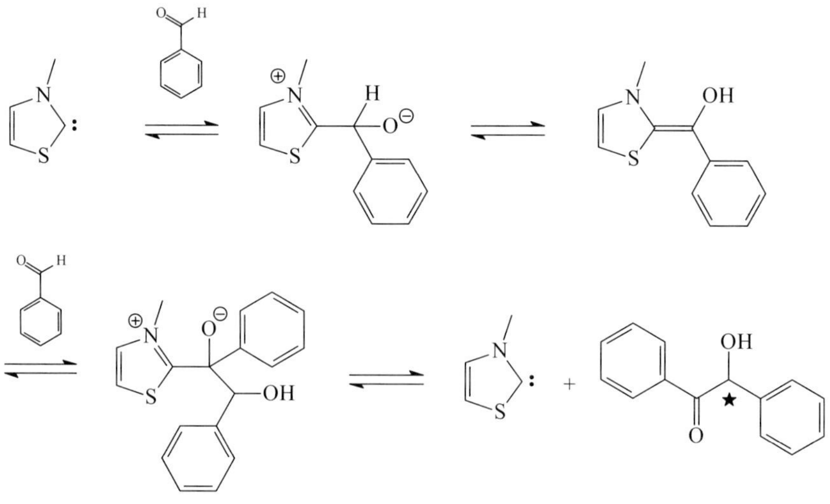

Scheme 12.3 Mechanism for the benzoin condensation using N-methylthiazol-2-ylidene [14].

the second benzaldehyde molecule. A second rearrangement gives rise to benzoin and the free carbene that is ready to participate in a subsequent turnover [2]. Notably, benzoin condensation is an equilibrium reaction and can be reversed.

The product of benzoin condensation bears a stereocenter and hence intensive research has focused on asymmetric catalyst design to obtain enantiopure product. Despite these efforts, starting as early as 1966 [48], only in recent years has significant progress been made in the preparation and application of enantioselective organocatalysts [34, 37, 49–55]. Increasing the steric bulk on the N-substituent and methylation of the 4- and 5-position of the early catalysts based on thiazole have been investigated. Unfortunately, the initial *ee*s were not very good, with a maximum reported value of 52% [49]. Bicyclic and polycyclic thiazoles were also investigated, but yielded only modest enantioselectivities [52–54]. Apart from 1,3-thiazole-based organocatalysts, N-heterocyclic carbenes derived from 1,2,4-triazole have also been reported for the benzoin condensation [34, 37, 55].

Triazole-based organocatalysts were shown to give much better *ee*s, because of more steric bulk at both sides of the carbene center: a bulkier phenyl-substituted N-atom instead of the single S-atom accounts for their success in the asymmetric benzoin condensation. Figure 12.2 and Table 12.1 present selected results on the organocatalysts that have been used up to date.

Table 12.1 Results of asymmetric benzoin condensation.

Compound	1	2	3	4	5	6	7	8	9	10	11	12	13	14	15
Yield (%)	10	6	20	6	34	20	50	100	2	18	45	47	22	66	83
ee (%)	22	52	35	12	20	11	21	26	28	30	80	48	63	75	90
References	48	49	50	51	52	52	52	53	54	54	55	55	55	33	56

The most successful catalyst is entry 15 in Table 12.1 [56]. This particular catalyst shields the Si-face of the Breslow intermediate, hence attack of the second aldehyde must come from the less-hindered Re-face. Moreover, the N-phenyl substituent causes a prealignment of the second benzaldehyde (Fig. 12.3). These observations have been confirmed by computational predictions [57]. Several catalysts mentioned in Table 12.1 have been used for the condensation of other aromatic and aliphatic aldehydes [34, 37, 55]. Also, intramolecular crossed aldehyde–ketone benzoin reactions have been reported [58]. In principle, these condensations are analogous to the benzoin condensation and follow the same mechanistic pathways.

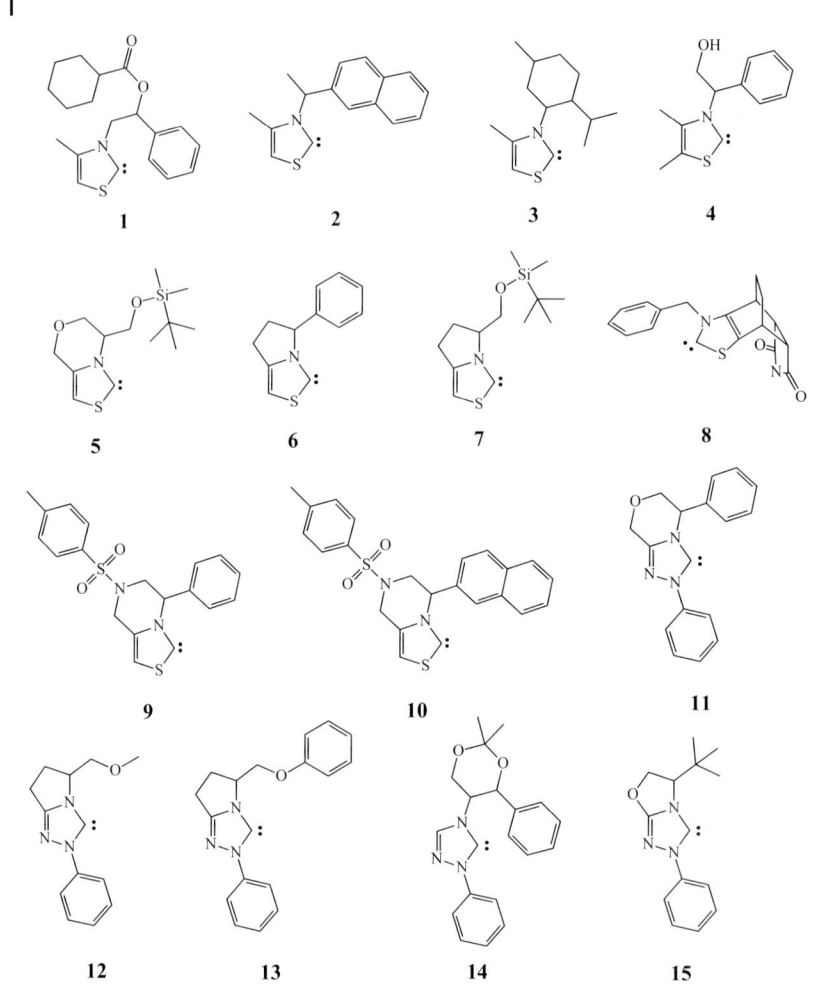

12 N-Heterocyclic Carbenes as Organic Catalysts

Fig. 12.2 Organocatalysts employed for asymmetric benzoin condensation.

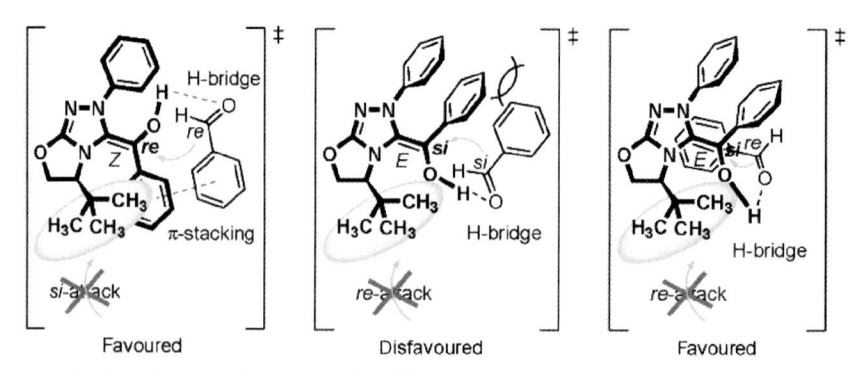

Fig. 12.3 Transition state involving catalyst **15**.

A related reaction that is catalyzed by heterocyclic carbenes is the formoin condensation [59–62]. The mechanism is similar to that of the benzoin condensation, One of the products of this reaction, glycolaldehyde, being the simplest monosaccharide, is important in industrial (bio)synthesis. Efforts have been made since the early 1980s to exploit the formoin condensation for the synthesis of glycolaldehyde, utilizing catalysts based on the thiazolium scaffold. Competition between the products that are also aldehydes and formaldehyde in benzoin-type condensations and/or base-catalyzed aldol condensations leads to a complicated mixture of C2, C3, C4 and even higher adducts (Fig. 12.4).

The triose was commonly found with thiazolium catalysts. Apparently the analogue of the Breslow intermediate is sufficiently reactive to react with a further equivalent of formaldehyde to give the triose [60, 61]. In 1996 important breakthroughs were reported by Enders [62]. After screening several thiazolium and imidazolium salts, he found that catalysts derived from triazolium salts afforded glycolaldehyde in 85% yield at 60% conversion of formaldehyde. Different substituents on the triazole ring influenced the reactivity of the catalyst but, unfortunately not the selectivity, suggesting that steric hindrance determines catalyst performance. Formoin condensation using triazole as the organocatalyst was shown to be under kinetic control, yielding a higher selectivity for glycolaldehyde than thiazole and imidazole catalysts.

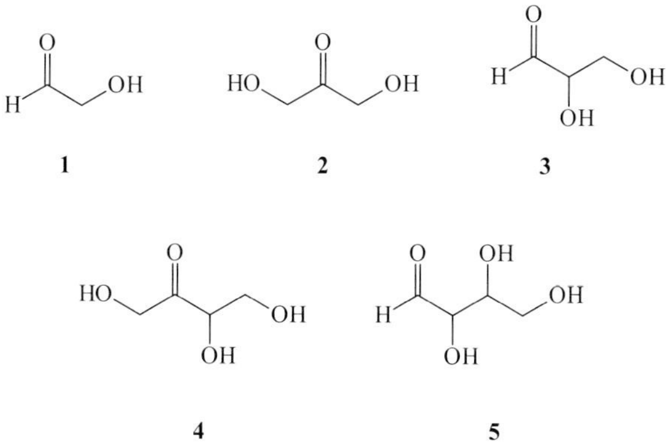

Fig. 12.4 Products found in the formoin condensation.

12.3.2
Michael-Stetter Reaction

12.3.2.1 Stetter Reaction: Addition of Acyl Intermediate to α,β-Unsaturated Aldehydes

The 1,4-addition of aldehydes to α,β-unsaturated ketones, esters, or nitriles to form 4-keto products is known as the Stetter reaction (Scheme 12.4) [63, 64]. Initially catalyzed using cyanide salts, azolium salts derived from thiazoles and imidazoles were quickly adopted as catalytic precursors based on their successful application for benzoin condensation. The use of these catalysts extended the range of substrates to aliphatic as well as aromatic and heterocyclic aldehydes. Catalysis of the Stetter reaction by azolium salts relies on the presence of base to form the active carbenes, which add to aldehydes and induce rearrangement to hydroxyenamine intermediates as in the benzoin condensation. Conjugate addition of this acyl anion equivalent to the α,β-unsaturated carbonyl compound followed by proton transfer and elimination from the tetrahedral intermediate gives the product and regenerates the NHC catalyst (Scheme 12.4). Products of benzoin condensation are likely formed simultaneously, but the reversibility of the benzoin condensation and the irreversibility of the conjugate addition–elimination to unsaturated carbonyl compounds allows the Stetter reaction products to be isolated in high yields.

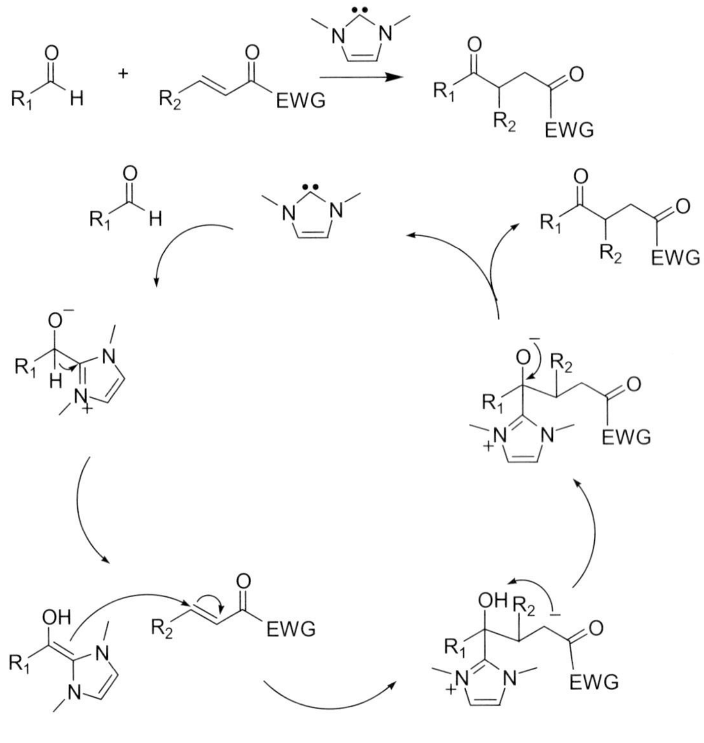

EWG = electron-withdrawing group, such as Cl-, OR-

Scheme 12.4 Stetter reaction.

For the better part of three decades, development of the Stetter reaction relied on thiazolium salts produced from intermediates in the industrial synthesis of vitamin B1. The field was rejuvenated in 1996 by the first report of chiral induction in a Stetter reaction. Based on their previous success with enantioselective benzoin condensations with chiral triazolium salts, Enders group found that chiral triazolium salts catalyzed the intramolecular Stetter reactions in good yield (20–70%) with *ee*s up to 74% [34].

Triazolium salts with chiral fused ring substituents were found by Rovis and co-workers to be very effective for enantioselective Stetter reactions [65–68]. Intramolecular reactions of benzene-bridged substrates can be accomplished using fused-ring triazolium catalysts in yields up to 95% with high diastereoselectivities (>10:1) and *ee*s frequently as high as 90%; alkyl-bridged substrates give only slightly lower yields while retaining high *ee*s. The generation of quaternary centers in benzo-fused products was demonstrated with consistently high *ee*s (>90%) and high yields for the formation of five-membered rings. The yields of analogous six-membered rings were lower (55%). Electron-donating substituents on the pendant phenyl ring led to higher yields for the six-membered rings, while electron-withdrawing substituents were more effective for the formation of five-membered rings with quaternary centers. Higher yields and better diastereoselectivities were generally obtained when isolated triazolylidenes were used as catalysts compared with catalysts generated *in situ* from the triazolium salts and added base.

More recently, cleverly designed thiazolium salts were applied to the Stetter reaction with promising results. Fused menthyl-substituted thiazoliums reported by Bach et al. catalyzed intramolecular Stetter reactions in 75% yield with 50% e.e. [69]. The artificial amino acid β-(N-benzylthiazolyl)alanine ("Taz") introduced by Miller et al. catalyzes the Stetter reaction in modest yield (40%) and good e.e. (80%), especially considering the distance of the chiral center from the nucleophilic carbon [70]. Furthermore, its formulation as an amino acid allows facile incorporation into small peptides, permitting combinatorial studies towards optimizing yield and selectivity. For instance, simple modification of Taz improves the yield (67%) at a slight cost to the e.e. (73%) [70].

12.3.3
a,β-Unsaturated Aldehydes as Homoenolate Equivalents

The use of a,β-unsaturated aldehydes in the Stetter reaction would be expected to lead to a mixture of products arising from the various possible combinations of benzoin condensations and 1,4-additions. Two groups have recently reported that judicious selection of the catalyst and conditions in the reaction of aromatic aldehydes with a,β-unsaturated aldehydes generates γ-butyrolactones as the major products [71, 72]. The steric bulk of IMes (1,3-dimesitylimidazol-2-ylidene) appears to favor attack of the aldehyde at the β-position of the hydroxyenamine intermediate. Intramolecular capture of the acyl intermediate by the newly formed alkoxide affords γ-butyrolactone products (Scheme 12.5). Despite the demonstrated nucleophilicity at the β-position, the reaction can be conducted in protic solvents without significant side reactions.

The nature of the carbene is critical for these reactions. IMes yields the best results for the formation of γ-butyrolactones. Thiazolium and triazolium salts are less effective, as are Icy (1,3-dicyclohexylimidazol-2-ylidene) or IPr (1,3-di-*iso*-propylimidazol-2-ylidene). The diastereoselectivity of the reaction shows a general preference for the cis product when IMes is used (cis/trans = 4:1–8:1). Modest enantioselectivities were observed when a fused tricyclic imidazolium salt was used as the catalyst precursor. Cinnamaldehydes with or without aromatic substituents are typical nucleophilic partners for electrophilic partners derived from substituted benzaldehydes or activated ketones such as $PhCOCF_3$.

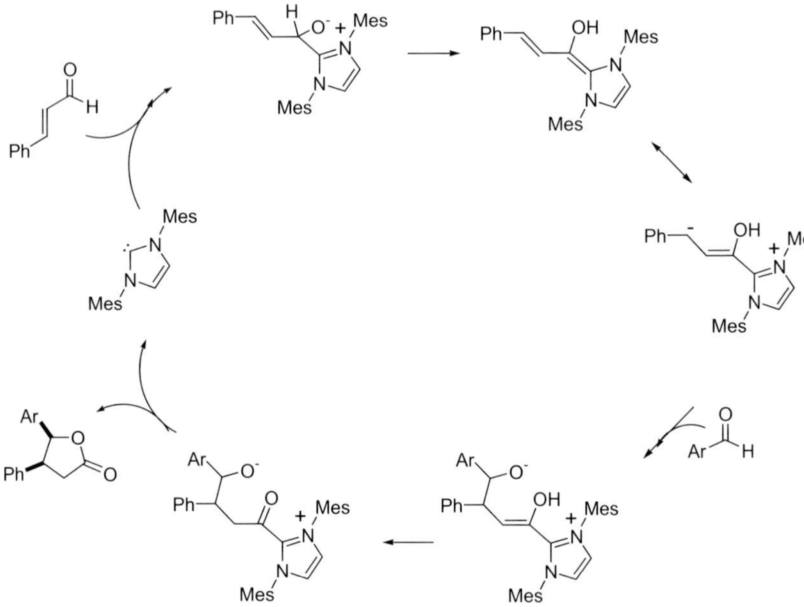

Scheme 12.5 Postulated catalytic cycle for lactone formation.

In addition, when N-sulfonylimines are used as the electrophile [73], γ-lactam products are formed.

Alternatively, the homoenolate intermediates can be protonated and then trapped by a separate alcohol to generate saturated esters [74]. The use of a chiral imidazoylidene carbene as a catalyst in the reaction allows for the kinetic resolution of chiral secondary alcohols. The average s factor [75] for this transformation is 4.8, thereby implicating a chiral activated ester as an intermediate in this process.

12.3.4
Conversion of α-Substituted Aldehydes into Esters

Several examples of NHC-catalyzed internal eliminations from α-substituted aldehydes have been described that result in unsubstituted or β-functional esters. Using triazolium or thiazolium salts as precursors, the groups of Bode and Rovis have shown that α-bromoaldehydes, α,β-epoxyaldehydes, and α,β-aziridinyl aldehydes can be transformed into aliphatic esters, β-hydroxyesters, and β-aminoesters, respectively, with retention of stereochemistry at the β-position for the latter two substrates (Scheme 12.6) [76, 77]. The mechanism is proposed to proceed via a hydroxyenamine intermediate analogous to that proposed in the benzoin condensation and Stetter reactions. Subsequent proton transfer and capture of the acyl group by added alcohols leads to the ester products. The use of α,β-aziridinyl aldehydes is particularly attractive as a means of synthesizing β-amino acids.

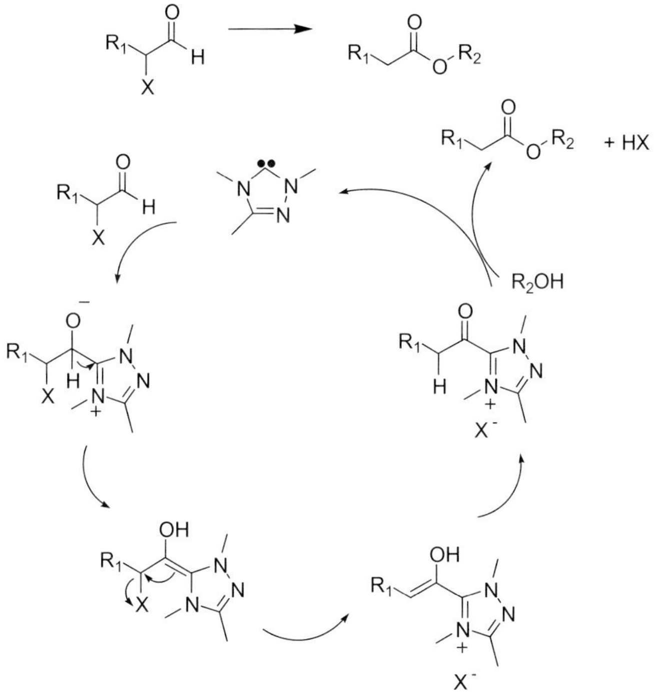

Scheme 12.6 Transformation of an α-aldehyde to an ester.

12.3.5
Transesterification

The nucleophilicity of NHCs is similar to that of the common nucleophilic transesterification agent 4-dimethylaminopyridine (DMAP). Reports from the groups of Hedrick and Nolan have demonstrated the efficiency of NHCs for this purpose (Scheme 12.7) [20, 22–24]. In the presence of an excess of ethanol (20 eq) the unsaturated carbene IMes (4 mol%) catalyzes the transesterification of methyl benzoate to ethyl benzoate; similarly, the reaction of benzyl alcohol with vinyl acetate (1.2 eq.) in the presence of catalytic IMes (0.5 mol%) led to near-quantitative formation of benzyl acetate within 5 min. As transesterifications are equilibrium reactions, high yields of products rely on using (a) an excess of the added alcohol, (b) molecular sieves to remove small alcoholic by-products (e.g., methanol), or (c) the use of vinyl esters that liberate vinyl alcohols that rapidly rearrange to alde-

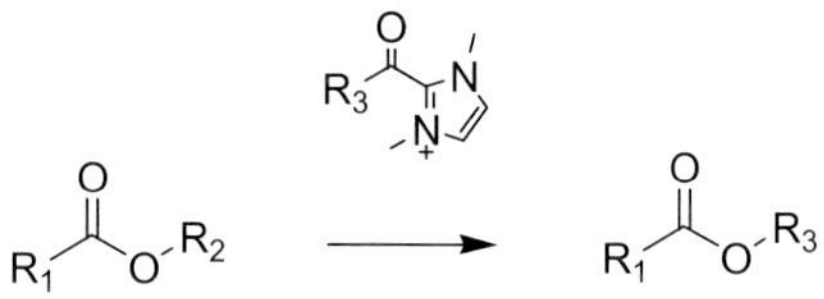

Scheme 12.7 Transesterification reaction.

hydes. NHC-catalyzed transesterification using 2-aminoalcohols provides a route for amidation of esters: following transesterification to the alcohol functionality, the amino group intramolecularly displaces the alcohol [78]. NHCs do not appear to activate amides, so amide formation is irreversible and provides the driving force for completion of the reaction.

Of the NHCs reported to catalyze transesterification, unsaturated imidazolylidenes such as IMes appear to be more active than the saturated imidazolinylidenes such as SIMes. Those NHCs bearing N-alkyl substituents are more active than those bearing N-aryl substituents, presumably due to greater electron density and nucleophilicity. Finally, those NHCs bearing small alkyl substituents (e.g., Me, Et) are more active transesterification catalysts than those with bulkier substituents (1-adamantyl, 1-cyclohexyl). NHC-catalyzed transesterifications are sensitive to steric effects; primary alcohols are readily esterified, while secondary alcohols are sluggish and tertiary alcohols only react at extended times. This allows for selective esterification in the case of polyol substrates.

The mechanism first proposed for NHC-catalyzed transesterification is analogous to that proposed for DMAP: nucleophilic attack of the NHC at the carboxylate leads to elimination of alkoxide and formation of an activated acyl imidazolium intermediate [20]. Acylation of the alkoxide generated from the added alcohol yields the ester and regenerates the carbene catalyst. Recently, spectroscopic (NMR) and crystallographic evidence revealed that unsaturated carbenes associate with alcohols by a hydrogen bond, raising the possibility that NHCs may be acting as general base catalysts for transesterification reactions [78]. The independent synthesis of acyl imidazolium salts from benzoyl chloride and its rapid reaction with alkoxide provides supporting evidence for a nucleophilic mechanism [79]; nevertheless, a general base-catalyzed mechanism cannot be ruled out.

Following the trend seen for benzoin condensation and Stetter reactions, the asymmetric version of NHC-catalyzed transesterification is under active investigation. Two groups have reported acylations for the kinetic resolution of secondary alcohols (Scheme 12.8). Chiral 1,3-bis(1-arylethyl)imidazolium salts were reported by Suzuki to effect the selective acylation of vinyl acetate (0.5 eq) at low temperatures to secondary alcohol substrates [80] with modest yields (up to 44% of a 50% theoretical maximum) and *ee*s (up to 58%). Maruoka's group used similar 1,3-bis(1-arylethyl)imidazolium salts as catalyst precursors with sterically demanding vinyl esters. Good results were achieved at low temperature (–78 °C) using vinyl diphenylacetate as the acylating agent, with *ee*s in the 90% range and 30–35% isolated yields of the acetates (50% theoretical maximum).

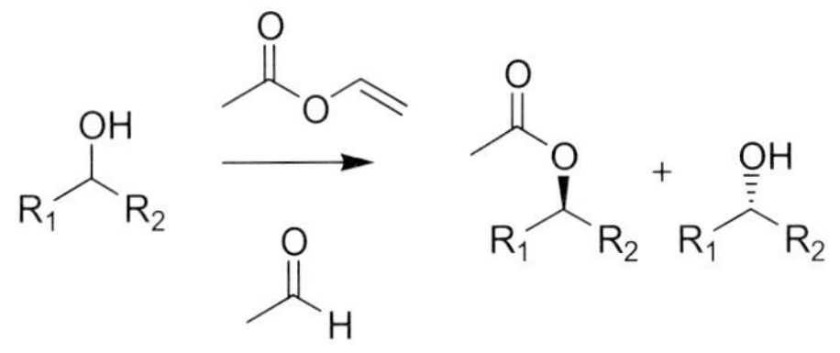

Scheme 12.8 Selective acylation reaction.

12.3.6
Nucleophilic Aromatic Substitution

Miyashita and coworkers have demonstrated the potential of N-heterocyclic carbenes to act as catalysts for nucleophilic aromatic substitution [81–84]. Aromatic aldehydes could be coupled to activated aromatic electrophiles (such as 4-fluoronitrobenzene) in the presence of a carbene catalyst to yield benzophenone derivatives. Nucleophilic attack of the hydroxyenamine intermediate on activated fluoroarenes was proposed as the key step in this reaction (Schemes 12.9 and 12.10) [84]. Several factors governed the rate and overall yield of the product benzophenone. As with traditional nucleophilic aromatic substitution, increasing the polarity of the solvent increased the yield and the reaction rate [83]. Activated haloarenes were most reactive: 4-fluoronitrobenzene was more reactive than 4-chloronitrobenzene [84], and a series of nitrogen heterocycles reacted in the order triazolopyrimidine > pyrazolopyrimidine > purine > pyrrolopyrimidine [83]. The electrophilic nature of the substrate could also be enhanced by the formation of methyl and phenyl sulfones *in situ*. The addition of methyl- or phenyl sodium sulfinate increased the yield of 2-benzoyl-3-phenylquinoxaline from 7% in 10 min to 96% and 87%, respectively [82].

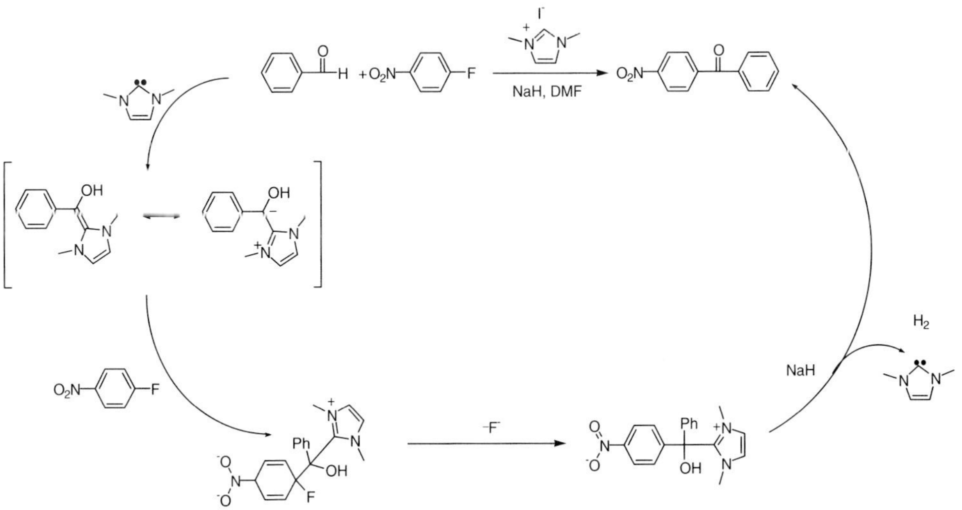

Scheme 12.9 Proposed mechanism of nucleophilic aromatic substitution.

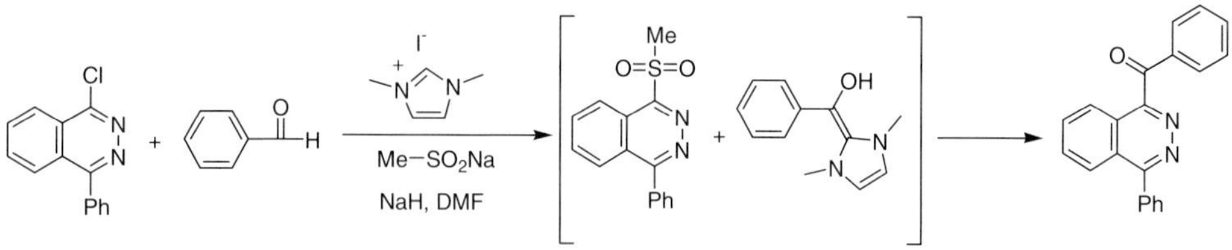

Scheme 12.10 Example of nucleophilic aromatic substitution.

These reactions were quite sensitive to the nature of the aldehyde. Aldehydes containing either highly donating or highly withdrawing substituents (e.g., *p*-methoxy and *p*-nitro) were ineffective. Aldehydes containing halogen substituents in the para position or those substituted in the meta position typically gave product in good to moderate yield [83, 84]. The structure in the carbene was also observed to play a role in the relative reactivity of the nucleophile and substrate. In the series of nitrogen heterocycles, triazolopyrimidine, pyrazolopyrimidine, purine, and pyrrolopyrimidine, the benzimidazolium salt was ineffective as a catalyst for the reaction of benzaldehyde with the less reactive electrophiles, purine and pyrrolopyrimidine. However, the imidazolium salt was an effective catalyst for the nucleophilic acyl substitution of halides for the whole series [83]. This observation was rationalized on the basis of the nucleophilicity of the hydroxyenamine intermediate.

In summary, N-heterocyclic carbenes are effective nucleophilic catalysts for various organic transformations and new methods and reactions are still being discovered. For example, a novel trifluoromethylation reaction of carbonyl compounds has been reported to be catalyzed by NHC catalysts [85]: both enolizable and non-enolizable aldehydes and α-keto esters undergo facile trifluoromethylation with TMSCF$_3$ to yield CF$_3$-substituted alcohols.

12.4
Living Ring-opening Polymerization

Living polymerization strategies provide a powerful means of generating well-defined polymers of predictable molecular weights, narrow molecular weight distributions, and controlled chain-end functionality. Living polymerization reactions also enable the synthesis of block copolymers and other novel topologies, which include a wide range of materials with unique physical and solution properties. The controlled ring-opening polymerization (ROP) of cyclic ester monomers such as lactide, ε-caprolactone and β-butyrolactone is an important synthetic route to well-defined polyesters, an important class of biologically relevant macromolecules. Many organometallic and co-ordination compounds mediate the controlled living polymerizations of cyclic ester monomers. Nevertheless, organocatalytic strategies offer novel mechanisms of enchainment and also provide a means of generating metal-free polymers, which can be advantageous in many microelectronics and biomedical applications.

Following the initial discovery that NHCs were highly active organic catalysts for ring-opening polymerization reactions [17], extensive work has been carried out to develop the scope and generality of these reactions and on the influence of structural diversity of the catalyst on polymerization behavior.

12.4.1
Imidazol-2-ylidenes

The unsaturated carbene IMes [1,3-bis(2,4,6-trimethylphenyl)imidazol-2-ylidene] is a very effective organocatalyst for the ring-opening polymerization of cyclic ester monomers to polyesters [17]. In the presence of an initiating species such as benzyl alcohol, the polymerization of lactide generated polylactides with narrow dispersities ($M_w/M_n < 1.15$) and molecular weights that closely tracked the monomer-to-initiator ratio. The linear relationship between molecular weight and conversion, the ability to extend the chains by further addition of monomer and the synthesis of block copolymers are all characteristic of a living polymerization reaction. The α- and ω-end group fidelity of the polymer chains were demonstrated by ^{1}H NMR with initiation from pyrenebutanol, resulting in observation of the pyrenebutyl ester and a secondary alcohol. Furthermore, gel-permeation chromatography measurements using both refractive index and UV measurements showed the distribution of pyrene throughout the polymer. While the polymerizations demonstrate remarkable control with narrow polydispersities, at high monomer conversions (>95%), the transesterification reactions lead to broadening of the polydispersities. The polymerization behavior is sensitive to the ratio of catalyst to initiator. Catalyst:initiator ratios between 0.25 and 1.5 produced narrowly dispersed PLAs in 1–2 M THF lactide solutions; higher concentrations or catalyst:initiator ratios led to broadened polydispersities. Oligomerization experiments revealed an optimal monomer:initiator:catalyst ratio of 1200:80:1 for the synthesis of narrowly dispersed oligomers [21].

A monomer-activated mechanism was proposed for the NHC-catalyzed ROP of cyclic esters, analogous to that proposed for enzymatic ROP [86]. Nucleophilic opening of the monomer by the carbene was proposed to generate a zwitterionic intermediate that undergoes a proton transfer with an alcohol to generate an alkoxide. Attack of the alkoxide on the acyl imidazolium intermediate liberates the ester and regenerates the carbene. Following acylation of the initiating alcohol, propagation entails acylation of the alcohol chain-ends by the activated monomer. As the activated monomer has an equal probability of acylating any chain-end, all chains grow with equal probability, leading to narrow polydispersities (Scheme 12.11) [17]. Other mechanisms are conceivable, such as an anionic mechanism initiated by deprotonation of the alcohol by the carbene, or activation of the alcohol nucleophile by hydrogen-bonding to the carbene [78]. However, the relative pK_as of the carbene (17–24 in DMSO) [87, 88] and alcohol (29 in DMSO) and the

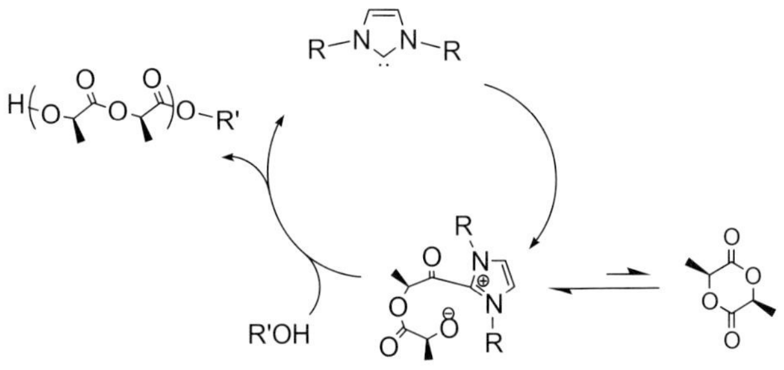

Scheme 12.11 Polymerization of lactide by imidazol-2-ylidenes.

mechanism of the widely studied carbene-activated benzoin condensation [14, 34, 64] support the monomer-activation mechanism.

The structural diversity of NHCs and the development of methods for generating active carbenes *in situ* has enabled studies on the influence of steric and electronic properties of the carbene in lactide polymerization. Convenient *in situ* generation of carbene from imidazolium salt precursors also allows rapid screening of the catalytic properties of the NHCs [21]. Carbenes with less sterically demanding groups were extremely active toward lactide polymerization; however, the polymerizations were less controlled than those utilizing IMes. The reaction of methylimidazole with Merrifield resin yields a solid-supported precatalyst, which is active for polymerization, although the polymerization is typically less controlled than its unsupported analogue. Electron-withdrawing groups attached to the carbene, studied by chlorination of the olefin backbone, significantly reduced the polymerization activity.

The presence of chiral center in lactide brings the possibility of stereo-control in poly-lactide. NHCs have been highly successful in enantioselective organic transformations and kinetic resolutions of amines and in transesterification reactions (Section 12.3.5). Recently, the mesityl substituted carbene IMes was reported to generate moderately isotactic PLA at low temperature [89].

While the use of imidazolium salts to deliver free carbenes has proved very efficient for lactide polymerization, Hedrick and Waymouth have also shown that imidazolium salt ionic liquids (ILs) [21] can also be used to generate free NHCs to mediate the controlled polymerizations of lactide [21]. Two strategies were demonstrated: (1) polymerization in neat ionic liquid, thus employing the IL as both solvent and catalyst reservoir; (2) biphasic polymerization in a THF/IL mixture in which the IL was employed solely as a catalyst reservoir. In the first instance, treatment of 1-ethyl-3-methylimidazolium tetrafluoroborate with initiator, lactide and base resulted in a monomer conversion of ca. 50% before polymer precipitation from the IL occurred. In contrast, the biphasic system, in which addition of potassium *tert*-butoxide results in the generation and subsequent phase migration of free carbene, led to a well-controlled polymerization, producing polymer with a relatively low polydispersity (M_w/M_n = < 1.4). Furthermore, once polymerization was complete, removal of the THF layer provided a facile method for isolating the polymer. The IL layer was reused for subsequent polymerization, demonstrating the repetitive reaction/recycle protocol available in this system.

The NHC platform has been shown to be general for other cyclic ester monomers. Hedrick and Waymouth showed that NHCs are efficient catalysts (Scheme 12.11) for the polymerization of ε-caprolactone, δ-valerolactone and β-butyrolactone [32]. These monomers typically require longer polymerization times, and generate slightly broader polymer polydispersities with the IMes catalyst, but these polymerizations have not been fully optimized with regard to matching the monomer with the optimal carbene catalyst. Less sterically demanding catalysts led to the well-controlled polymerization of caprolactone to yield polymers with narrower polydispersities. Prolonged reaction times led to increased polydispersity due to transesterification of the polymer chains. NHCs are also useful in the synthesis of block copolymers such as the diblock copolymer of ε-caprolactone and lactide. Norbornene macromonomers have also been reported by using bicyclohept[2.2.1]-2-ene-5-methanol as an initiator to form the norbornene-g-lactide graft copolymer by using Grubbs' catalyst [21].

12.4.2
Imidazolin-2-ylidenes

Saturation of the NHC backbone gives strikingly different reactivity for both free carbene and carbene-ligated metal complexes [90]. For example, saturated NHCs with less sterically hindered N-substituents lead to ready dimerization of the carbenes [15, 91, 92]. Polymerization of lactide by imidazolin-2-ylidenes (SIMes) generated *in situ* from the respective hydrochloride salt proceeds rapidly to produce narrowly dispersed polymers with molecular weights that closely track the monomer-to-initiator ratio (Scheme 12.12) [21]. The "Wanzlick" dimer, bis(1,3-diphenyl-2-imidazolidinylidene), was shown to be a potent lactide polymerization catalyst, though with slightly broader polydispersities. Application of SIMes and SIPh (from the Wanzlick dimer) as catalysts for ε-caprolactone polymerization resulted

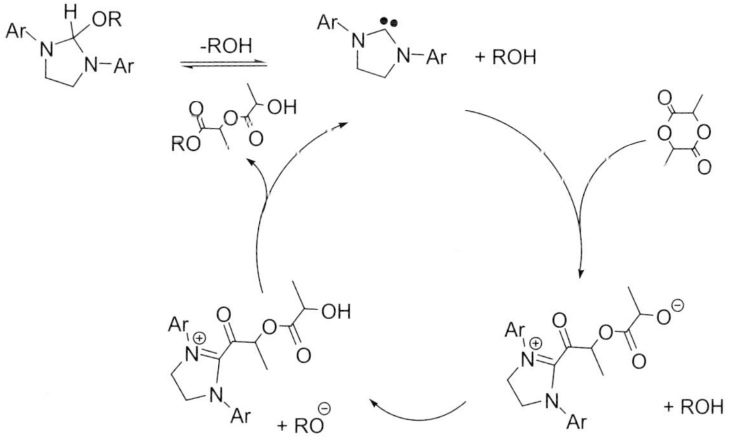

Scheme 12.12 Polymerization of lactide by an imidazolin-2-ylidene adduct.

in different polymerization behavior: SIMes gives no polymerization activity, while less hindered SIPh was active for the polymerization of ε-caprolactone and δ-valerolactone.

The use of protected carbenes to provide a stable delivery platform for imidazolin-2-ylidenes has also received attention (Fig. 12.5) [19, 39]. These saturated carbenes react with organic compounds containing acidic proton to generate carbene adducts that serve as efficient pre-catalysts for living ring-opening polymerization. Elevated temperatures were required to thermally cleave the carbene-adduct bond and release free carbene into solution. In the presence of a benzyl alcohol initiator and 1.5 eq of carbene adduct in a 1–2 M THF or toluene solution of lactide, a well-controlled polymerization was observed whereby the molecular weights closely track the monomer:initiator ratio. End-group fidelity was demonstrated by both ^{1}H NMR and UV-GPC. At 65 °C both C_6F_5-Mes and CCl_3-Mes (Fig. 12.5) resulted in high monomer conversion within 3 h, while C_6F_4-Mes only produced modest monomer conversion within 24 h under the same conditions. The difference was attributed to the less favorable formation of the free carbene in the latter case. Use of the less sterically hindered C_6F_5-Ph led to a less well-controlled polymerization and broader polymer polydispersities, which are attributed to the formation of the Wanzlick dimer [93, 94].

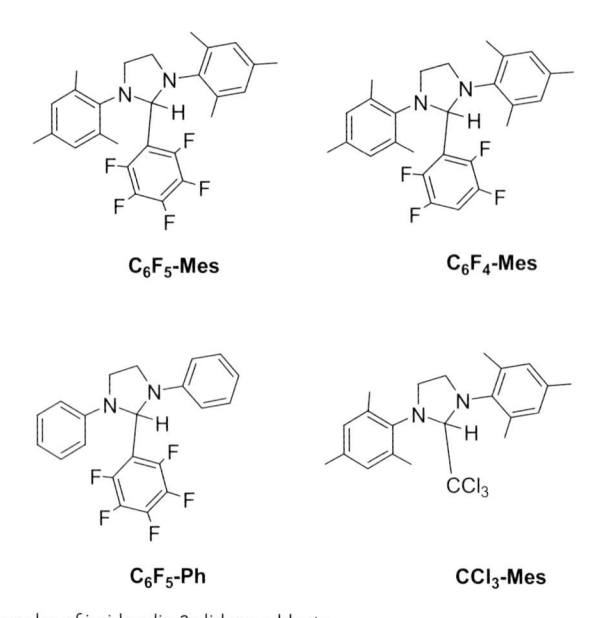

Fig. 12.5 Examples of imidazolin-2-ylidene adducts.

The use of alcohol adducts has also been recently reported to effect the efficient polymerization of lactide. The application of alcohol adducts provides a simple unimolecular strategy for the polymerization of lactones [19]. Alcohol adducts of imidazolin-2-ylidenes form reversibly at room temperature, providing a facile strategy for generation of both the free carbene (catalyst) and the alcohol (initiator/propagating species). Polymerization proceeded in a rapid, well-controlled manner to produce polymers of predicted molecular weight and narrow polydispersity. Various alcohol adducts were reported to provide various polymer end-groups, confirmed by ^{1}H NMR and UV-GPC. Furthermore, to demonstrate the versatility of this system, multifunctional carbene adducts were prepared (Fig. 12.6) and used to polymerize lactide to produce block copolymers, telechelic polymers and star polymers. The reversible nature of the alcohol adduct formation is partially responsible for the exquisite control observed in this system. The reversible formation of a "dormant" alcohol adduct by combination of free carbene with the propagating alcohol chain end leads to reversible deactivation, thus maintaining a low concentration of catalyst in solution, analogous to modern controlled radical polymerization processes.

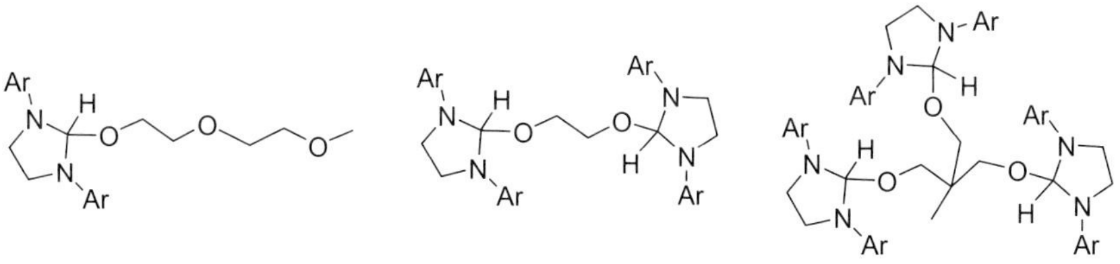

Fig. 12.6 Examples of multifunctional carbene adducts.

12.4.3
1,2,4-Triazol-5-ylidenes

The commercially available 1,3,5-triazol-2-ylidenes have also recently been shown by Hedrick, Waymouth and Dubois to be active polymerization catalysts [18]. As with the saturated imidazolin-2-ylidenes, triazolylidenes are able to reversibly form alkoxytriazolylidenes, but in the latter case elimination of the alcohol to generate the free carbene only occurs readily at above 90 °C. Attempted polymerization of lactide using 1,3,4-triphenyl-4,5-dihydro-1H-1,2,4-triazol-5-ylidene (triphenyl triazole carbene) with methanol as an initiator at ambient temperature and 50 °C resulted in poorly controlled polymerizations, attributed to formation of alkoxytriazolylidene at these temperatures. However, polymerization at 90 °C afforded extremely well-controlled polymerizations to yield polymers with molecular weights close to those predicted from the monomer:initiator ratio, with polydisperities of <1.10 and end-group fidelity. Furthermore, on completion of polymerization the system proved extraordinarily resistant to transesterification, suggesting that formation of the carbene–alcohol adduct reduced the concentration of active transesterification agent. The polymerization could also be stopped and started on demand by simply changing the polymerization temperature. Study of the polymerization kinetics at 90 °C revealed that the system was first order with respect to [lactide], [alcohol] and [carbene]. The versatility of the system was demonstrated by the preparation of block copolymers using a hydroxypoly(ethylene oxide) macroinitiator and a 24-arm hydroxyl end-capped third-generation dendrimer derived from 2,2′-bis(hydroxymethyl)propionic acid. The polymerization resulted in narrowly dispersed block copolymers without degradation of the poly(ester) macroinitiators.

12.4.4
Thiazol-2-ylidenes

Thiazolium carbenes have also been shown to be efficient catalysts for the living polymerization of lactones [21]. Thiazolium salts were investigated as precatalysts (Fig. 12.7) in the presence of 2–5 eq triethylamine (triethylamine does not initiate polymerization by itself) in the presence of benzyl alcohol initiator in CH_2Cl_2. Compared with imidazol- and imidazolin-ylidenes, thiazol-ylidenes polymerize lactide at a lower rate. The molecular weights of the polymers produced displayed excellent correlation with those predicted from the monomer:initiator ratio, with polydispersities below 1.10. The molecular weight was shown to rise linearly with monomer conversion, and end-group fidelity was demonstrated by both ^{1}H NMR and GPC measurements. Nevertheless, high molecular weight polymers were difficult to prepare, even at extended reaction times and elevated temperatures.

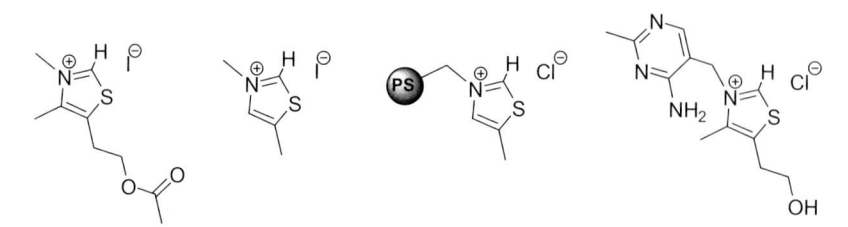

Fig. 12.7 Examples of thiazol-2-ylidene precursors.

References

1. Wöhler, F., Liebig, J. *Ann. Pharm.* **1832**, *3*, 249–282.
2. Enders, D., Balensiefer, T. *Acc. Chem. Res.* **2004**, *37*, 534–541.
3. Dalko, P.I., Moisan, L. *Angew. Chem. Int. Ed. Engl.* **2004**, *43*, 5138–5175.
4. Spivey, A.C., Arseniyadis, S. *Angew. Chem. Int. Ed.* **2004**, *43*, 5436–5441.
5. Wilson, R.M., Jen, W.S., MacMillan, D.W.C. *J. Am. Chem. Soc.* **2005**, *127*, 11 616–11 617.
6. Fu, G.C. *Acc. Chem. Res.* **2000**, 412–420.
7. Ahrendt, K.A., Borths, C.J., MacMillan, D.W.C. *J. Am. Chem. Soc.* **2000**, *122*, 4243–4244.
8. Cordova, A., Notz, W., Zhong, G., Betancort, J.M., Barbas, C.F. *J. Am. Chem. Soc.* **2002**, *124*, 1866–1868.
9. Qiao, S., Fu, G.C. *J. Org. Chem.* **1998**, *63*, 4168–4170.
10. Vedejs, E., Daugulis, O., Diver, S.T. *J. Org. Chem.* **1996**, *61*, 430.
11. Vedejs, E., Rozners, E. *J. Am. Chem. Soc.* **2001**, *123*, 2428–2429.

12 Tao, B., Lo, M.C., Fu, G.C. *J. Am. Chem. Soc.* **2001**, *123*, 353–354.
13 Liu, L., Breslow, R. *J. Am. Chem. Soc.* **2003**, *125*, 12 110–12 111.
14 Breslow, R. *J. Am. Chem. Soc.* **1958**, *80*, 3719–3726.
15 Wanzlick, H.W., Kleiner, H.J. *Angew. Chem.* **1961**, *73*, 493.
16 Nair, V., Bindu, S., Sreekumar, V. *Angew. Chem. Int. Ed.* **2004**, *43*, 5130–5135.
17 Connor, E.F., Nyce, G.W., Myers, M., Mock, A., Hedrick, J.L. *J. Am. Chem. Soc.* **2002**, *124*, 914–915.
18 Coulembier, O., Dove, A.R., Pratt, R.C., Sentman, A.C., Culkin, D.A., Mespouille, L., Dubois, P., Waymouth, R.M., Hedrick, J.L. *Angew. Chem. Int. Ed.* **2005**, *44*, 4964–4968.
19 Csihony, S., Culkin, D.A., Sentman, A.C., Dove, A.P., Waymouth, R.M., Hedrick, J.L. *J. Am. Chem. Soc.* **2005**, *127*, 9079–9084.
20 Nyce, G.W., Lamboy, J.A., Connor, E.F., Waymouth, R.M., Hedrick, J.L. *Org. Lett.* **2002**, *4*, 3587–3590.
21 Nyce, G.W., Glauser, T., Connor, E.F., Mock, A., Waymouth, R.M., Hedrick, J.L. *J. Am. Chem. Soc.* **2003**, *125*, 3046–3056.
22 Grasa, G., Kissling, R.M., Nolan, S.P. *Org. Lett.* **2002**, 3583–3586.
23 Grasa, G.A., Guveli, T., Singh, R., Nolan, S.P. *J. Org. Chem.* **2003**, *68*, 2812–2819.
24 Singh, R., Kissling, R.M., Letellier, M.A., Nolan, S.P. *J. Org. Chem.* **2004**, *69*, 209–212.
25 Viciu, M.S., Zinn, F.K., Stevens, E.D., Nolan, S.P. *Organometallics* **2003**, *22*, 3175–3177.
26 Trnka, T.M., Morgan, J.P., Sanford, M.S., Wilhelm, T.E., Scholl, M., Choi, T.L., Ding, S., Day, M.W., Grubbs, R.H. *J. Am. Chem. Soc.* **2003**, *125*, 2546–2558.
27 Moss, R.A. *Acc. Chem. Res.* **1989**, *22*, 15–21.
28 Bourissou, D., Guerret, O., Gabbai, F.P., Bertrand, G. *Chem. Rev.* **2000**, *100*, 39–91.
29 Arduengo, A.J. *Acc. Chem. Res.* **1999**, *32*, 913–921.
30 Herrmann, W.A., Kocher, C. *Angew. Chem. Int. Ed. Engl.* **1997**, *36*, 2163–2187.
31 Alder, R.W., Blake, M.E., Oliva, J.M. *J. Phys. Chem. A* **1999**, *103*, 11200–11211.
32 Sentman, A.C., Csihony, S., Waymouth, R.M., Hedrick, J.L. *J. Org. Chem.* **2005**, *70*, 2391–2393.
33 Arduengo, A.J., Calabrese, J.C., Davidson, F., Dias, H.V.R., Goerlich, J.R., Krafczyk, R., Marshall, W.J., Tamm, M., Schmutzler, R. *Helv. Chim. Acta* **1999**, *82*, 2348–2364.
34 Enders, D., Breuer, K., Teles, J.H. *Helv. Chim. Acta* **1996**, *79*, 1217–1221.
35 Stefan, R., Simon, G., Hideaki, W., Siegfried, B. *Synlett* **2001**, *3*, 430–432.
36 Scholl, M., Ding, S., Lee, C.W., Grubbs, R.H. *Org. Lett.* **1999**, *1*, 953–956.
37 Enders, D., Breuer, K., Raabe, G., Runsink, J., Teles, J.H., Melder, J.P., Ebel, K., Brode, S. *Angew. Chem. Int. Ed. Engl.* **1995**, *34*, 1021–1023.
38 Csihony, S.C., Culkin, D.A., Sentman, A.C., Dove, A.P., Waymouth, R.M., Hedrick, J.L. *J. Am. Chem. Soc.* **2005**, *127*, 9079–9084.
39 Nyce, G.W., Csihony, S., Waymouth, R.M., Hedrick, J.L. *Chem-Eur. J.* **2004**, *10*, 4073–4079.
40 Lapworth, A. *J. Chem. Soc.* **1903**, *83*, 995–1005.
41 Ukai, T., Tanaka, R., Dokawa, T.J. *Pharm. Soc. Jpn.* **1943**, *63*, 296–300.
42 Castells, J., López-Calahorra, F., Domingo, L. *J. Org. Chem.* **1988**, *53*, 4433–4436.
43 van den Berg, H.J., Challa, G., Pandit, U.K. *J. Mol. Cat.* **1989**, *51*, 1–12.
44 Castells, J., Domingo, L., López-Calahorra, F., Marti, J. *Tetrahedron Lett.* **1993**, *34*, 517–520.
45 Breslow, R., Schmuck, C. *Tetrahedron Lett.* **1996**, *37*, 8241–8242.
46 López-Calahorra, F., Rubires, R. *Tetrahedron Lett.* **1995**, *35*, 9713–9728.
47 López-Calahorra, F., Castro, E.O., A., Marti, J. *Tetrahedron Lett.* **1996**, 5019–5022.
48 Sheehan, J.C., Hunneman, D.H. *J. Am. Chem. Soc.* **1966**, *88*, 3666–3667.
49 Sheehan, J.C., Hara, T. *J. Org. Chem.* **1974**, *39*, 1196–1199.
50 Takagi, W., Mamura, Y., Yano, Y. *Bull. Chem. Soc. Jpn.* **1980**, 478–480.

51 Marti, J., Castells, J., López-Calahorra, F. Tetrahedron Lett. **1993**, 521–524.
52 Knight, R.L., Leeper, F.J. Tetrahedron Lett. **1997**, 3611–3614.
53 Gerhard, A.U., Leeper, F.J. Tetrahedron Lett. **1997**, 3615–3618.
54 Dvorak, C.A., Rawal, V.H. Tetrahedron Lett. **1998**, 2925–2928.
55 Knight, R.L., Leeper, F.J. J. Chem. Soc., Perkin Trans. I **1998**, 1891–1893.
56 Enders, D., Kallfass, U. Angew. Chem. Int. Ed. **2002**, 41, 1743–1745.
57 Dudding, T., Houk, K.N. Proc. Natl. Acad. Sci. U.S.A. **2004**, 101, 5770–5775.
58 Hachisu, Y., Bode, J.W., Suzuki, K. J. Am. Chem. Soc. **2003**, 125, 8432–8433.
59 Castells, J., Geijo, F., López-Calahorra, F. Tetrahedron Lett. **1980**, 4517–4520.
60 Matsumoto, T., Inoue, S. J. Chem. Soc., Chem. Commun. **1983**, 171–172.
61 Matsumoto, T., Yamamoto, H., Inoue, S. J. Am. Chem. Soc. **1984**, 106, 4829–4832.
62 Teles, J.H., Melder, J.P., Ebel, K., Schneider, R., Gehrer, E., Harder, W., Brode, S., Enders, D., Breuer, K., Raabe, G. Helv. Chim. Acta **1996**, 79, 61–83.
63 Stetter, H., Kuhlmann, H. Org. React. **1991**, 40, 407–496.
64 Stetter, H. Angew. Chem. Int. Ed. Engl. **1976**, 15, 639–647.
65 Kerr, M.S., de Alaniz, J.R., Rovis, T. J. Am. Chem. Soc. **2002**, 124, 10 298–10 299.
66 Kerr, M.S., Rovis, T. Synlett **2003**, 1934–1936.
67 Kerr, M.S., Rovis, T. J. Am. Chem. Soc. **2004**, 126, 8876–8877.
68 Read de Alaniz, J., Rovis, T. J. Am. Chem. Soc. **2005**, 127, 6284–6289.
69 Pesch, J., Harms, K., Bach, T. Eur. J. Org. Chem. **2004**, 9, 2025–2035.
70 Mennen, S.M., Blank, J.T., Tran-Dube, M.B., Imbriglio, J.E., Miller, S.J. Chem. Commun. **2005**, 195–197.
71 Sohn, S.S., Rosen, E.L., Bode, J.W. J. Am. Chem. Soc. **2004**, 126, 14 370–14 371.
72 Burstein, C., Glorius, F. Angew. Chem. Int. Ed. **2004**, 6205–6208.
73 He, M., Bode, J.W. Org. Lett. **2005**, 7, 3131–3134.
74 Chan, A., Scheidt, K.A. Org. Lett. **2005**, 7, 905–908.
75 Chen, C.S., Fujimoto, Y., Girdaukas, G., Sih, C.J. J. Am. Chem. Soc. **1982**, 104, 7294–7299.
76 Read de Alaniz, J., Rovis, T. J. Am. Chem. Soc. **2004**, 126, 9518–9519.
77 Chow, K.Y.K., Bode, J.W. J. Am. Chem. Soc. **2004**, 126, 8126–8127.
78 Movassaghi, M., Schmidt, M.A. Org. Lett. **2005**, 7, 2453–2456.
79 Csihony, S., Nyce, G.W., Sentman, A.C., Waymouth, R.M., Hedrick, J.L. Polym. Prepr. (Am. Chem. Soc., Div. Polym. Chem.) **2004**, 45, 319–320.
80 Yumiko Suzuki, Y., Yamauchi, K., Muramatsu, K., Sato, M. Chem. Commun. **2004**, 2770–2771.
81 Miyashita, A., Obae, K., Suzuki, Y., Oishi, E., Iwamoto, K., Higashimo, T. Heterocycles **1997**, 45, 2159.
82 Miyashita, A., Suzuki, Y., Iwamoto, K., Oishi, E., Higashimo, T. Heterocycles **1998**, 49, 405.
83 Miyashita, A., Suzuki, Y., Iwamoto, K., Higashimo, T. Chem. Pharm. Bull. **1998**, 46, 390.
84 Suzuki, Y., Toyota, T., Imada, F., Sato, M., Miyashita, A. Chem. Commun. **2003**, 1314.
85 Song, J.J., Tan, Z., Reeves, J.T., Gallou, F., Yee, N.K., Senanayake, C.H. Org. Lett. **2005**, 7, 2193–2196.
86 Gross, R.A., Kumar, A., Kalra, B. Chem. Rev. **2001**, 101, 2097–2124.
87 McGuinness, D.S., Green, M.J., Cavell, K.J., Skeleton, B.W., White, A.H. J. Organomet. Chem. **1998**, 565, 165.
88 Kim, Y.-J., Streitwieser, A. J. Am. Chem. Soc. **2002**, 124, 5757.
89 Jensen, T.R., Breyfogle, L.E., Hillmyer, M.A., Tolman, W.B. Chem. Commun. **2004**, 2504–2505.
90 Trnka, T.M., Grubbs, R.H. Acc. Chem. Res. **2001**, 18–29.
91 Arduengo, A.J., Goerlich, J.R., Marshall, W.J. J. Am. Chem. Soc. **1995**, 117, 11 027–11 028.
92 Wanzlick, H.W., Esser, F., Kleiner, H.J. Chem. Ber.-Recueil **1963**, 96, 1208.
93 Wanzlick, H.W., Lachmann, B., Schikora, E. Chem. Ber.-Recueil **1965**, 98, 3170.
94 Wanzlick, H.W. Angew. Chem. **1962**, 74, 129.

Subject Index

a

abnormal binding 58
π-accepting character 183
acyclic diene metathesis (ADMET) 8
acylation of vinyl acetate 286
addition of C–H bond 234
ADMET (acyclic diene metathesis) 34, 36
aerobic oxidation 110, 111, 218
– of alcohols 103
Ag-NHC 180
– complexes 257
alcohol adducts 293
alcohol oxidation 108
alcohol surrogate 111
alcohols into olefins 111
aliphatic esters 284
alkene binding 28
alkoxide base 61
alkoxy-imidazolylidene ligand 201
alkoxytriazolylidene 293
alkylidene 2, 19
alkyne dimerization 41
allenylidene 48
allyl complex intermediate 45
allylic alkylation 216, 267
allylic phosphate 266
allylic substitution 214
amidation of esters 285
ammonia 78, 80
anionic ligands 12, 269
anionic mechanism 289
antimicrobials 236
AROM/CM reaction 214
aryl amination 55, 59, 73
aryl chlorides 92, 97
aryl triflates 97
α-arylation of carbonyl compounds 217
arylation of amides 63
arylation of ester 63
arylation of ketones 55, 63, 64
α-asarone 10
asymmetric 1,2-addition 189
– secondary alcohols 205
asymmetric 1,4-addition 189, 199
asymmetric allylic alkylation 269
asymmetric allylic substitution 189
asymmetric amide α-arylation 189
asymmetric catalysis 183, 185, 258
asymmetric conjugate addition 210
asymmetric hetero-Diels–Alder reaction 275
asymmetric hydrogenation 189, 193, 194
– dimethylitaconate 195
asymmetric hydrosilylation 189, 206, 215
asymmetric olefins metathesis 189
asymmetric ring-closing metathesis (ARCM) 212
asymmetry 266
atom transfer radical polymerization (ATRP) 40
atropisomer complex 214
atropisomers 199
attachment through the alkylidene moiety 19
attachment through the anionic ligand 18
Au-NHC complexes 257
axial chirality 189, 193
axially chiral binaphthalene 213
aza-Michael reaction 204

b

π back donation 125, 257
basicity 91
benzimidazole 223
benzimidazolium salts 88, 134
(benzimidazolyl-NHC)Pt(dvtms) 134
benzoin condensation 275, 278
benzylidene 8, 10, 15, 40
BINAP 208
binaphthyl axial chirality 216

N-Heterocyclic Carbenes in Synthesis. Edited by Steven P. Nolan
Copyright © 2006 WILEY-VCH Verlag GmbH & Co. KGaA, Weinheim
ISBN: 3-527-31400-8

BINOL 10
biphasic 38
block copolymers 288, 293, 294
Buchwald–Hartwig amination
– 2- and 3-aminopyridines 62
– imines 92
– indole 92
– mechanism 93
– tetra-substituted 62
Buchwald–Hartwig reaction 59, 73
Büchner reaction 273
bulky ligands 114
γ-butyrolactones 283

c

C_2-symmetric 187, 188, 200
C_2-symmetry strategy 211
C5 metallation 229
C–C bond-forming reactions 63
C–C cross-coupling 183
C–H oxidative addition 229
C–N bond-formation 60
C–N cross-coupling 183
carbene adducts 291
– acetylene 277
– alcohol 277
– chloroform 277
– hemiacetal 277
– methyl phenyl sulfone 277
– pentafluorobenzene 277
carbene transfer 268
– agent 259
– reaction 258
β-carbon elimination 174
catalyst deactivation 234
catalyst immobilization 90
catalytic kinetic resolution 189
Chalk–Harrod mechanism 146
chelate complex 232
chelate formation 233
chelating imidazole-2-ylidene NHC 224
chemoselectivity 91, 273
chiral alkylation catalysts 216
chiral backbone 259
chiral bidentate anionic N-heterocyclic
    carbene 212
chiral carbene 109
chiral diamines 267
chiral imidazoylidene carbene 284
chiral induction 282
chiral NHC 184
chiral NHC-complex 188, 197
chiral NHC-ruthenium complexes

– asymmetric ring-opening/cross
    metathesis 14
– bidentate chiral imidazolidene 14
– binaphthyl 15
– phosphine-free 14
chiral N-heterocyclic carbenes 217, 218
chiral N-substituents 200
chiral phosphine 188
chiral substituents at nitrogen 259
cis-disubstituted alkenes 242
classical synthetic methods 185
CM 6, 22
– acrylonitrile 11
– E-selectivity 9
– electron-deficient olefin 9
– erogorgiaene 9, 10
colloidal platinum 120
commercial viability 75
1,4-conjugate addition 199, 261
conjugate addition 201, 258
conjugate addition–elimination 281
1,4-conjugate Friedel–Crafts reaction 275
copper-NHC complexes, 1,4-addition of
    dialkylzinc 199
counter-ion 229
coupling of allenes 178
Crabtree's catalyst 241, 242
– inactive hydride-bridged trimers 243
– sterically hindered alkene 243
– tetra-substituted alkenes 243
cross coupling 257
cross-coupling reactions 55, 56, 151, 236
cross-metathesis (CM)
– E-selectivity 5
– tri-substituted alkenes 5
crossover 177
– reaction 50
Cu-NHC complexes 257
cyclic enones 202
cycloaddition, nitriles 172
cycloaddition, reactions
– isocyanates 170
– isocyanurates 172
– pyridones 170
cycloisomerization 46, 47, 48
cyclopentanes 166
cyclopropanation 39, 40
– of styrene 273

d

dehalogenation 67, 68
dehydrogenation 34
deprotonation of imidazolium 276

diastereoselectivities 283
diazo decomposition 272
Diels–Alder reaction 275
differential scanning calorimetry 277
dimerization of alkynes 41
4-dimethylaminopyridine (DMAP) 285
1,3-dipolar cycloadditions 275
dissymmetry 188
divinyltetramethylsiloxane 121
σ-donating 125
σ-donor ligands 123
σ-donors, N-heterocyclic carbenes (NHC) 2

e

electrochemical reduction 57
14-electron active species 10
14-electron NHC-ruthenium complexes 13
electron density 142
electron-donating 127
electronic effects 263
electronic flexibility 230
enantiocontrol 209
enantioselective catalysis 183
enantioselective conjugate addition 215
enantioselective ruthenium olefin catalysts
– asymmetric induction 13
– desymmetrization 13
– enantioselectivity 14
enantioselective Stetter reaction 282
enantioselectivity 185, 202, 212, 253, 279, 283
enyne metathesis, erogorgiaene 9, 10
enzymatic ROP 289
epothilones 5
ethyl diazoacetate 271
ethylene 180
ethyne 47

f

ferrocenyl based-NHC ligands 195
Fischer-type carbenes 8
fluorous solid-phase extraction 21
fluxional behavior 235
fluxionality 234
formoin condensation 278, 280
free carbene 278
Friedel–Crafts reaction 275
functional group tolerance 120
functionalization of C–H bonds 268
furan formation 37, 39

g

glycolaldehyde 280
Grubbs I catalyst 9
Grubbs II catalyst 9

h

H_2 oxidative addition 28
Heck 55, 243
– coupling 236
– reactions 58, 90, 236
hemilabile behavior 235
N-heterocyclic carbenes (NHC) 55
– strong σ-donor 3
heterodimerization 91
Hg(0) 236
high basicity 234
high performance polyethylenes 83
high selectivities 219
hindered alkenes 195
Hiyama coupling 119
homoallylic silyl alcohol 175
homogeneous hydrogenation 241
Hoveyda–Grubbs catalyst 9
hydride 48
β-hydride elimination 93, 111, 115, 166
hydride ligands 248
hydroarylation 69
– of alkynes 70
hydroformylation 243
β-hydrogen
– elimination 81
– transfer 68
hydrogenation 27, 29, 42, 122
– and hydrosilylation reactions 27
– imines 31
– ketones 31
– of olefins 242, 247
hydrosilylation 27, 32, 33, 119, 139, 144, 148, 243, 265
– activation 138
– anti-Markovnikov 129
– chemoselectivity 131
– concentration of active catalyst 144
– concentration of alkene 143
– concerted mechanism 149
– deactivation pathway 142
– decoordination of the alkene 146
– dimeric hydride 141
– high rate 134
– homogen 138
– initiation rate constant 145
– kinetic 137
– kinetic isotope effects 137, 149, 150

- kinetic model   146
- mechanistic studies   137
- of acetophenone   207
- of carbonyl compounds   206, 262
- proposed catalytic cycle   147
- rate-determining step   136, 138, 139
- reactivity   132
- regiocontrol   129
- selectivity   132
- temperature variation   146
- terminal olefins   129
- variation of silane   143

hydrosilylation of alkynes
- dimethyl(pyridyl)silane   151
- economically viable   151
- electronic nature alkyne   156, 157
- enhanced reactivity   152
- H_2PtCl_6 151
- high regiocontrol   157
- high selectivity   152
- low catalyst loading   152
- propargylic alcohols   151
- rhodium   151
- ruthenium   151
- selectivity   152
- vinylsilanes   150

hydroxyenamine intermediate   284, 288

i

IAd   95
ICy   41, 95, 163
ICyCuCl   263
IMes   8, 41, 57, 64, 83, 84, 91, 95, 98, 112, 176, 181, 245, 285, 286, 289, 290
IMesCuCl   265
IMes.HCl   64, 85
imidazolinium salts   185
imidazolium salts   58, 68, 83, 84, 88, 90, 99, 217, 259, 261
- nucleophilic acyl substitution of halides   288

imidazolyl ligands   134
1,2 insertion   146
insertion into C–H bonds   271
intramolecular   100
- cyclization   37
- cycloaddition   271
- cyclopropanation   271
- cyclopropanation of dienynes   271
- hydroamination   235
- Stetter reaction   283
ion pair structure   230
ionic liquids   21, 73, 90, 179, 269, 290

IPr   63, 83, 95, 98, 108, 163, 167, 174, 178, 273
[(IPr)$_2$Cu]X   265
IPrCuCl   261, 265
IPrHCl   61, 99
(IPr)Pd(acac)Cl   64
(IPr)Pd(allyl)Cl   67
(IPr)Pd(palladacycle)Cl   66, 67
IR frequencies   233
Ir–hydride species   115
Ir–NHC-catalyzed Oppenauer oxidation   115
Ir–NHC hydrogenation mechanism
- dihydride mechanism   248
- para-hydrogen induced polarization   248
Ir-allyl intermediates   252
Ir-catalyzed hydrogenation of arenes   188
[Ir(COD)(PCy$_3$)(py)]PF$_6$ 242
[Ir(COD)(phosphine)(NHC)]PF$_6$ 246
[Ir(COD)(py)$_2$]PF$_6$ 244
[Ir(COD)(SIMes)(py)]PF$_6$ 244
iridium-based hydrogenation catalyst
- longevity   241
- stability   241
iridium N-heterocyclic catalysts
- aliphatic alkenes   250
- chiral   250
Ir(py)(PCy$_3$)(COD)PF$_6$ 195
isolated Pd(0)/(II)(NHC) complexes   95
isomerization   27, 28, 34, 44, 45, 47, 49, 139, 144, 150, 241
- benzyl alcohol   36
- N-allyl   35
- O-allyl   35
- O-homoallyl   35
- of olefins   164
isomerized alkenes   122, 129
(–)-isopinocamphenyl   217
isotactic PLA   290
ItBu   95

k

Karstedt catalyst   119, 121, 122, 137
kinetic resolution   108, 109
- of secondary alcohols   218

l

lactide polymerization   289
ligand bite angle   227
linkers   225
living ring-opening polymerization   288
LNi(CO)$_3$ 233

m

macrocyclization of ynals   177
Mannich reaction   275
MAO   180
mechanism   36
- ruthenocyclobutane   13
mechanistic studies, phosphine dissociation   6
metallation   227, 228
metathesis   45
- and hydrogenation   42
- and isomerization   44
- reaction   183
metathesis catalysts   1
- decomposition   1
- release–return mechanism   1
- thermal stability   1
methane oxidation, methanol   108
methyl methacrylate (MMA)   40
methylaluminoxane (MAO)   180
methylidene   37
Michael acceptors   200
microwave   98, 99
molecular oxygen   103, 104, 107
molecular sieves   285
molecular weight distribution   40
monocarbene palladium(II)   61, 70
monoligated Pd(0)   73
monomer-activation mechanism   289
monosaccharide   280
Montmorillonite K-10   236

n

nanoparticulate Pd(0)   236
NHC   78, 81
- π-acceptor   82
- A_H angles   154
- A_L angles   154
- buried volume   154
- electronic   155
- electronic factors   80
- electronic properties   136
- high basicity   77
  nucleophilic aromatic substitution   287
- steric   77, 136
- steric effect   155
- steric factors   80
- tilt angle   152
NHC-catalyzed internal eliminations   284
NHC-iridium complexes   196
NHC-JM-Phos   188
NHC ligands   7, 28, 163
NHC-oxazoline ligands   194

NHC precursor   186
[NHC-Pt]   139, 148
NHC-Pt complexes   119, 158
NHC-Ru complexes   8
- continuous flow reactors   16
- immobilization   17
- Merrifield-supported   17
- metal contamination   16
- monolithic systems   18
- non-porous supports   18
- poly-divinylbenzene   19
- reuse   16
- solid supported   16
NHC-ruthenium   3, 7, 9
- four-membered cyclic NHC   12
- initiator   11
- second-generation complexes   5, 6
- triazol-5-ylidene   7
(NHC)Au$^+$ 270
(NHC)AuCl   270
(NHC)Au(OAc)   270
(NHC)CuCl   262
(NHC)Pd species, steric environment   67
(NHC)Pd(0)   64
(NHC)Pd(allyl)Cl   63, 65, 67, 68
(NHC)Pd(OAc)$_2$, acidic conditions   69
(NHC)Pt(dvtms)   125, 127, 129, 142, 151
(NHC)Pt(dvtms) complexes   123
Ni hydrides   179
Ni-catalyzed reductive coupling   176
Ni-mediated catalysis
- catalytic activity   169
- cycloaddition   167
- cyclopropylen-ynes   164
- nitrogen-containing heterocycles   167
- oxygen-containing heterocycles   167
- proposed mechanism   165
- pyrones   167
- rearrangement reactions   163
- vinylcyclopropanes   163
Ni-NHC complexes   179
nickel hydride   177
nickel silyl   177
nickel-cycloaddition of CO_2 168
Norbornene macromonomers   291
nucleophilic addition   278
nucleophilic catalysts   275
nucleophilicity   286

o

olefin metathesis   1, 243, 257
- asymmetric   211
- 14-electron species   3

– N-heterocyclic carbenes (NHC)   2
– mechanism   2
– molybdenum   1
– phosphine   2
– phosphine ligands   2
– ruthenium   1
– tungsten   1
oligomerization   178
oligomers   179
organic catalysts   275
organocatalysis   275
– enantioselective   275
organolithium reagents   176
Osborn–Schrock catalyst   241
oxazoline   197
oxazoline-NHC iridium complex   193
oxidation   41, 57, 103
– Cu   105
– of alcohols   31
– Pd   104, 105
– Pd-hydroperoxo   107
oxidative addition   55, 56, 60, 91, 93, 146
oxidative C–C cleavage   230
oxidative carbonylation   113
oxidative catalysis   117
oxidative coupling   166

p

palladium   55, 56
– allyl   61
– catalysis   214
palladium-NHC   77
palladium-NHC complexes   204
palladium(II) N-heterocyclic carbene complexes   58
paracyclophane   196, 197, 203
paracyclophane-NHC ligands   210
Pt-carbene   123
P(Cyp)$_3$ 178
Pd-hydride   107, 111
Pd-(NHC)   86
Pd-NHC complexes   73, 114
Pd(0) species   62
Pd(0)-L compound   92
[Pd^0(NHC)]   93
Pd^0(NHC) species   93
Pd^0(NHC)$_2$ 98
Pd^0(NHC)(dvds)   85
PdCl$_2$(NHC)$_2$ 97
[PdCl$_4$]$^-$ 92
Pd(dba)$_2$ 98
Pd(OAc)$_2$ 99
peroxo species   104

[(Ph$_3$P)CuH]$_6$ 261
phenyl boronic acid   203
phosphine   10, 56, 103, 121, 122, 124, 164, 169, 172, 174, 178, 245, 257
– steric and electronic tunability   223
tris(*tert*-butyl)phosphine   122
phosphine-free   9
– catalysts   22
phosphine-oxazoline   250
phosphoramidite ligand   258
pincer carbene   30
– complexes   223
pincer ligands, planarity   227
planar chiral NHC ligands   205
planar chiral stable carbene   208
planar chirality   192, 204, 210
planarity of the backbone   259
platinum carbene   129
– reactivity   134
– selectivity   134
platinum colloids   122
polycarbonate   114
polyesters   289
polymerization   8, 12, 28, 178
– of ethylene   180
– of lactide   269
– of lactones   292
– of norbornene   180
polyol substrates   286
polysiloxane   120
potassium trifluoroborates   203
preactivation   60
preformed complexes   88
preparation of 1,2-bisboronate esters   268
primary C–H activation   273
princers   234
prochiral dienes   252
prochiral unsubstituted alkenes   252
L-proline amino acid   205
protected carbenes   291
proton transfer   281
^{195}Pt NMR   124
pyridine   172, 173

q

quaternary carbons   267
quaternary stereogenic centers   216

r

radical polymerization   293
RCM   8, 18, 21
– and isomerization products   46
RCM/hydrogenation   44

RCM/isomerization 46
ReactOp 145
recyclability 236
reductive coupling reactions
– nickelapentenacycle 174
– oxidative coupling 174, 175
– regioselectivity 174
reductive elimination 55, 93, 169, 178, 224, 234
reductive hydrogenation 241
regiocontrol 158
regioselectivity 171, 173, 265, 273
Rh-NHC complexes 185
Rhodium-NHC complexes 203, 207
Rh(PPh$_3$)$_3$Cl 241
ring rearrangement metathesis (RRM) 20
ring-closing metathesis (RCM) 2, 44
– asymmetric 211
– E-geometry 3
– E/Z selectivity 3
– tetra-substituted double bonds 7
– Z-geometry 3
ring-opening polymerization (ROP) 277
– β-butyrolactone 288
– ε-caprolactone 288
– cyclic ester monomers 288
– lactide 288
Rochow process 119
ROM-CM 20, 21
ROMP 8
RRM 20, 21
Ru hydride species 48
Ru-vinylidene 48
ruthenium 1
– hydride 31
ruthenium NHC complexes 210
– hydrosilylation 32

s

saturated esters 284
selectivity 29, 129, 199, 209, 214, 281
silicon polymers 119
silicon-hydride 156
silver-based catalyst 268
silver carbene 200
silver NHC 225
– ligands 200
silver-NHC complex 186
SIMes 12, 95, 244, 258, 286
– ligand 8, 34
single-bond character 125
SIPr 63, 95, 163, 169, 172
SIPr.HCl 99

(SIPr)Pd(allyl)Cl 63
solution calorimetry 2
solvent 45
Sonogashira 99, 243
– coupling 236
(–)-sparteine 108, 218
Speier's catalyst 119
stable catalysts 246
star polymers 293
stereocontrol 207, 215
stereogenic centers 189, 190, 217
stereoselective hydrosilylation 185
stereoselective induction 187, 197
stereoselectivity 200, 201
steric 91
– bulk 129
– properties 196
sterically hindered NHC 181
Stetter reaction 281
Stryker's reagent 263
substitution 234
Suzuki–Miyaura reaction 69, 243
– arylboronic acids 64
– boronic acid 65
– isopropanol 65
– mechanism 65
– room temperature 65
– tri-ortho-substituted biaryls 65, 66
synthetic strategies
– alkylation 228
– cyclometallation of the imidazolium salt 228
– de novo 228
– direct metallation 228
– oxidative addition 228

t

tandem dehydrogenation/hydrogenation 29
tandem processes 27
tandem RCM/hydrogenation 42, 43
tandem RCM/isomerization 48, 49
tandem reactions 42, 50
telechelic polymers 293
telomerization 58, 73, 78, 86, 90, 91, 171
– bmim 90
– bmiy 90
– buta-1,3-diene 74, 76
– cationic complexes 89
– chemoselectivity 82, 88
– conversion 88
– DFT calculations 78
– isoprene 74, 76
– Kuraray company 76

- linear dimerization product 87
- mechanism 77
- MeOh 74
- MeOH 76
- NHC ligands 75
- of alkenes 55
- Pd-(NHC) complexes 83
- phenol 75
- plasticizer production 76
- rearrangement 79
- regioselectivity 78, 83
- selectivity 84, 85, 88
- taxogen 74
- telogen 74
- telomers 74
terminal alkenes 28
tetrakis[3,5-bis(trifluoromethylphenyl)]borate 248
tetra-substituted alkene 21, 22, 245, 248
thermal stability 145
thermogravimetric analysis 277
thermolysis 277
thiamine cofactors 275
thiazolium carbenes 294
thiourea 135
transesterification 277, 285, 289, 293
transfer hydrogenation 42, 243
- aldimines 181
- catalysis 235
- imines 30, 181
- ketimines 181
- ketones 30

transmetallation 55, 225, 226, 231
- reaction 258
tri-substituted double bonds 21
1,3,5-triazol-2-ylidenes 293
triazolium salts 283
tridentate NHC ligands 227
triflates 60
trifluoromethylation of carbonyl compounds 288
tripod ligands 232
tripodal NHC ligand 103
triscarbene 227
trisubstituted olefins 164

u
α,β-unsaturated carbonyl 281
unsaturated NHC ligand 246

v
o-vanillin 10
vinylidene 8

w
Wacker cyclization 112
Wacker oxidation 113
Wacker-type oxidation 112
"Wanzlick" dimer 291, 292
Wilkinson's catalyst 241
Wittig reaction 50

z
zwitterionic intermediate 289